MATLAB Roadmap to Applications

Yi Chen · Long Huang

MATLAB Roadmap to Applications

Volume I Fundamental

 Springer

Yi Chen
Xi'an Jiaotong-Liverpool University
Suzhou, Jiangsu, China

Long Huang
Xi'an Jiaotong-Liverpool University
Suzhou, Jiangsu, China

ISBN 978-981-97-8787-6   ISBN 978-981-97-8788-3 (eBook)
https://doi.org/10.1007/978-981-97-8788-3

This Springer imprint is published by the registered company Springer Nature Singapore Pte Ltd.
The registered company address is: 152 Beach Road, #21-01/04 Gateway East, Singapore 189721, Singapore

If disposing of this product, please recycle the paper.

For My Family

—Yi Chen

[踏春行]

初踏雨痕寸寸新，

暖日杨花点点晴，

水镜月盈山歌隐，

一枕春晓落叶静。

For My Family

—Long Huang

Foreword

MATLAB, a powerful mathematical software package, has become an indispensable tool for engineers, scientists, and researchers in various academic and industry domains. Its versatility in data analysis, visualization, computing, and simulation has made it a cornerstone of many fields, including energy, manufacturing, transportation, healthcare, supply chain, information technologies, aerospace, business, and finance, just name a few.

It is my pleasure to present this comprehensive guide, "*MATLAB from Fundamentals to Applications—I Fundamentals*," which caters to students, engineers, and researchers seeking to utilize MATLAB in their daily work. What sets this book apart is its cross-disciplinary approaches, showcasing the wide range of applications that MATLAB offers in fields such as engineering, science, and finance. This makes the book suitable for a diverse audience, including undergraduate and postgraduate students, university professors, as well as industry professionals.

The key features of this book are:

- **Comprehensive Coverage**: This book covers all the fundamental concepts of MATLAB, serving as a one-stop shop for beginners starting from scratch.
- **Multidisciplinary Applications**: With a focus on real-world applications, this book demonstrates how MATLAB can be utilized to analyze and solve complex problems in various fields, catering to the needs of a wide range of readers.
- **Hands-on Approach**: The book emphasizes practical learning by including hands-on experience and course works, allowing readers to apply the concepts they learn in real-world scenarios.
- **Solution-Oriented**: To enhance and assess the learning progress, this book provides solutions to problems and labs, giving readers the opportunity to test their understanding of the materials.
- **Online Resources**: Supplementary online resources and modules are also available to further support the learning experience, providing additional guidance and assistance to readers.
- **Author's Experience**: The lead author of this book is a distinguished professor, IET Fellow, and IMechE Fellow, with extensive experience in MATLAB. This

ensures that the book drew reliable and cutting-edge sources of knowledge, offering valuable insights and expertise to readers.

In the era of **Artificial Intelligence**, we firmly believe that this book is an essential resource for anyone seeking to learn and become a mastery of MATLAB. Whether you are a student, a researcher, or a professional, this book will equip you with the necessary skills to harness the power of MATLAB in your work. It is our hope that readers will find this book not only informative but also practical, enabling them to achieve their goals and excel in their respective fields.

With great enthusiasm, we invite you to embark on this MATLAB journey and enjoy the numerous findings that await you to explore.

30 August 2024

Prof. Tongdan Jin
Texas State University
San Marcos, USA
tj17@txstate.edu

Preface

Welcome to the first edition of "MATLAB from Fundamentals to Applications—
I Fundamentals." As esteemed professors with extensive experience in MATLAB,
we are delighted to present this comprehensive guide to students, engineers, and
researchers who are seeking to master MATLAB as a powerful tool for their everyday
work.

MATLAB is a widely used programming language in various domains, including
engineering, science, finance, and many others. With its extensive range of features,
MATLAB can sometimes appear overwhelming to beginners. Thus, we have taken
the initiative to write this book, which systematically covers all the fundamental
concepts of MATLAB. Through a step-by-step approach, supported by numerous
examples and exercises, readers will develop a solid understanding of MATLAB.

This book has been carefully designed to provide readers with a roadmap for
learning MATLAB, starting from the basics and progressing to advanced concepts.
Each chapter is structured in a manner that facilitates step-by-step learning, with
clear explanations, illustrative examples, and exercises to reinforce comprehension.
In addition, the book incorporates practical lab works and courseworks, enabling
readers to apply their knowledge in real-world scenarios.

Key Features

One of the standout features of this book is its emphasis on a hands-on approach. Lab
works and course works are included to enable readers to apply the concepts they
learn to practical situations. Moreover, solutions to problems and labs are provided to
help readers assess their understanding and track their progress. To further enhance
the learning experience, this book is accompanied by online resources and modules,
including video lectures, quizzes, and supplementary exercises.

Audience

This book is tailored to cater to a wide range of readers, including:

- Undergraduate and postgraduate students studying engineering, physics, mathematics, computer science, and related disciplines.
- Researchers who utilize MATLAB for data analysis and visualization in fields such as biology, chemistry, economics, and social sciences.
- Engineers who employ MATLAB for modeling, simulation, and control system design.
- Professionals who wish to acquire MATLAB as a valuable tool for their daily work, including data analysts, statisticians, and financial analysts.

Organization

The book is organized as follows:

- Chapter 1: Introduction to MATLAB, encompassing its history, features, and installation.
- Chapters 2 and 3: Fundamental concepts of MATLAB programming, covering topics such as data types, operators, expressions, vectors, arrays, matrices, and data structures.
- Chapters 4–6: Conditional statements, loop statements, scripts, and functions.
- Chapters 7–9: Inputs and outputs, data visualisation, programming, and algorithm development.
- Chapter 10: Object-oriented programming.
- Appendices A and B: Solutions to problems and FAQs.

Pedagogical Features

To ensure an effective learning experience, this book incorporates various pedagogical features, including:

- Step-by-step explanations and examples to facilitate comprehension.
- Exercises at strategic points to reinforce understanding.
- Lab works and examples to enable practical application.
- Solutions to problems and labs to assess progress in each chapter.

Prerequisites

The only prerequisite for this book is a basic understanding of mathematics and computer science.

This book is an indispensable resource for undergraduate students, postgraduate students, and professionals across a range of disciplines, including engineering, physics, mathematics, and computer science. It is our sincere hope that this book will serve as a valuable companion for anyone seeking to master MATLAB.

Suzhou, China

Prof. Yi Chen
IET Fellow
IMechE Fellow
leo.chen@ieee.org

Prof. Long Huang

Acknowledgements

We express our sincere gratitude for the financial support received in part from the Research Development Fund (RDF-21-02-019), which facilitated the development of this book.

We would like to extend our heartfelt appreciation to our esteemed colleagues from various institutions for their unwavering support and encouragement throughout this endeavor. These institutions include Xi'an Jiaotong-Liverpool University, Newcastle University, University of Glasgow, University of Strathclyde, Northumbria University, University of Liverpool, University of Cranfield, University of Edinburgh, Edinburgh Napier University, Cardiff University, Queen's University Belfast, Massachusetts Institute of Technology, California Institute of Technology, University College London, University of Bath, University of Bristol, University of Manchester, University of British Columbia, University of Alberta, and Texas State University.

Our sincere thanks go to the reviewers who generously provided valuable feedback and insightful comments, which significantly contributed to enhancing the quality and clarity of the manuscript.

We are deeply grateful to the editorial team at Springer Nature for their guidance, professionalism, and support in publishing this book. Their expertise and dedication have been instrumental in bringing this project to fruition.

On a personal note, one of the authors would like to extend heartfelt gratitude to his wife for her unconditional love, support, and patience throughout the duration of this project. Her understanding and encouragement were invaluable sources of motivation.

Contents

Acronyms

AI	Artificial Intelligence
CI	Continuous Integration
CPU	Central Processing Unit
EISPACK	Matrix Eigensystem Package
GPU	Graphics Processing Unit
GUI	Graphical User Interface
GUIs	Graphical User Interfaces
IDE	Integrated Development Environment
LINPACK	Linear Equation Package
ODE	Ordinary Differential Equation
OOP	Object-Oriented Programming

List of Figures

List of Tables

Chapter 1
Introduction

Chapter Learning Outcomes

- Explain the **purpose** and **applications** of MATLAB in various fields
- Gain an **overview** of the MATLAB **environment** and **user interface**
- Understand the **structure** and **organisation** of this book
- Appreciate the **importance** of MATLAB as a **programming language** and **numerical computing tool**
- Describe the **topics** covered in this book
- Discover the **benefits** of using MATLAB.
- Identify the different types of **users** who can benefit from using MATLAB.

Chapter Key Words

- **MATLAB**: A high-performance numerical computing environment and programming language widely used in various fields, including engineering, science, and finance, for data analysis, algorithm development, and Visualisation.
- **Programming Language**: A formal language designed to instruct computers and computing devices to perform specific tasks by writing code or programs, following a set of rules and syntax.
- **User Interface**: The means by which users interact with a computer system or application, typically consisting of graphical elements like menus, toolbars, and windows.
- **Environment**: In the context of MATLAB, it refers to the integrated development environment (IDE) where users can write, edit, and execute MATLAB code, as well as access various tools and features.
- **Mathematical Operations**: Fundamental arithmetic operations, such as addition, subtraction, multiplication, and division, as well as more advanced operations like matrix calculations and trigonometric functions, performed using MATLAB.

© The Author(s) 2025

Y. Chen and L. Huang, *MATLAB Roadmap to Applications*,
https://doi.org/10.1007/978-981-97-8788-3_1

- **Variables**: Named storage locations in computer memory used to store and manipulate data values, which can be assigned, modified, and referenced within MATLAB programs.
- **Data Types**: The classification of data based on the type of values they can represent, such as numbers, characters, or logical values, which determines the operations and functions that can be applied to them in MATLAB.
- **Arrays**: Ordered collections of elements, often of the same data type, which can be one-dimensional (vectors) or multi-dimensional (matrices) in MATLAB, and are widely used for numerical computations and data manipulation.

1.1 What is MATLAB

MATLAB (Matrix Laboratory) is a high-performance, **numerical computing environment** and **programming language** developed by MathWorks. It is widely used in various fields, including engineering, science, finance, and academia, for data analysis, algorithm development, and Visualisation [2]. MATLAB stands out for its ease of use, powerful matrix and array manipulation capabilities, and extensive toolboxes for diverse applications.

MATLAB provides an IDE that combines a code editor, debugger, and Visualisation tools, facilitating the entire workflow from initial prototyping to final implementation. The IDE offers a user-friendly interface, making it accessible to both novice and experienced users. One of MATLAB's key strengths is its **high-level programming language** that leverages matrix and array operations, enabling concise and efficient code development for scientific and engineering computations.

The MATLAB language is designed to perform **numerical computations** efficiently, with built-in support for linear algebra, signal processing, image processing, and other mathematical functions. It also provides powerful **data Visualisation** capabilities, allowing users to create high-quality 2D and 3D plots, graphs, and animations to explore and communicate their data effectively.

MATLAB is highly **extensible** through the use of toolboxes, which are collections of specialised functions and applications tailored to specific domains, such as control systems, machine learning, optimisation, and finance. These toolboxes expand MATLAB's functionality, enabling users to tackle a wide range of problems and leverage cutting-edge algorithms and techniques.

Furthermore, MATLAB seamlessly integrates with other programming languages and technologies, including C, C++, Fortran, Java, .NET, and Python, allowing for **interoperability** and code reuse across different platforms and systems. This flexibility makes MATLAB an invaluable tool for developing and deploying applications in diverse environments, from embedded systems to cloud computing platforms.

As shown in Fig. 1.1, the MATLAB product tree has different components [1].

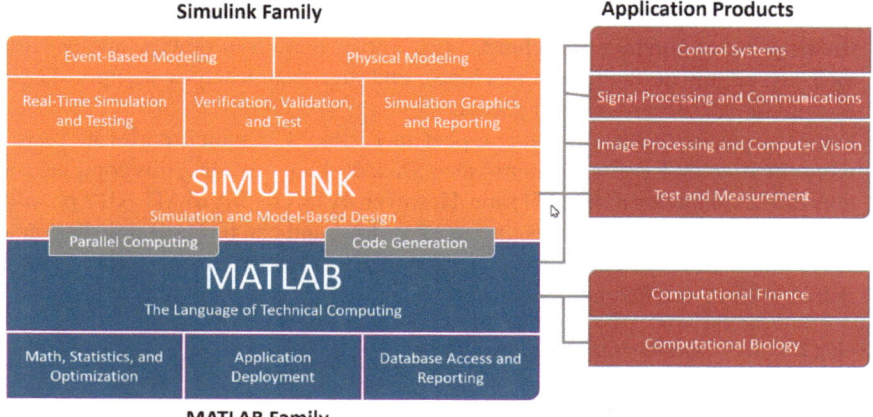

Fig. 1.1 MATLAB family tree

1.2 Why MATLAB?

- **Ease of Use** One of the primary reasons to choose MATLAB is its user-friendly environment, which promotes accessibility and productivity. MATLAB's intuitive syntax, combined with its interactive desktop, makes it easy for beginners to get started with programming and numerical computations. Experienced programmers also appreciate MATLAB's simplicity, which allows them to focus on problem-solving rather than dealing with complex language constructs.
Example:

Listing 1.1 Simple arithmetic in MATLAB

```
% Simple arithmetic operations
a = 5;
b = 3;
c = a + b
d = a * b
```

Listing 1.2 Results: Simple arithmetic in MATLAB

```
% Results:
c = 8
d = 15
```

- **High-Performance Computing** MATLAB's architecture is optimised for efficient numerical computations, making it a powerful tool for handling large data sets and complex calculations. It leverages optimised libraries for linear algebra, signal processing, and other mathematical operations, enabling researchers and engineers to solve computationally intensive problems.

One of the key features of MATLAB is its ability to **scale computations from the desktop to clusters and clouds**. This allows users to prototype and develop their algorithms on their local machines, and then seamlessly scale up to larger computational resources when needed.

For example, consider the code snippet in Listing 1.3, which performs matrix multiplication on large random matrices. On a desktop machine, this computation may take a significant amount of time. However, by leveraging MATLAB's parallel computing capabilities, such as the `parfor` loop and batch processing, the same computation can be distributed across multiple cores or cluster nodes, greatly reducing the overall computation time.

Listing 1.3 Matrix operations in MATLAB

```
% Create large random matrices
A = rand(1000, 1000);
B = rand(1000, 1000);

% Matrix multiplication
C = A * B;
```

MATLAB also provides a rich set of tools for **parallel computing**, including the Parallel Computing Toolbox, which enables users to harness the power of multi-core processors, GPUs, and computer clusters. This toolbox provides high-level constructs, such as parallel for-loops (`parfor`), distributed arrays, and parallel numerical algorithms, which allow users to parallelise their code with minimal changes to their existing MATLAB programs.

In addition to the Parallel Computing Toolbox, MATLAB also integrates with popular cluster and cloud platforms, such as Amazon Web Services (AWS), Microsoft Azure, and Google Cloud Platform (GCP). This enables users to **scale their computations to the cloud**, taking advantage of the virtually unlimited computational resources available in these platforms.

- **Vast Collection of Toolboxes** MATLAB offers a comprehensive range of toolboxes that extend its capabilities to various domains, including signal processing, control systems, machine learning, computer vision, and many more. These toolboxes provide pre-built functions, algorithms, and applications, enabling users to solve complex problems without starting from scratch.
- **Integrated Development Environment** MATLAB's IDE provides a seamless environment for developing, debugging, and visualizing code. Features like live code editing, debugging tools, and integrated documentation make the development process more efficient and productive.
- **Cross-Platform Compatibility** MATLAB is available on multiple platforms, including Windows, macOS, and Linux, ensuring consistent performance and compatibility across different operating systems. This cross-platform compatibility facilitates collaboration and code sharing among teams and organisations.

By highlighting the ease of use, high-performance computing capabilities, extensive toolbox collection, integrated development environment, and cross-platform compatibility, this section effectively communicates the key advantages of using MATLAB for a wide range of technical computing tasks.

1.3 Who Should Use MATLAB?

MATLAB can be used by a wide range of users:

- Undergraduate engineering students for numerical computing courses.
- Graduate students and researchers for developing and prototyping algorithms.
- Data scientists for data analytics and visualisations.
- Professional engineers across various domains communications, signal processing, control systems, computer vision, etc.

  Specifically,

- **Engineers and Scientists** MATLAB is widely used by engineers and scientists across various disciplines, such as mechanical engineering, electrical engineering, aerospace engineering, and physics. Its powerful computational capabilities, combined with its intuitive programming environment, make it an ideal tool for numerical analysis, data processing, and system modeling.
  Example:

Listing 1.4 Solving a system of linear equations

```
% Define the coefficient matrix and constants
A = [1 2 3; 4 5 6; 7 8 10];
b = [6; 15; 33];

% Solve the system of linear equations
x = A \ b
```

- **Researchers and Academics** MATLAB is extensively used in research and academic institutions for data analysis, algorithm development, and simulation. Its rich ecosystem of toolboxes and libraries enables researchers to tackle complex problems in fields such as signal processing, machine learning, control systems, and computational biology [3].
  Example:

Listing 1.5 Plotting a sine wave

```
% Generate a sine wave
t = 0:0.01:2*pi;
y = sin(t);

% Plot the sine wave
figure;
plot(t, y);
xlabel('Time');
```

```
ylabel('Amplitude');
title('Sine Wave');
```

- **Students and Educators** MATLAB's intuitive syntax and interactive environment make it a popular choice for teaching and learning programming, numerical methods, and computational mathematics in educational settings. Many universities and colleges incorporate MATLAB into their curricula across various STEM fields.
- **Industry Professionals** MATLAB is widely adopted in various industries, including automotive, aerospace, finance, and biotechnology. Its ability to handle large datasets, perform complex simulations, and integrate with other programming languages makes it a valuable tool for professionals working in these domains.

  By highlighting the diverse user base, including engineers, scientists, researchers, academics, students, educators, and industry professionals, this section effectively communicates the broad applicability and versatility of MATLAB across various domains and professions.

1.4 What is Covered in this Book

- Introduction to MATLAB: These topics will provide an overview of MATLAB, its history, and its significance in the field of technical computing. It will explain the core features of MATLAB, such as its matrix-based language, built-in graphics capabilities, and integrated development environment.
- MATLAB Programming Fundamentals: These topics will cover the fundamental concepts of MATLAB programming, including **variables**, **data types**, **operators**, and **control structures**. It will introduce the syntax and structure of MATLAB scripts and functions, as well as provide examples to illustrate these concepts. Example:

Listing 1.6 Variable assignment and arithmetic operations

```
% Variable assignment
a = 5;
b = 3;

% Arithmetic operations
sum = a + b;
difference = a - b;
product = a * b;
quotient = a / b;
```

- Data Handling and Visualisation: These topics will delve into techniques for importing, preprocessing, and analysing data in MATLAB. It will cover topics such as **reading and writing data files**, **data manipulation**, and **data visualisation** using various plotting functions and techniques.

Example:

Listing 1.7 Plotting a scatter plot

```
% Generate sample data
x = rand(100, 1);
y = 2 * x + randn(100, 1);

% Create a scatter plot
figure;
scatter(x, y);
xlabel('X');
ylabel('Y');
title('Scatter Plot');
```

- Advanced MATLAB Programming: These topics will explore more advanced programming concepts in MATLAB, such as **object-oriented programming**, **functional programming**, and **parallel computing**. It will also introduce techniques for developing and deploying MATLAB applications.
- MATLAB Toolboxes and Problem Solving This section will provide an overview of the various toolboxes available in MATLAB, which extend its functionality to specific domains, such as **signal processing**, **control systems**, **optimisation**, and **machine learning**. It will also showcase real-world applications of MATLAB in various fields.

  By covering a comprehensive range of topics, from introductory concepts to advanced programming techniques and domain-specific applications, this section outlines the scope and depth of the material covered in the book, catering to readers with varying levels of expertise and interests.

1.5 What You Will Learn by the Book

This book aims to provide a comprehensive introduction to MATLAB, a powerful programming language and environment widely used by engineers and scientists for technical computing. From the basics of MATLAB syntax and command-line operations to advanced topics in numerical computation and data visualisation, you will gain the knowledge necessary to excel in your academic or professional pursuits. This section will outline the key skills and concepts you will master by reading and working through the examples in this book.

By the end of this book, you will:

- Understand what is MATLAB and its key capabilities
- Get an overview of MATLAB's history and evolution
- Learn about MATLAB's programming environment and interface
- Understand basic MATLAB concepts like arrays, matrices, data types, etc.
- Programming in MATLAB and Problem Solving
- Data Analysis and Visualisation.

By covering these topics, the book will provide you with a comprehensive understanding of MATLAB's capabilities and equip you with the skills to leverage its power for various computational tasks and applications.

1.6   MATLAB History and Timeline

This section chronicles the remarkable journey of MATLAB, highlighting the pivotal milestones that have shaped its evolution from its inception to its current state as a powerful computational platform. The timeline unfolds, tracing MATLAB's origins, major updates, and groundbreaking advancements that have propelled its widespread adoption across diverse fields [4].

Comprehending MATLAB's historical development is essential for grasping its capabilities and profound impact. The narrative begins with its mathematical roots in solving matrix equations and eigenvalue problems using Algol 60, a foundational step that paved the way for subsequent breakthroughs. The Matrix Eigensystem Package(EISPACK) and Linear System Package(LINPACK) projects played catalytic roles in advancing mathematical software, culminating in Cleve Moler's pioneering development of the first MATLAB version in Fortran, which demonstrated its potential as an interactive matrix calculator.

A seminal moment occurred with MATLAB's transition to C and its commercial release in 1984, marking a significant milestone that facilitated the expansion of its functionality and widespread adoption across diverse fields. It is noteworthy that MATLAB's continuous evolution, manifested through the introduction of new features and toolboxes, has significantly contributed to its versatility and extensive utilization in engineering, mathematics, and scientific research endeavors.

For students, professional researchers, and engineers, understanding MATLAB's historical context and evolution offers valuable insights into its capabilities and the profound impact it has exerted across various domains.

- **Founders**
  MATLAB was co-founded by **Cleve Moler** [5], a mathematician and computer scientist, and **Jack Little** [6], an electrical engineer and computer scientist [7]. **Moler** was a professor at several universities, including the University of Michigan, Stanford University, and the University of New Mexico, before co-founding MathWorks. **Little** earned degrees from MIT and Stanford before meeting Moler and recognising the potential for MATLAB as a commercial product.
- **Mathematical Origins (1965–1970):**
  The genesis of MATLAB is deeply intertwined with the pioneering work of **J. H. Wilkinson** and his associates, whose research papers on algorithms for addressing matrix linear equations and eigenvalue problems established the bedrock upon which MATLAB was built [8]. These seminal works were encapsulated in the authoritative volume *"Handbook for Automatic Computation, Volume*

II: Linear Algebra", serving as the mathematical and computational foundation for the development of MATLAB.

Drawing inspiration from these scholarly articles, the inaugural rendition of MAT-LAB was crafted in Algol 60. This early prototype was not a programming language by contemporary standards but functioned as a rudimentary interactive matrix calculator. It was devoid of the advanced features that define modern programming languages, such as user-defined programs, toolboxes, graphic capabilities, ordinary differential equations (ODEs) solvers, and fast Fourier transform (FFT) utilities. Despite its simplicity, it was a potent mathematical instrument specifically tailored for solving matrix linear equations and eigenvalue problems. In the late 1960s, **Cleve Moler** embarked on the development of numerical linear algebra software using Fortran, a programming language renowned for its scientific computing applications [9]. This initiative marked a significant step towards the evolution of MATLAB as we know it today.

- **EISPACK and LINPACK (1970–1976):**
  During the 1970s, **Cleve Moler** played a pivotal role in authoring the scientific subroutine libraries known as **EISPACK** and **LINPACK**, which served as the bedrock for MATLAB's mathematical prowess [9]. The inception of these libraries represented a significant juncture in the evolution of numerical computing software.
  EISPACK, which emerged in the early 1970s, offered a suite of routines dedicated to eigenvalue computations. Subsequently, **LINPACK**, introduced later in the decade, provided comprehensive solutions for linear equations and linear least-squares problems. These libraries collectively formed the foundational capabilities of MATLAB.

  - **EISPACK** The EISPACK initiative was launched by a consortium of researchers at Argonne National Laboratory. It was designed as a package for tackling eigenvalue problems by translating Algol 60 algorithms into Fortran. With an aim to develop and rigorously test high-calibre mathematical software, the first iteration of EISPACK was unveiled in 1971. This project was seminal in setting the stage for subsequent advancements in mathematical software [10].
  - **LINPACK** The LINPACK project, spearheaded by **Cleve Moler** [5] and his collaborators, was proposed to the U.S. National Science Foundation with the objective of exploring methodologies for the development of mathematical software. Unlike EISPACK, LINPACK was developed directly in Fortran, eschewing the need for Algol translation. This project culminated in the creation of LINPACK, a package specifically designed to address linear equation problems. Comprising 44 subroutines across various numeric precisions, LINPACK was initially penned in Fortran and represented a landmark development in the domain of mathematical software [11].

- **Historic MATLAB (1970s–1980s):**
  MATLAB traces its origins back to the late 1970s at the University of New Mexico, where **Cleve Moler**, a computer science professor, was working on developing a programming environment for his students. Inspired by the LINPACK and EIS-PACK projects, which provided Fortran libraries for numerical linear algebra, Moler aimed to create a user-friendly interface that would make these libraries more accessible to students and researchers [9]. Initially called MATLAB, the software was designed to provide a more intuitive and interactive environment for working with matrices and numerical computations, serving as an interactive matrix calculator to provide easy access to LINPACK and EISPACK functionalities.
  MATLAB allowed users to perform matrix operations and numerical computations in an interactive and user-friendly manner, providing an accessible interface for the LINPACK and EISPACK libraries. Initially used in academia, MATLAB's ease of use and powerful computational abilities quickly gained popularity among students and researchers. This period saw the transformation of MATLAB from a simple matrix calculator into a comprehensive computational environment, offering a simple syntax and powerful matrix operations that gained popularity among engineers and scientists.

- **Commercial MATLAB (1980s–2020s):**

  – The first commercial version of MATLAB (MATLAB 1.0) was released in 1984, Moler collaborated with **Jack Little**, to further develop and refine MATLAB. Jack Little wrote a new version in C, which debuted as PC-MATLAB at the IEEE Conference on Decision and Control in 1984 [12]. MATLAB was rewritten in C and commercially released for the first time. This version introduced a more extensive function library and improved performance, expanding its capabilities. It quickly gained popularity among researchers and engineers due to its ease of use and powerful computational capabilities, which included features such as matrix manipulation, plotting, and numerical analysis functions. This release marked the beginning of MATLAB's widespread use and popularity
  – In 1984, **Steve Bangert**, a computer engineer, joined the project, and the three co-founders established MathWorks, a company dedicated to commercialising MATLAB.
  – MATLAB's user-friendly interface, marked by the introduction of a graphical user interface (GUI) in 1987, has made it accessible to users with varying levels of technical expertise. The GUI allows users to visually navigate through MAT-LAB's features using menus and buttons, simplifying complex computations and enabling users to focus on their specific tasks.
  – In 1992, MATLAB expanded its reach by becoming available on multiple platforms, including Windows and Macintosh, making it accessible to a broader audience of researchers, engineers, and students from diverse backgrounds.

- One of the key advancements in MATLAB's capabilities occurred with the integration of Simulink in 2000. Simulink provided researchers and engineers with a graphical simulation and model-based design environment, enabling them to Analyse and simulate complex dynamic systems.
- MATLAB's specialiszation in various domains is evident through its specialised toolboxes. For instance, the Image Processing Toolbox, introduced in 2004, equipped MATLAB users with a comprehensive set of functions and algorithms for image analysis and processing.
- Similarly, the Parallel Computing Toolbox, introduced in 2011, enabled users to leverage parallel computing capabilities for faster and more efficient computation.

Nowadays, MATLAB boasts a global user base of over one million, highlighting its widespread adoption and profound influence in academia, industry, and research. Its intuitive interface, extensive library of functions, and powerful computational capabilities render it an invaluable tool for users across various disciplines and industries.

MATLAB continued to evolve with regular updates and the introduction of new features and toolboxes, becoming a widely used software tool in engineering, mathematics, and scientific research, enabling tasks such as numerical analysis and data visualisation. A list of toolboxes is provided below:

- **ODEs (1980s):** MATLAB's commercial version included numerical solutions for ordinary differential equations (ODEs), which are central to Simulink, MATLAB's companion product for simulation and model-based design.
- **1987:** MATLAB introduced a graphical user interface (**GUI**), making it easier for users to interact with the software. The GUI allowed users to visually navigate through MATLAB's features and perform operations using menus and buttons.
- **Simulink (1990s):** In the early 1990s, Simulink, a block diagram environment for modeling and simulating dynamic systems, was integrated into MATLAB, allowing engineers and scientists to design and simulate complex systems.
- **Sparse Matrices (1992):** MATLAB 4 introduced sparse matrices, a memory-efficient representation for large arrays with few non-zero values.
- **Cell Arrays (1996):** MATLAB 5 introduced cell arrays, allowing for indexed collections of heterogeneous MATLAB objects.
- **Structures (1996):** Structures were added, providing a way to create complex data structures with associated methods.
- **Objects (2008):** MATLAB's object-oriented programming capabilities were enhanced, simplifying tasks involving specialised data structures.
- **Desktop and Live Editor (2000, 2016):** The MATLAB desktop and Live Editor significantly improved usability, especially for users without prior programming experience.
- **Parallel Computing (2004):** MATLAB introduced the Parallel Computing Toolbox, supporting both coarse-grained and fine-grained parallelism, unlocking new realms of computational power.

– **Toolboxes (2018):** As of release 2018a, MATLAB offers an impressive arsenal of 63 specialised toolboxes, catering to diverse applications across numerous domains.
– **Data Types (2000s):** To accommodate larger datasets and diverse applications, MATLAB introduced single precision, integer, and logical data types, expanding its versatility and applicability.

- **MATLAB in the 21st Century**
  As the new millennium dawned, MATLAB expanded its reach into various industries, solidifying its position as a pivotal tool in engineering, finance, biotech, and numerous other domains. It embraced cutting-edge technologies such as **cloud computing**, **parallel computing**, and **automatic code generation** [13], further enhancing its capabilities and versatility.

  MATLAB's burgeoning popularity paved the way for the development of numerous toolboxes, broadening its functionality to encompass realms such as machine learning, deep learning, data analytics, and control systems. Toolboxes like the Statistics and Machine Learning Toolbox, Neural Network Toolbox, and Control System Toolbox furnished users with specialised functions and algorithms tailored for specific applications, extending MATLAB's prowess into novel frontiers.

  Today, MATLAB stands as one of the most widely adopted programming languages and environments for scientific and engineering computations. Its evolutionary journey continues unabated, with regular releases introducing new features and enhancements, ensuring its enduring relevance and preeminence. MATLAB's extensive ecosystem, comprising online communities, comprehensive documentation, and robust support, empowers users to leverage its capabilities effectively, fostering innovation and driving advancements across myriad disciplines.

1.7   MATLAB Products and Services (2024a)

- **Product Families** [1]

  – MATLAB

    · Parallel Computing

       · Parallel Computing Toolbox
       · MATLAB Parallel Server

    · AI, Data Science, and Statistics

       · Deep Learning Toolbox
       · Statistics and Machine Learning Toolbox
       · Curve Fitting Toolbox
       · Text Analytics Toolbox

- · Math and Optimization

  - · Optimization Toolbox
  - · Global Optimization Toolbox
  - · Symbolic Math Toolbox
  - · Mapping Toolbox
  - · Partial Differential Equation Toolbox

- · Reporting and Database Access

  - · Database Toolbox
  - · MATLAB Report Generator

- · Code Generation

  - · MATLAB Coder
  - · Embedded Coder
  - · HDL Coder
  - · HDL Verifier
  - · Filter Design HDL Coder
  - · Fixed-Point Designer
  - · GPU Coder

- · Application Deployment

  - · MATLAB Compiler
  - · MATLAB Compiler SDK
  - · MATLAB Production Server
  - · MATLAB Web App Server

- · Verification, Validation, and Test

  - · Requirements Toolbox
  - · MATLAB Test

- Simulink

  - · Event-Based Modeling

    - · Stateflow
    - · SimEvents

  - · Physical Modeling

    - · Simscape
    - · Simscape Battery
    - · Simscape Driveline
    - · Simscape Electrical
    - · Simscape Fluids
    - · Simscape Multibody

- Real-Time Simulation and Testing

  - Simulink Real-Time
  - Simulink Desktop Real-Time

- Reporting

  - Simulink Report Generator

- Systems Engineering

  - System Composer
  - Requirements Toolbox

- Code Generation

  - Simulink Coder
  - Embedded Coder
  - DDS Blockset
  - AUTOSAR Blockset
  - C2000 Microcontroller Blockset
  - Fixed-Point Designer
  - Simulink PLC Coder
  - Simulink Code Inspector
  - DO Qualification Kit (for DO-178)
  - IEC Certification Kit (for ISO 26262 and IEC 61508)
  - HDL Coder
  - HDL Verifier

- Application Deployment

  - Simulink Compiler

- Verification, Validation, and Test

  - Requirements Toolbox
  - Simulink Check
  - Simulink Coverage
  - Simulink Design Verifier
  - Simulink Fault Analyzer
  - Simulink Test
  - Polyspace Bug Finder
  - Polyspace Bug Finder Server
  - Polyspace Code Prover
  - Polyspace Test
  - Polyspace Access
  - Polyspace Code Prover Server
  - Polyspace Client for Ada
  - Polyspace Server for Ada

- **Application Products**

  - Signal Processing

    - Signal Processing Toolbox
    - DSP System Toolbox
    - Audio Toolbox
    - Wavelet Toolbox
    - DSP HDL Toolbox

  - Image Processing and Computer Vision

    - Image Processing Toolbox
    - Computer Vision Toolbox
    - Lidar Toolbox
    - Medical Imaging Toolbox
    - Vision HDL Toolbox
    - Image Acquisition Toolbox

  - Control Systems

    - Control System Toolbox
    - System Identification Toolbox
    - Predictive Maintenance Toolbox
    - Robust Control Toolbox
    - Model Predictive Control Toolbox
    - Fuzzy Logic Toolbox
    - Simulink Control Design
    - Simulink Design Optimization
    - Reinforcement Learning Toolbox
    - C2000 Microcontroller Blockset
    - Motor Control Blockset

  - Test and Measurement

    - Data Acquisition Toolbox
    - Instrument Control Toolbox
    - Image Acquisition Toolbox
    - Industrial Communication Toolbox
    - Vehicle Network Toolbox
    - ThingSpeak

  - RF and Mixed Signal

    - Antenna Toolbox
    - RF Toolbox
    - RF PCB Toolbox
    - RF Blockset
    - Mixed-Signal Blockset
    - SerDes Toolbox
    - Signal Integrity Toolbox

– Wireless Communications

  · Communications Toolbox
  · 5G Toolbox
  · LTE Toolbox
  · WLAN Toolbox
  · Bluetooth Toolbox
  · Satellite Communications Toolbox
  · Wireless HDL Toolbox
  · Wireless Testbench

– Radar

  · Radar Toolbox
  · Phased Array System Toolbox
  · Sensor Fusion and Tracking Toolbox
  · Mapping Toolbox

– Robotics and Autonomous Systems

  · Automated Driving Toolbox
  · Robotics System Toolbox
  · Navigation Toolbox
  · ROS Toolbox
  · Sensor Fusion and Tracking Toolbox
  · RoadRunner
  · RoadRunner Asset Library
  · RoadRunner Scenario
  · RoadRunner Scene Builder
  · Simulink 3D Animation

– FPGA, ASIC, and SoC Development

  · HDL Coder
  · HDL Verifier
  · Deep Learning HDL Toolbox
  · Wireless HDL Toolbox
  · Vision HDL Toolbox

– Embedded Systems

  · Embedded Coder
  · HDL Coder
  · HDL Verifier
  · Filter Design HDL Coder
  · Fixed-Point Designer
  · GPU Coder
  · C2000 Microcontroller Blockset
  · PLC Coder

- · DO Qualification Kit (for DO-178)
- · IEC Certification Kit (for ISO 26262 and IEC 61508)

- – Industrial Automation

  - · Industrial Communication Toolbox
  - · Vehicle Network Toolbox
  - · Simulink PLC Coder
  - · Simulink Test
  - · Simulink Requirements
  - · Simulink Coverage
  - · Motor Control Blockset

- – Aerospace

  - · Aerospace Blockset
  - · Aerospace Toolbox
  - · UAV Toolbox
  - · DO Qualification Kit (for DO-178)
  - · Simulink 3D Animation

- – Computational Finance

  - · Datafeed Toolbox
  - · Database Toolbox
  - · Econometrics Toolbox
  - · Financial Toolbox
  - · Financial Instruments Toolbox
  - · Risk Management Toolbox
  - · Spreadsheet Link (for Microsoft Excel)

- – Computational Biology

  - · Bioinformatics Toolbox
  - · SimBiology

- – Code Verification

  - · Polyspace Bug Finder
  - · Polyspace Bug Finder Server
  - · Polyspace Code Prover
  - · Polyspace Test
  - · Polyspace Access
  - · Polyspace Code Prover Server
  - · Polyspace Client for Ada
  - · Polyspace Server for Ada.

- **Services**

  - – Software Maintenance
  - – Training
  - – Consulting.

- **License Types**

  - Industry Use
  - Student Use
  - University Use
  - Academic Teaching Use
  - Primary and Secondary School Use
  - Startup Use
  - Home Use.

- **Cloud Solutions**

  - MATLAB Online
  - MATLAB Online Server
  - Simulink Online
  - MATLAB Drive
  - ThingSpeak
  - MATLAB Mobile
  - MATLAB Grader

- **Community and Third-Party**

  - File Exchange
  - Hardware Support Packages and Services
  - Third-Party Products and Services
  - MATLAB and Simulink Books

1.8   How to Use this Book

This book is designed to be used as:

- A textbook for undergraduate and postgraduate students.
- A reference manual for researchers and professional engineers.
- A self-learning guide for enthusiasts.

The chapters are organised logically building up from fundamentals to advanced concepts. Multiple illustrative examples are provided throughout. While no strict prerequisites are required, some background in the following areas will help readers get the most out of this textbook:

- Programming experience–Prior experience in any programming language like C, Python, Java, etc. will be helpful to understand basic programming constructs and data structures.
- Mathematics–Foundational knowledge in mathematics including calculus, linear algebra, probability and statistics will enable better understanding of examples and applications.

- Engineering/Science basics–Some familiarity with basic engineering or science concepts will provide context for many of the examples. However, the book covers fundamentals as well.
- Computer skills–Basic computer skills including proficiency with an operating system, file management, office productivity tools etc. will be useful.

The book is designed in a modular fashion allowing even beginners with no prior experience to pick up MATLAB skills systematically. The programming aspects are built up gradually with abundant examples. Necessary mathematical and scientific context is provided along the way.

Readers with some amount of prior experience in programming, mathematics or an engineering/science discipline will likely be able to progress through the material more quickly. However, the book can be used even by complete beginners starting from first principles.

1.9 MATLAB Environment and Settings

The MATLAB Environment and Settings play a crucial role in optimising the user experience and streamlining the workflow. This section will delve into the various aspects of the MATLAB Desktop, preferences, and platform-specific considerations [14].

- **MATLAB Desktop**
  The **MATLAB Desktop** serves as the primary user interface, providing a unified environment for managing files, variables, code development, debugging, and visualisation [14].
  This subsection will cover:

  – **Desktop Layout**: This section provides an overview of the various components of the MATLAB Desktop, including the **Command Window**, **Workspace Browser**, **Current Folder Browser**, **Editor Window**, and plotting areas, as shown in Fig. 1.2. The **Command Window** is where users can enter commands and run scripts, while the **Workspace Browser** displays all the variables in the current workspace. The **Current Folder Browser** allows users to navigate and manage files within the current directory, and the **Editor Window** is used for writing, editing, and debugging code. The plotting areas enable users to visualise data in various forms, such as graphs and charts.
  – **Toolstrip and Toolbars**: This section explains the **Toolstrip** and **Toolbars**, which provide quick access to commonly used functions and tools. The **Toolstrip** is a ribbon-like interface that organises tools into tabs and sections, facilitating easy access to different functionalities such as file operations, plotting tools, and code execution. The **Toolbars** offer shortcuts to frequently used commands, enhancing the efficiency of the user interface.

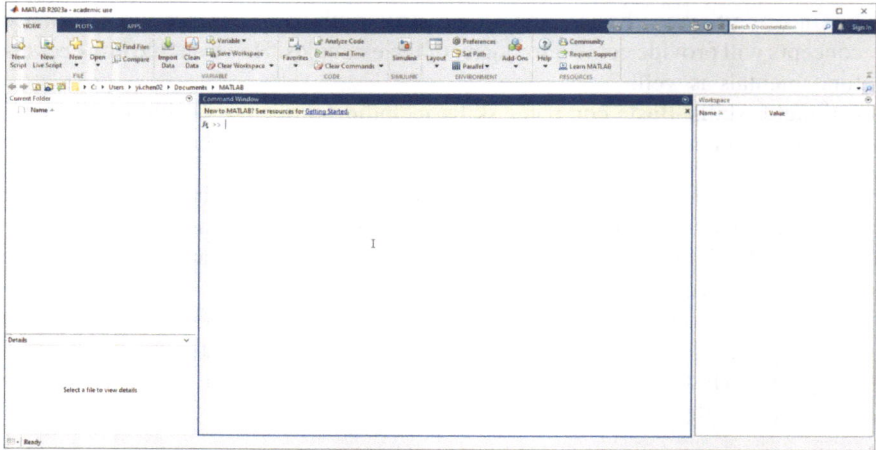

Fig. 1.2 MATLAB main window

- **Customisation**: This section discusses techniques for customising the desktop
  layout, including docking windows, creating custom desktop layouts, and setting
  preferences. Users can dock and undock windows, arrange them in different
  configurations, and save these configurations as custom layouts. Preferences
  such as font size, colour schemes, and keyboard shortcuts can be personalised
  to suit individual needs, thereby

Listing 1.8 Customising the MATLAB Desktop

```
% Change desktop layout
desktop('-layout', 'dock');
% Create a custom desktop layout
desktop.layout.LayoutConfig('Layout Name', 'Custom');
```

- **Preferences and Settings**

  MATLAB offers a wide range of **preferences and settings** to tailor the environment to individual needs and preferences. This subsection will explore:

  - **General Preferences**: Configuring the MATLAB environment, including font
    size, colour scheme, and keyboard shortcuts.
  - **Language-Specific Preferences**: Setting preferences for the MATLAB lan-
    guage, such as indentation, code folding, and auto-completion.
  - **Hardware and Performance Settings**: Optimising MATLAB's performance
    by configuring memory usage, parallel computing, and GPU acceleration.

Listing 1.9 Accessing MATLAB Preferences

```
% Open MATLAB Preferences
preferences;

% Set a specific preference programmatically
setpref('EditorPage', 'DefaultFontSize', 14);
```

- **Platform-Specific Considerations**

  MATLAB supports various platforms, including Windows, macOS, and Linux.
  While the core functionality remains consistent across platforms, there are some
  platform-specific considerations to keep in mind:

  - **Installation and Licensing**: Platform-specific installation procedures and
    licensing requirements.
  - **System Integration**: Integrating MATLAB with platform-specific tools and
    utilities, such as system libraries and compilers.
  - **Performance Optimisation**: Platform-specific performance tuning and optimi-
    sation techniques.

Listing 1.10 Checking the Current Platform

```
% Get the current platform
platform = computer;

% Perform platform-specific actions
if strcmp(platform, 'PCWIN64')
% Windows-specific code
elseif strcmp(platform, 'GLNXA64')
% Linux-specific code
elseif strcmp(platform, 'MACI64')
% macOS-specific code
end
```

- **Command Window**

  The Command Window is a vital component of the MATLAB Desktop, allowing users to interact with MATLAB by entering commands, executing code, and
  viewing results. This subsection will cover:

  - **Command Syntax**: Explanation of the syntax for entering commands, including
    the use of semicolons, newlines, and comment lines.
  - **Command History**: Techniques for recalling and re-executing previous com-
    mands using the command history.
  - **Formatting Output**: Methods for controlling the formatting of output in the
    Command Window.

Listing 1.11 Using the Command Window

```
% Enter a command and display output
x = 1:10;
disp(x)

% Suppress output using a semicolon
y = 2:2:20;

% Recall and execute a previous command
hist % View command history
rehash 5 % Execute the 5th command in the history
```

- **Editor**

  The MATLAB Editor is a powerful tool for creating, editing, and debugging MAT-LAB code, as shown in Fig. 1.3. This section will cover:

  - **Creating and Opening Files**: Techniques for creating new MATLAB files (scripts, functions, classes) and opening existing files.
  - **Code Editing Features**: An overview of code editing features such as syntax highlighting, code folding, auto-indentation, and code completion.
  - **Debugging Tools**: Explanation of debugging tools available within the Editor, including setting breakpoints, stepping through code, and monitoring variables.

Listing 1.12 Debugging in the MATLAB Editor

```
% Set a breakpoint
editor.breakpoint('set', 'file.m', 10)
```

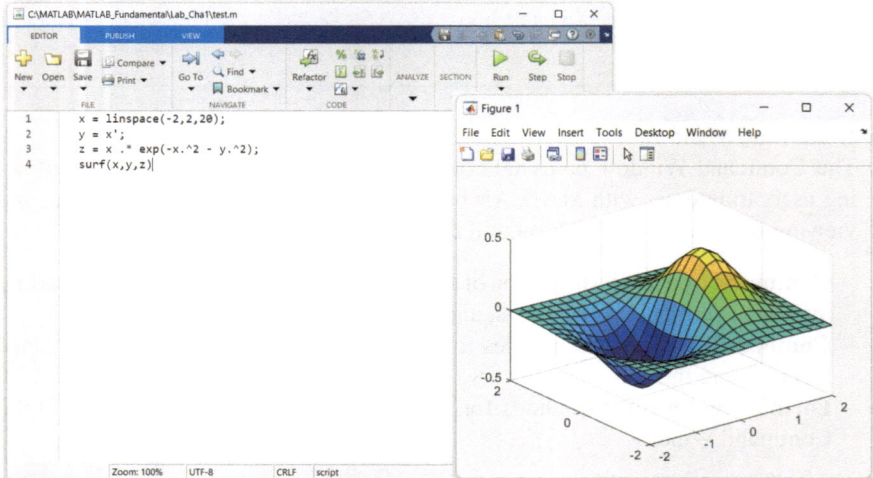

Fig. 1.3 MATLAB online with a MathWorks account

```
% Step through code
editor.stepInto()
editor.stepOver()

% Monitor variables
editor.addWatchpoint('x')
```

- **Live Editor**

  The Live Editor provides a versatile and user-friendly environment for developing and presenting computational narratives. By combining executable code, rich text formatting, and inline visualisations, Live Scripts facilitate the creation of interactive and self-explanatory documents. This enables users to effectively communicate their ideas, share results, and collaborate with others, as shown in Fig. 1.4.

  One of the key advantages of the Live Editor is its ability to support **interactive execution**. Users can execute code sections individually or run the entire script, allowing for incremental development and immediate feedback. This feature enhances the exploratory nature of MATLAB programming and enables users to quickly iterate on their code and see the results in real-time.

  In addition to its interactive capabilities, the Live Editor also provides seamless **exporting and sharing** options. Live Scripts can be exported to various formats, including HTML, PDF, and LaTeX, making it easy to share the documents with colleagues, incorporate them into reports or presentations, or publish them online. This flexibility ensures that the insights and results captured in Live Scripts can be effectively disseminated and communicated to a wider audience.

Listing 1.13 Creating a Live Script

```
% Create a new Live Script
edit('live')

% Execute a code section
run(1) % Run the first code section
% Export to HTML
export('example.mlx', 'html')
```

1. **MATLAB Online** provides a convenient way to access the Live Editor from anywhere with an internet connection. To use MATLAB Online:

   a. Visit https://matlab.mathworks.com in a web browser.
   b. Sign in with a MathWorks account or create a new account if needed, as shown in Fig. 1.5.
   c. Once logged in, users can create, edit, and run Live Scripts directly in the browser, as shown in Fig. 1.6.

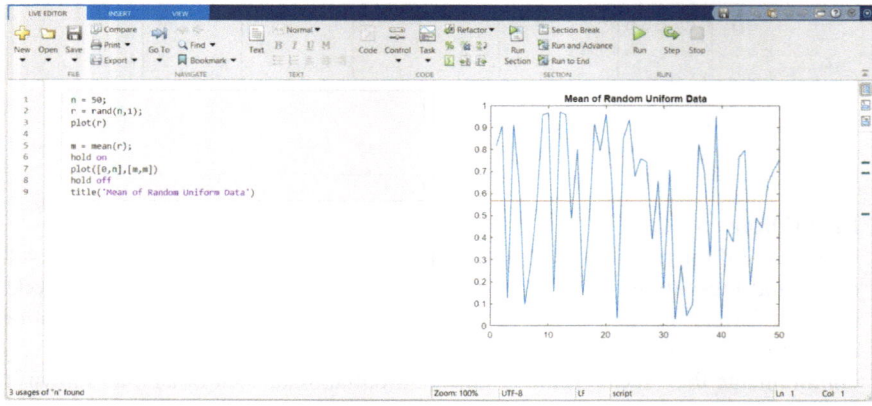

Fig. 1.4   MATLAB online main page

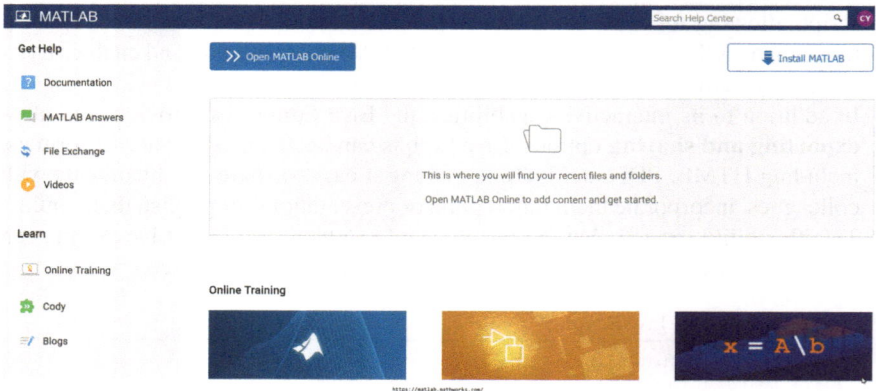

Fig. 1.5   MATLAB editor and figure

MATLAB Online offers a seamless experience, allowing users to access their files, collaborate with others, and utilise the Live Editor's features without the need for a local installation.

2. **Accessing Live Editor in Local MATLAB Installation**

For users with a local installation of MATLAB on their computer, accessing the Live Editor is straightforward [15]:

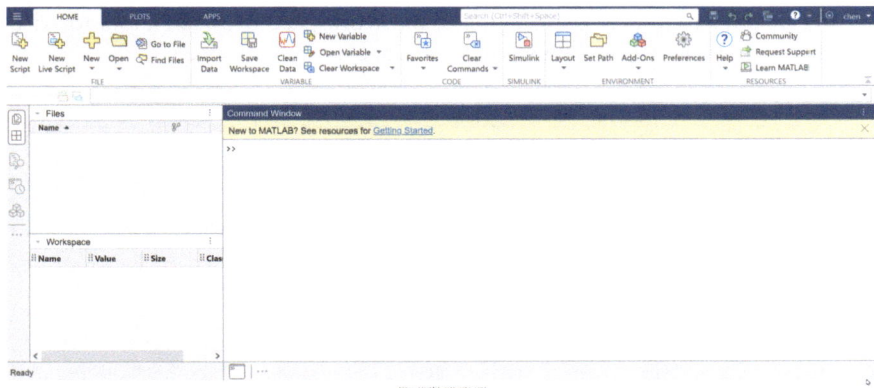

Fig. 1.6 MATLAB LiveEditor

a. Open the MATLAB software on the computer.
b. In the MATLAB toolstrip, click on the New Live Script" button or select Live Script" from the "New" menu.
c. A new Live Script will open in the MATLAB Editor, ready for editing and execution.

Local installation of MATLAB provides the full range of features and capabilities, including the Live Editor, and allows users to work offline without the need for an internet connection.

Listing 1.14 Accessing Live Editor in Local MATLAB

```
% Open a new Live Script in local MATLAB
livescript

% Alternatively, use the "New" menu
uiopen('live')
```

Regardless of whether users choose to access the Live Editor online or through a local installation, they will have access to the same powerful features and capabilities for creating interactive and engaging computational narratives.

- **Variables and Workspace**

MATLAB's workspace is a crucial component, storing variables and data structures created during a session. This section will cover:

- **Variable Types**: An overview of the different variable types in MATLAB, including arrays, structures, and objects.
- **Managing Variables**: Techniques for creating, modifying, and deleting variables in the workspace, as well as techniques for saving and loading variables to/from files.

– **Workspace Browser**: Explanation of the Workspace Browser, a tool for visualizing and interacting with variables in the current workspace.

Listing 1.15 Working with Variables

```
% Create a variable
x = 1:10;

% Save variables to a file
save('data.mat', 'x')

% Load variables from a file
load('data.mat')
```

- **Debugging**

  Debugging is an essential part of the programming process, and MATLAB provides
  powerful tools to help identify and fix errors. This section will discuss:

  – **Types of Errors and Warnings**: An overview of the different types of errors
  and warnings in MATLAB, such as syntax errors, runtime errors, and warnings.
  – **Identifying and Fixing Errors**: Techniques for identifying the cause of errors
  and warnings, including methods for tracking down and fixing bugs.
  – **Debugging Tools**: Explanation of MATLAB's debugging tools, such as setting
  breakpoints, stepping through code, monitoring variables, and inspecting the
  call stack.

- **File Types**

  MATLAB supports various file types for different purposes, such as code, data,
  models, and documentation. This subsection will provide an overview of the most
  common file types:

  – .m: MATLAB code file that contains scripts or functions. These files can include
  commands that you would run in the command window or define functions that
  can be used within other scripts or functions.
  – .mat: Binary data file that stores variables that are created during a MAT-
  LAB session. This format is specific to MATLAB and can be loaded into the
  workspace with the load function.
  – .mlx: MATLAB Live Script file that combines code, output, and formatted text
  in an interactive document. It's useful for sharing and presenting workflows,
  algorithms, and analyses.
  – .mdl: File that contains a model created with an older version of Simulink, a
  graphical programming environment for modeling, simulating, and analysing
  multidomain dynamical systems.
  – .mex: MEX-file (short for MATLAB Executable) that contains a program
  intended to be called from MATLAB that is written in C, C++, or Fortran and
  compiled into binary form. MEX-files can provide high-performance functions
  that are callable directly from MATLAB.

- .fig: MATLAB Figure file that stores the data needed to recreate a plot or graphical user interface. These files are generated by the figure window's Save option and can be reopened in MATLAB for further manipulation or editing.
- .p: MATLAB P-code file, which contains MATLAB code that has been obfuscated into a form of bytecode. P-code files can be executed just like .m files, but their contents cannot be easily viewed, providing a measure of intellectual property protection.
- .slx: File that represents a model created with the latest versions of Simulink. This format is XML-based and is the successor to the older .mdl file format.
- .mldatx: File that contains simulation data from MATLAB. It is often used with Simulink to store simulation input and output data.
- .sfx: MATLAB Shared library or toolbox file that is intended to be shared across different machines or platforms. This file type is used for add-on applications that extend MATLAB's capabilities.
- .mupad: File associated with MuPAD, which used to be MATLAB's symbolic math engine before it was replaced by MATLAB's Symbolic Math Toolbox. Although MuPAD notebooks are deprecated, some users might still encounter these files.
- .asv: MATLAB AutoSave file that is created automatically by MATLAB as a backup of an unsaved script or function file (.m). This feature helps in recovering data in case of an unexpected interruption or crash.
- .mexw64: MEX-file for 64-bit Windows operating systems. Similar to .mex files, but specifically compiled for 64-bit Windows environments.
- .mexa64: MEX-file for 64-bit Linux operating systems. Similar to .mex files, but specifically compiled for 64-bit Linux environments.
- .mexmaci64: MEX-file for 64-bit macOS. Similar to .mex files, but specifically compiled for 64-bit macOS environments.
- .mlapp: MATLAB App Designer file, which is used for building MATLAB apps with the App Designer tool. These can include both the design and code of the app.
- .mlappinstall: MATLAB App installation file, which is created when you package an app using MATLAB App Designer. This file type is used for distributing or sharing MATLAB apps.
- .mltbx: MATLAB Toolbox installation file. It is a package of MATLAB code files, data files, apps, examples, and documentation, used to share or distribute toolboxes.
- .mlprj: MATLAB Project file used by MATLAB's Project tool to manage and share all the files associated with a project, track their status, and integrate with source control systems.
- .mn: MATLAB Notebook file, which is an interactive document that can integrate MATLAB code, output, and formatted text in a notebook interface (primarily used in earlier versions of MATLAB).
- .rpt: MATLAB Report Generator file, which contains information to generate reports from MATLAB applications and can include code, results, graphs, and formatted text.

- .cdf: Common Data Format file, which is used for storing multidimensional data. MATLAB can import and export data in CDF format using the appropriate functions.
- .nc: NetCDF (Network Common Data Form) file, which is a set of software libraries and machine-independent data formats that support the creation, access, and sharing of array-oriented scientific data.
- .hdf, .h5, .hdf5: Hierarchical Data Format files, which are designed to store and organise large amounts of data. MATLAB supports reading from and writing to these formats, often used in high-performance computing and scientific research.
- .tgz, .zip: Compressed file archives that MATLAB can create or extract. These formats are used to package multiple files into a single archive that's easier to distribute or transfer.
- .java: Java class files that can be called from within MATLAB. MATLAB integrates with Java, allowing users to write Java code and create Java objects within the MATLAB environment.
- .jar: Java ARchive files that can be added to MATLAB's Java classpath, enabling the use of the Java classes they contain from within MATLAB.
- .csv: Comma-Separated Values file, a common data exchange format that can be read into and written from MATLAB as a table or matrix.
- .txt, .dat: Text or data files that contain plain text and can be imported into MATLAB. They are often used to store numerical data in a simple format that can be read by various software programs.
- .xml: Extensible Markup Language file, which is a markup language that defines a set of rules for encoding documents in a format that is both human-readable and machine-readable. MATLAB can read and write XML files.
- .json: JavaScript Object Notation file, a lightweight data-interchange format that is easy for humans to read and write. MATLAB provides functions to encode and decode JSON data.
- .ini: Initialisation file, a configuration file for Windows programs. MATLAB can read and write INI files, though it's not a common practice for MATLAB-specific applications.
- .xlsx, .xls: Excel spreadsheet files. MATLAB can read from and write to these files directly, allowing for data exchange between MATLAB and Microsoft Excel.
- .sim: Simulink model file used in older versions of Simulink prior to the introduction of the .slx format.
- .sldd: Simulink Data Dictionary file, which stores design data, such as definitions of bus objects and data type objects, separately from Simulink models.
- .slxp: Simulink protected model file, which is a compiled version of a Simulink model that protects the intellectual property of the model's design.
- .req: Requirements file, which can be used in conjunction with MATLAB's Requirements Management Interface to link requirements to MATLAB models and code.

- .sldvdata: Simulink Design Verifier data file, which contains analysis results from the Simulink Design Verifier tool used for formal verification and validation of Simulink models.
- .sfx: MATLAB Shared library or toolbox file that is intended to be shared across different machines or platforms. This file type is used for add-on applications that extend MATLAB's capabilities.

- **Help and Documentation** MATLAB provides extensive help and documentation resources to assist users in learning and troubleshooting. This subsection will cover:

  - **Accessing Help**: Methods for accessing MATLAB's built-in help system, including keyword searches, browsing by topic, and accessing demo examples.
  - **Online Resources**: An overview of online resources, such as the MATLAB documentation, community forums, and file exchange.
  - **Contextual Help**: Techniques for accessing contextual help within the MATLAB environment, such as using the "Help" button in dialog boxes or the "Help on Selection" feature in the Editor.

Listing 1.16 Accessing MATLAB Help

```
% Search for help on a topic
doc mean

% Open the MATLAB documentation browser
doc

% Access contextual help
help editor % Help on the Editor
```

- **MATLAB Grader**
  The MATLAB Grader is a tool for automatically grading MATLAB assignments and providing feedback to students, as shown in Fig. 1.7. This subsection will provide an overview of its features and capabilities.

  - **Automated Grading**: Explanation of how the MATLAB Grader can automatically grade MATLAB assignments by running test cases and comparing outputs.
  - **Feedback and Reporting**: Discussion of the feedback and reporting capabilities of the MATLAB Grader, including the generation of detailed reports for students and instructors.
  - **Integration with Learning Environments**: Overview of how the MATLAB Grader can be integrated with learning management systems (LMS) and other educational platforms.

- **Startup and Shutdown** Startup command line flags, startup and shutdown files
- **Add-Ons** Find, run, and install add-ons, including optional features, apps, toolboxes, and support packages

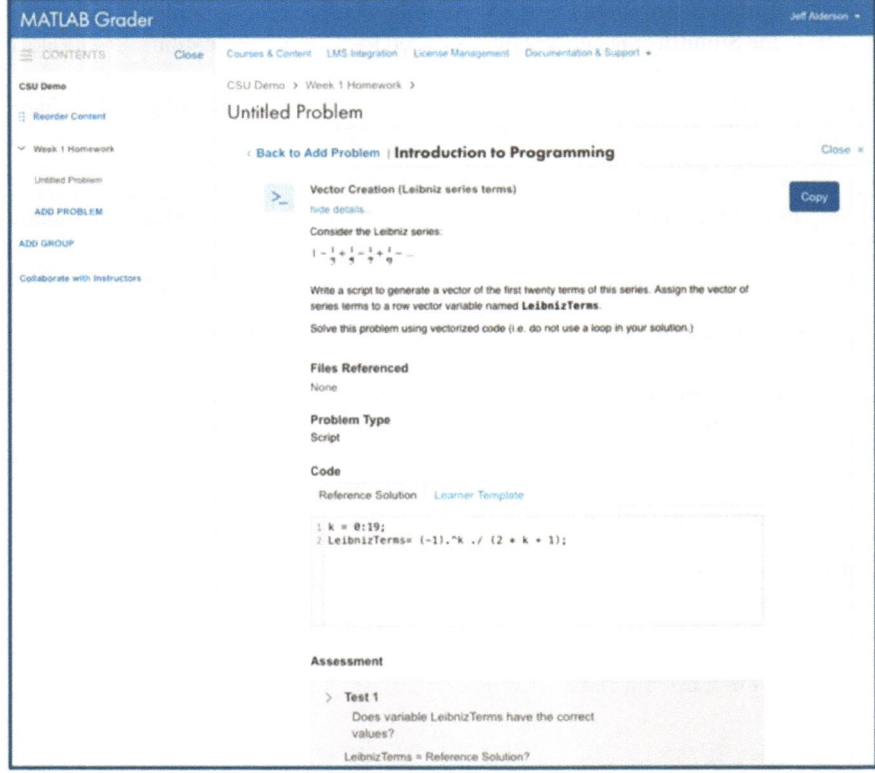

Fig. 1.7 MATLAB grader

- **System Commands** Interact programmatically with operating system and MAT-LAB environment
- **Internationalisation** Locale settings and messages.

1.10   MATLAB Basic Concepts

- **Current Directory**:
  - The **current directory** refers to the folder location where MATLAB is currently operating.
  - It is the default folder where MATLAB looks for files and where it saves newly created files.
  - **Example:** `cd` command to see current directory

Listing 1.17 Displaying the current directory.

```
cd
```

• **Defined Path**:

– The defined path in MATLAB is a set of folders that MATLAB searches for files and functions.

– By adding folders to the defined path, you can access files and functions from any location without specifying the full file path.

– The `addpath` function is used to add a folder to the defined path, and the `savepath` function saves the current path for future MATLAB sessions.

Listing 1.18 Adding a folder to the defined path

```
addpath('C:\My_MATLAB_Files')
savepath
```

• **Saving Files**:

– MATLAB allows you to save your work as scripts, functions, or data files.

– MATLAB code can be saved in script files with a .m extension, while data can be saved in various formats such as .mat, .txt, or .csv. The `save` function is used to save data to a file.

– **Example:** `save filename.mat` to save .mat file

Listing 1.19 Saving data to a .mat file

```
save('data.mat', 'variable1', 'variable2')
```

• **Loading Files**:

– **Loading files** in MATLAB involves reading data from external files into the MATLAB workspace.

– MATLAB provides functions to load different types of files such as load for .mat files, readtable for .txt or .csv files, and imread for image files.

– **Example:** `load('data.mat')` to load .mat file

Listing 1.20 Loading data from a .mat file

```
data = load('data.mat')
```

• **File Naming Constraints**:

MATLAB has certain **naming constraints** for files and variables. File names must start with a letter and can contain letters, numbers, or underscores. Variable names must also start with a letter, can contain letters, numbers, or underscores, and must not be a keyword or function name. When working with MATLAB, it is crucial to adhere to specific file naming constraints to ensure compatibility and avoid errors. This section outlines the primary constraints and best practices for naming files.

- **Length Limitations** The maximum length of a file name is 63 characters. Exceeding this limit may result in errors or unexpected behaviour.

Listing 1.21 File name length limit

```
% Avoid excessively long file names
thisIsAVeryLongFileName.m   % May exceed the length
    limit
```

- **Valid Characters**

    · **Start with a Letter**: MATLAB file names must begin with a letter (A-Z, a-z).
    · **Followed by Letters, Digits, or Underscores**: After the initial letter, the file name can include letters, digits (0-9), and underscores (_).
    · **No Special Characters**: Spaces, punctuation marks, and non-ASCII characters are not allowed in file names. For example, `my_script.m` is valid, but `my script!.m` is not.

Listing 1.22 Valid file names

```
% Valid file names
myFile.m
data_analysis.m
project1.mat
my_Script1.m
```

- **Case Sensitivity:** MATLAB is case-insensitive on Windows, meaning `MyScript.m` and `myscript.m` are considered different files, but case-sensitive on UNIX and Linux systems. It is advisable to maintain consistent casing conventions to avoid confusion.

Listing 1.23 Case-sensitive file names

```
% Case-sensitive file names
myFile.m
myfile.m   % Considered a different file
```

- **Reserved Words:** File names should not be the same as MATLAB reserved words, such as `for`, `while`, `if`, etc. Avoid using names that conflict with MATLAB built-in functions or keywords. For instance, `sum.m` would conflict with MATLAB's sum function, leading to potential issues during execution. For example, using names like 'plot.m' or 'for.m' can lead to unexpected behaviour.
- **Spaces:** File names should not contain spaces. If spaces are necessary, they can be replaced with underscores (_) or hyphens (-).
- **Extension:** File names for MATLAB functions and scripts should end with the `.m` extension. For example, `myFunction.m` is a valid script file name.
- **No Leading Dots:** File names should not start with a dot (.), as this is reserved for hidden files in UNIX and Linux systems.
- **No Special Characters:** Avoid using characters like < > ? * | as they can cause issues with file operations in different operating systems.

- **No Trailing Periods:** File names should not end with a period, as this can be interpreted as an extension.

To illustrate these constraints, consider the following MATLAB code snippet that demonstrates valid and invalid file names:

Listing 1.24 Examples of Valid and Invalid File Names

```
% Valid file names
validFileName1 = 'myScript1.m';
validFileName2 = 'data_analysis.m';

% Invalid file names
invalidFileName1 = 'my script!.m'; % Contains space
    and exclamation mark
invalidFileName2 = '123file.m';      % Starts with a
    digit
invalidFileName3 = 'sum.m';          % Conflicts with
    built-in function
```

By adhering to these file naming constraints, users can ensure that their MATLAB scripts and functions are easily identifiable, maintainable, and free from naming conflicts.

- **Variables**:

  - Variables in MATLAB are used to store and manipulate data.
  - They can be assigned values of different data types such as numbers, strings, arrays, or structures.
  - Variables are dynamically typed, meaning their data type can change during program execution.
  - **Example:** $x = 5$; to assign a value to variable x

- **Operators**:

  - MATLAB supports various operators such as arithmetic, relational, logical, and assignment operators.
  - Arithmetic operators perform mathematical calculations, relational operators compare values, logical operators perform Boolean operations, and assignment operators assign values to variables.
  - **Example:** $a + b, a == b, a \& b, a = 5$

- **Expressions**:

  - In MATLAB, expressions are combinations of variables, operators, and functions that are evaluated to produce a result.
  - Expressions can be simple arithmetic calculations or complex mathematical operations involving arrays and matrices.
  - **Example:** $3 + 4, \sin(x), A*x$

- **Arrays**:
  - Arrays in MATLAB are used to store collections of data.
  - They can be one-dimensional (vectors), two-dimensional (matrices), or multi-dimensional.
  - MATLAB provides powerful built-in functions for creating, accessing, and manipulating arrays.
  - **Example:** `a = [1 2 3], A = rand(3,5)`

- **Matrix Manipulations**:
  - MATLAB is renowned for its matrix manipulation capabilities.
  - It provides numerous functions for matrix operations such as addition, subtraction, multiplication, inversion, transposition, and solving linear equations.
  - **Example:** `A*B, inv(A), A'`

- **Data Types**:
  - MATLAB supports various data types including numeric, character, logical, cell, and structure arrays.
  - Each data type has its own properties and functions for manipulation.
  - **Example:** `double, char, logical, cell, struct`

- **Arithmetic Operations**:
  - MATLAB allows performing arithmetic operations on variables and arrays.
  - These operations include addition, subtraction, multiplication, division, exponentiation, and more.
  - **Example:** `a + b, a - b, a * b, a / b, a^2`

- **Built-in Functions**:
  - MATLAB offers a vast collection of built-in functions for a wide range of applications.
  - These functions perform tasks such as mathematical calculations, data analysis, signal processing, image processing, and more.
  - **Example:** sum, mean, fft, image

- **Plotting**:
  - MATLAB provides powerful tools for creating plots and visualisations.
  - It provides functions for creating 2D and 3D plots, histograms, scatter plots, bar graphs, and more.
  - Plots can be customized with various options such as colours, markers, line styles, and labels.
  - **Example:**

Listing 1.25 Plotting

```
x = 0:0.1:2*pi;
y = sin(x);
plot(x, y, 'r-o') % red line with circle markers
```

```
title('Sine Wave')
xlabel('x')
ylabel('sin(x)')
```

- **Programming Constructs**:

  - MATLAB supports various programming constructs such as conditional statements (if-else), loops (for and while), switch-case statements, and function definitions.
  - These constructs enable you to write structured and modular MATLAB code.
  - **Example:**

Listing 1.26 Programming Constructs

```
if x > 0
disp('Positive')
else
disp('Non-positive')
end
```

- **Control Flow**:

  - Control flow refers to the order in which statements and instructions are executed in a program.
  - MATLAB provides various control flow structures such as loops (for, while), conditional statements (if-else, switch), and function calls.
  - These structures allow you to control the flow of execution based on certain conditions or for repetitive tasks.
  - **Example:**

Listing 1.27 Display index number through for-loop

```
for i = 1:10
disp(i)
end
```

- **Functions**:

  - Functions in MATLAB are reusable blocks of code that perform a specific task.
  - They allow you to modularize your code and make it more organised and easier to maintain.
  - MATLAB provides built-in functions as well as the ability to create user-defined functions.
  - **Example:**

Listing 1.28 Function example

```
function y = square(x)
y = x^2;
end
```

- **File Input/Output**:

  – File input/output (I/O) refers to reading data from files and writing data to files.
  – MATLAB provides functions for reading and writing data in various formats such as text files, spreadsheets, images, and sound files.
  – This allows you to work with external data and save your results for future use.
  – **Example:**

Listing 1.29 Load and Write Data

```
% Read data from text file
data = load('data.txt');

% Write matrix to Excel file
xlswrite('output.xlsx', A)

% Save plot as image
print('figure.png')
```

- **Debugging and Error Handling**:

  – Debugging is the process of finding and fixing errors or bugs in your code.
  – MATLAB provides tools and techniques for debugging, such as setting breakpoints, stepping through code, and inspecting variables.
  – Error handling involves anticipating and handling errors that may occur during program execution.
  – MATLAB provides mechanisms for catching and handling errors, ensuring that your program continues to run smoothly.
  – **Example:**

Listing 1.30 Try to capture error

```
try
error_prone_code
catch err
fprintf('Error: %s\n', err.message)
fprintf 'Error identifier:` %d\n', err.identifier
end
```

- **Symbolic Math**:

  – Symbolic math in MATLAB allows you to work with mathematical expressions symbolically, rather than numerically.
  – You can perform operations such as differentiation, integration, simplification, and equation solving symbolically.
  – This is useful in fields such as mathematics, engineering, and physics, where exact symbolic solutions are desired.
  – **Example:**

Listing 1.31 Symbolic example

```
syms x
f = x^2 + 2*x + 1;
diff(f) % Differentiate f
int(f) % Integrate f
```

- **Toolboxes and Applications**:
  - MATLAB offers a wide range of toolboxes and applications for specific domains and applications.
  - These toolboxes provide additional functions and features tailored to specific fields such as image processing, control systems, optimisation, and more.
  - Understanding how to use and leverage these toolboxes can greatly enhance your MATLAB skills and expand your capabilities.
  - **Example:**

Listing 1.32 Read and process image

```
image = imread('image.png');
edges = edge(image, 'canny'); % Image Processing
    Toolbox

sys = tf(1, [1 2 1]);
step(sys) % Control System Toolbox
```

- **Logical Operations**

  Logical operations in MATLAB allow evaluating conditions and performing operations based on logical values (true or false). Logical operators such as **AND** (&&), **OR** (||), and **NOT** ($\sim$) are used to combine or negate logical conditions. Logical operations are often used in conditional statements and for selecting elements from arrays based on specific criteria:

  - **AND** (&&): The && operator performs a logical **AND** operation. It returns `true` if both operands are true; otherwise, it returns `false`. This operator is used for short-circuit evaluation, meaning the second operand is evaluated only if the first operand is true.
  - **OR** (||): The || operator performs a logical **OR** operation. It returns `true` if at least one of the operands is true; otherwise, it returns `false`. Similar to &&, this operator also uses short-circuit evaluation, where the second operand is evaluated only if the first operand is false.
  - **NOT** ($\sim$): The $\sim$ operator performs a logical **NOT** operation. It inverts the logical state of its operand. If the operand is `true`, the result is `false`, and vice versa.

Listing 1.33 Logical operations in MATLAB.

```
x = 5;
y = 10;
if x > 0 && y > 0
disp('Both x and y are positive')
```

```
end

A = [1 2; 3 4];
B = A > 2; % Logical indexing
```

For example:

```
The logical expression $a && b$ represents the AND operation between
$a$ and $b$.
The logical expression $a \lor (\sim b)$ represents the OR operation
between $a$ and the negation of $b$.
```

- **String Manipulation** MATLAB provides functions for **string manipulation**, allowing you to work with and modify character arrays and strings [2]. Common string operations include concatenation, substring extraction, searching, replacing, and converting between character arrays and strings:

Listing 1.34 String manipulation in MATLAB.

```
str1 = 'Hello';
str2 = 'World';
concatenated = [str1, ' ', str2]; % Concatenation
substring = str1(1:3); % Substring extraction
found = contains(str1, 'el'); % Searching for a
    substring
```

- **Cell Arrays** **Cell arrays** in MATLAB are containers that can hold elements of different data types and sizes . They provide a flexible way to store and organize heterogeneous data. Cell arrays are created using curly braces and can be accessed and manipulated using indexing:

Listing 1.35 Cell arrays in MATLAB.

```
cell_array = {'Hello', 42, [1 2; 3 4]};
element = cell_array{1}; % Accessing elements
cell_array{2} = 'World'; % Modifying elements
```

- **Structures** **Structures** in MATLAB are data types that group related data using named fields [2]. They allow organizing and accessing data in a more meaningful and intuitive way. Structures are created using the struct() function or by directly assigning values to named fields:

Listing 1.36 Structures in MATLAB.

```
student.name = 'John';
student.age = 20;
student.grades = [85, 90, 92];
name = student.name; % Accessing fields
```

1.11 Laboratory

1. **Installing and Launching MATLAB**

   a. Download and install **MATLAB** on your computer.
   b. Launch **MATLAB** and explore the desktop environment.

   Solution:

   a. Follow the installation instructions provided by MathWorks for your specific operating system.
   b. After installation, launch **MATLAB** from the Start menu (Windows) or Applications folder (macOS/Linux).
   c. Familiarise yourself with the various components of the desktop environment, such as the **Command Window, Workspace, Editor**, and others.

2. **Get familiar with MATLAB Environment, create and view a matrix**

   a. Create a matrix **A** with 2 rows and 3 columns.
   b. Locate the **MATLAB Workspace** and double-click to view **A**.

   Solution:

   a. In the **Command Window**, enter the following command to create a 2x3 matrix:

Listing 1.37 Creating a matrix.

```
A = [1 2 3; 4 5 6]
```

   b. In the **Workspace** window, locate the variable **A** and double-click on it to view its contents.

3. **Working with the Command Window**

   a. Open the **Command Window** in **MATLAB**.
   b. Perform basic arithmetic operations (addition, subtraction, multiplication, division) using the **Command Window**.
   c. Assign values to variables and display their contents.

   Solution:

   a. To open the **Command Window**, click on the "Command Window" icon in the **MATLAB** desktop environment or press the "Ctrl+1" keyboard shortcut.
   b. Enter arithmetic expressions like "2 + 3" or "4 * 5" and press Enter to see the results.
   c. Assign values to variables using the assignment operator (=). For example, "x = 10" assigns the value 10 to the variable x. Display the variable's contents by typing its name and pressing Enter.

4. **Familiarising with Command Window, "doc" Function, and Figure Operations**

In this lab work, you will become familiar with the **Command Window**, the **"doc"** function, and various operations related to **figures** in **MATLAB**. Please follow the step-by-step guide below:

a. On the command line, create an array **"x"** with the natural numbers from 1 to 100.

The basic syntax to create the array **"x"** is:

Listing 1.38 Creating array x.

```
x = 1:100;
```

b. Create an array **"y"** where each element is twice the corresponding element in **"x"** (i.e., y = 2*x).

The basic syntax to create the array **"y"** is:

Listing 1.39 Creating array y.

```
y = 2*x;
```

c. Use the **"doc plot"** command on the command line to access the documentation and learn how the **plot** function is used.

On the command line, enter the following command:

Listing 1.40 Accessing plot function documentation.

```
doc plot
```

This will open the **MATLAB** documentation for the **plot** function, providing information on its usage and options.

d. Draw a **figure** using **"x"** as the horizontal coordinate and **"y"** as the vertical coordinate.

The basic syntax to draw the **figure** is:

Listing 1.41 Drawing the figure.

```
plot(x, y);
```

This will create a plot with **"x"** as the horizontal axis and **"y"** as the vertical axis.

e. In the **Figure** interface, adjust the **LineWidth** of the curve to 2 by selecting **"Edit" -> "Current Object Properties"**.
   Follow these steps:
   i. Click on the curve of the image in the **Figure** window.
   ii. Navigate to the **"Edit"** menu.
   iii. Select **"Current Object Properties"**.
   iv. Adjust the **LineWidth** to 2.
   This will change the thickness of the curve to 2.
f. In the command history, select the commands used in the previous steps, copy them to a new **MATLAB** script (m-script), and save the script.

Below is the complete solution for your reference:

Listing 1.42 Complete solution.

```
% Lab Work 4: Familiarizing with Command Window, "doc"
    Function, and Figure Operations

% Step 1: Create the array x with natural numbers from 1
    to 100
x = 1:100;

% Step 2: Create the array y by multiplying each element
    of x by 2
y = 2*x;

% Step 3: Access the documentation for the plot function
doc plot;

% Step 4: Draw the figure using plot function
plot(x, y);

% Step 5: Adjust the LineWidth of the curve to 2 in the
    Figure interface

% Step 6: Copy the commands to a new MATLAB script and
    save the script
```

Make sure to follow the steps carefully and review the provided solution. This lab work will help you become more comfortable with the **Command Window**, the **"doc"** function, and **figure** operations in **MATLAB**.

5. **Creating and Running a Script**

Solution:

a. Open a new script file in the **MATLAB Editor**.
b. Write a simple script that performs a series of calculations and displays the results.
c. Save and run the script.

Solution:

a. To open a new script file, click on the "New Script" icon in the **MATLAB** desktop environment or press the "Ctrl+N" keyboard shortcut.
b. Write a series of **MATLAB** statements in the script, such as variable assignments, arithmetic operations, and output statements using the `disp()` function.
c. Save the script with a .m extension (e.g., myScript.m).
d. Run the script by clicking the "Run" icon in the Editor or by pressing the "F5" key.

6. **Importing and Exporting Data**

   a. Import a sample data file (e.g., a CSV file) into **MATLAB** using the appropriate function.
   b. Perform some basic data manipulation or analysis on the imported data.
   c. Export the processed data to a new file format (e.g., Excel spreadsheet).

Solution:

a. Use the `readtable()` or `readmatrix()` function to import a CSV file into **MATLAB**.

   The basic syntax to import a CSV file is:

Listing 1.43 Importing a CSV file.

```
data = readtable('filename.csv');
```

b. Perform data manipulation or analysis operations on the imported data, such as filtering, sorting, or computing summary statistics.
c. Use the `writematrix()` or `writetable()` function to export the processed data to a new file format like an Excel spreadsheet.

   The basic syntax to export data to an Excel file is:

Listing 1.44 Exporting data to Excel.

```
writetable(data, 'output.xlsx');
```

7. **Exploring Built-in Functions and Documentation**

   a. Explore **MATLAB's** built-in functions by using the **help** and **doc** commands.
   b. Find and read the documentation for a specific function (e.g., `sin()`, `plot()`, or any other function of your choice).
   c. Use the function in a simple script or command to observe its functionality.

Solution:

a. In the **Command Window**, type "help" followed by a function name (e.g., "help sin") to get a brief description and syntax of the function.

b. Type "doc" followed by a function name (e.g., "doc sin") to open the full documentation for that function in the **MATLAB Help Browser**.

c. After reading the documentation, use the function in a script or command to observe its functionality. For example, create a script that generates a sine wave plot using the `sin()` and `plot()` functions.

The basic syntax to plot a sine wave is:

Listing 1.45 Plotting a sine wave.

```
x = 0:0.1:10;
y = sin(x);
plot(x, y);
```

8. **Get familiar with creating 'favorite' folder**

   Step 1: Based on Lab 1.2, input `close all` in the command line to observe the change of the **Figure** window.

   Step 2: Enter `clear` in the command line and observe the change of variables in the **Workspace**.

   Step 3: Enter `clc` in the command line and observe the change of the **Command Window**.

   Step 4: In the home page -> Favorites, create a new favorite item (the label is set to "clear up"), and enter the above commands into the code area, and save.

   Step 5: After repeating Lab 1.2, click "clear up" in the favorites and observe the changes of **Figure**, **Workspace**, and **Command Window** again.

9. **To solve the system of equations:**

$$
\begin{aligned}
2x + 3y - 6z - 12w &= 4 \\
5x - 7y + 4z + 2w &= -3 \\
x + 8z - 2w &= 9 \\
-6x + 5y - 4z + 10w &= -8
\end{aligned}
\tag{1.1}
$$

We can use the `syms` function in **MATLAB** to define the variables and the `solve` function to find the solution:

Listing 1.46 Solving the system of equations.

```
syms x y z w

eq1 = 2x + 3y - 6z - 12w == 4;
eq2 = 5x - 7y + 4z + 2w == -3;
eq3 = x + 8z - 2w == 9;
eq4 = -6x + 5y - 4z + 10w == -8;

sol = solve([eq1, eq2, eq3, eq4], [x, y, z, w]);

sol.x
sol.y
sol.z
sol.w
```

The solution to the system of equations is:

$$x = -\frac{47}{97}$$
$$y = \frac{2}{97}$$
$$z = \frac{349}{291}$$
$$w = -\frac{136}{291}$$

(1.2)

Therefore, the values of **x**, **y**, **z**, and **w** that satisfy the equations are as shown above.

1.12 Problems

1. **Creating a Simple Calculator**

   - Write a MATLAB script that prompts the user to enter two numbers and an operation (addition, subtraction, multiplication, or division).
   - Perform the requested operation on the two numbers and display the result.
   - Include error handling to gracefully handle invalid input or division by zero.

2. **Data Analysis and Visualisation**

   - Load a dataset from a provided file (e.g., a CSV file containing weather data or stock prices).
   - Perform data cleaning and preprocessing steps as necessary (e.g., handling missing values, removing outliers).
   - Analyse the data by computing summary statistics and visualizing the results using appropriate plots (e.g., line plots, histograms, scatter plots).

3. **Implementing a Simple Algorithm**

   - Implement a sorting algorithm (e.g., bubble sort, insertion sort) in MATLAB as a function.
   - Write a script that generates a random array of numbers and calls your sorting function to sort the array.
   - Verify the correctness of your implementation by comparing the sorted array with the expected output.

4. **Solving Systems of Equations**

   - Write a MATLAB function that takes a system of linear equations as input (in the form of coefficient matrices and constant vectors).

- Use MATLAB's built-in functions to solve the system of equations and return the solution vector.
- Test your function with multiple sets of linear equations, including cases with unique solutions, no solutions, and infinitely many solutions.

5. **Creating a Simple Game**

- Design and implement a simple game using MATLAB's graphical capabilities (e.g., a number guessing game, a simple version of Tic-Tac-Toe or Hangman).
- Create a graphical user interface (GUI) for the game, with components for user input, displaying game state, and providing feedback.
- Implement the game logic and rules within MATLAB functions and callbacks.

1.13 Summary

This chapter provided a comprehensive introduction to the MATLAB environment, covering various aspects essential for effective use and understanding of this powerful computational tool. The key points discussed in this chapter are summarised below:

- **MATLAB Overview**: MATLAB is a high-level programming language and numerical computing environment widely used in academia, research, and industry for data analysis, algorithm development, and visualisation.
- **MATLAB Interface**: The MATLAB interface consists of several components, including the Command Window, Editor, Workspace, and various toolboxes and apps. Understanding the functionality and purpose of each component is crucial for efficient workflow.
- **Data Types and Variables**: MATLAB supports various data types, such as scalars, vectors, matrices, and structures. Variables in MATLAB do not require explicit declaration of data types, allowing for flexible and dynamic programming.
- **Array Indexing and Operations**: MATLAB provides powerful array indexing and manipulation capabilities, enabling efficient handling of large datasets and matrix operations, which are fundamental in many scientific and engineering applications.
- **Plotting and Visualisation**: MATLAB offers extensive plotting and visualisation tools, allowing users to create high-quality 2D and 3D plots, charts, and graphical user interfaces (GUIs) for effective data representation and analysis.
- **Programming Constructs**: MATLAB supports various programming constructs, such as loops, conditional statements, and functions, enabling the development of complex algorithms and applications.
- **File Types and Data Import/Export**: MATLAB supports a wide range of file types for code, data, models, and documentation, facilitating seamless integration with other software and data sources.

- **Help and Documentation**: MATLAB provides extensive help and documentation resources, including built-in help, online resources, and contextual help, empowering users to learn and troubleshoot effectively.
- **MATLAB Grader**: The MATLAB Grader is a powerful tool for automatically grading MATLAB assignments and providing feedback to students, making it an invaluable resource in educational settings.

For undergraduate students, this chapter serves as a foundational introduction to MATLAB, equipping them with the essential skills and knowledge required for various academic and research endeavors. By understanding the core concepts, data types, programming constructs, and visualisation capabilities of MATLAB, students can effectively utilise this powerful tool for numerical computations, data analysis, and problem-solving in their respective fields of study.

Postgraduate students and researchers will find this chapter beneficial for its comprehensive coverage of MATLAB's advanced features and functionalities. The in-depth discussion on array operations, toolboxes, and integration with other software and data sources empowers them to tackle complex research problems, develop sophisticated algorithms, and conduct data-driven analyses across various domains, such as engineering, sciences, and computational fields.

For professional engineers and practitioners, this chapter serves as a valuable resource for leveraging MATLAB's powerful capabilities in industry and real-world applications. The sections on graphical user interfaces (GUIs), file handling, and integration with other software systems provide a solid foundation for developing user-friendly applications, automating processes, and streamlining workflows. Additionally, the extensive help and documentation resources ensure that professionals can effectively navigate and utilise MATLAB's vast array of features and toolboxes for their specific domains and projects.

By combining theoretical concepts with practical examples, exercises, and real-world applications, this chapter equips readers from diverse backgrounds with the knowledge and skills necessary to harness the full potential of MATLAB, enabling them to solve complex problems, Analyse and visualise data, and drive innovation in their respective fields.

References

1. MathWorks, "MATLAB Product Family," https://www.mathworks.com/products.html
2. MathWorks, "MATLAB Fundamentals," [Online]. Available: https://www.mathworks.com/help/matlab/, accessed on Feb. 17, 2024
3. Hanselman D, Littlefield B (2003) Mastering MATLAB 7. Pearson Education
4. Cleve M, "A Brief History of MATLAB," [Online]. Available: https://www.mathworks.com/company/newsletters/articles/a-brief-history-of-matlab.html, accessed on Feb. 17, 2024
5. MathWorks, "Founders," [Online]. Available: https://uk.mathworks.com/company/aboutus/founders/clevemoler.html, accessed on Feb. 17, 2024
6. MathWorks, "Founders–Jack Little," [Online]. Available: https://uk.mathworks.com/company/aboutus/founders/jacklittle.html, accessed on Feb. 17, 2024

7. Moler C, Little J (2020) A history of MATLAB. In: Proc. ACM Program. Lang., vcl. 4, no. HOPL, pp 1–67, Art. no. 81. [Online]. Available: https://dl.acm.org/doi/pdf/10.1145/3386331

8. Wilkinson JH, Reinsch C (1965) Handbook for automatic computation, volume II: linear algebra. Springer-Verlag, Berlin

9. Moler C, Numerical Linear Algebra Software Development, [Online]. Available: https://www.mathworks.com/company/newsletters/articles/numerical-linear-algebra-software-development.html, accessed on Feb. 17, 2024

10. Burton S (1974) Garbow, EISPACK-A package of matrix eigensystem routines. Comput Phys Commun 7(4):179–184

11. Stewart GW (1977) Research, development, and LINPACK, mathematical software, pp 1–14 Proceedings of a Symposium Conducted by the Mathematics Research Center, the University of Wisconsin-Madison, March 28–30

12. Little J (1984) PC-MATLAB: A Matrix Laboratory for the IBM PC. MathWorks, Natick, MA

13. MathWorks, "MATLAB and Simulink in the Cloud," [Online]. Available: https://www.mathworks.com/solutions/cloud.html, accessed on Feb. 17, 2024

14. MathWorks, "Environment and Settings," https://ww2.mathworks.cn/help/matlab/desktop-tools-and-development-environment.html

15. MathWorks, "Live Scripts and Functions," [Online]. Available: https://uk.mathworks.com/help/matlab/live-scripts-and-functions.html, accessed on Feb. 17, 2024

Chapter 2
Data Types, Operators, and Expressions

Chapter Learning Outcomes

- Comprehend the concept of **data types** and their significance in MATLAB programming.
- Distinguish between the various **data types** available in MATLAB, such as **numeric types**, **characters**, and **logical values**.
- Utilise appropriate **type conversion** techniques to manipulate and transform data as required.
- Apply **arithmetic operators**, such as **addition**, **subtraction**, **multiplication**, and **division**, to perform basic mathematical computations.
- Employ **relational operators**, including **equality** and **inequality operators**, to compare values and evaluate conditions.
- Utilise **logical operators**, such as **AND**, **OR**, and **NOT**, to create **logical expressions** and make decisions based on conditions.
- Grasp the concept of **operator precedence** and utilise **parentheses** to control the order of operations.
- Utilise MATLAB's built-in functions and operators to work with **strings** and perform **string manipulations**.
- Perform arithmetic and logical **operations** using various **operators** and construct complex **expressions**.

Chapter Key Words

- **Data Types:** The different types of data that can be stored and manipulated in MATLAB, such as numeric types, characters, and logical values.

© The Author(s) 2025
Y. Chen and L. Huang, *MATLAB Roadmap to Applications*,
https://doi.org/10.1007/978-981-97-8788-3_2

- **Type Conversion:** The process of converting data from one type to another in MATLAB.
- **Arithmetic Operators:** Mathematical operators used to perform basic arithmetic operations in MATLAB, including addition, subtraction, multiplication, and division.
- **Relational Operators:** Operators used to compare values and evaluate conditions in MATLAB, such as equality and inequality operators.
- **Logical Operators:** Operators used to create logical expressions and make decisions based on conditions in MATLAB, such as AND, OR, and NOT.
- **Strings:** Sequences of characters in MATLAB, used to represent and manipulate text data.
- **Variables**: Learn how to create, manipulate, and use variables in MATLAB code.
- **Constants**: Recognise and use constants in MATLAB programs.
- **Literals**: Understand the concept of literals and how to use them in expressions.
- **Operators**: Grasp the different types of operators in MATLAB, such as arithmetic, relational, logical, and assignment operators.
- **Expressions**: Construct and evaluate expressions using variables, operators, and functions in MATLAB.
- **Operator Precedence**: Apply the rules of operator precedence to ensure correct order of operations in expressions.

2.1 MATLAB Built-in Data Types

In MATLAB, there are several built-in data types that provide the foundation for storing and manipulating data. These data types are designed to handle different kinds of information and support various operations. Here is a list of MATLAB's built-in data types as shown in the Fig. 2.1.

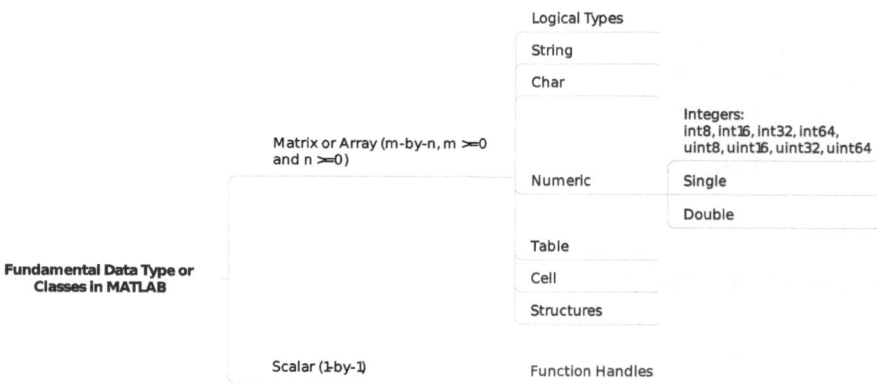

Fig. 2.1 Fundamental data types of classes in MATLAB

- **Matrix or Array (m-by-n, m ≥ 0 and n ≥ 0)**

  MATLAB has 17 fundamental data types (or classes) [1, 2]. Each of these classes is in the form of a matrix or array.

  - **Numeric Data Type**

    By default, MATLAB stores all numeric variables as double-precision floating-point values. Additional data types store text, integer or single-precision values, or a combination of related data in a single variable. More information on data types can be found in the Help index [3].

    MATLAB also supports signed and unsigned integer types and single-precision floating-point, by means of functions such as `int8`, `uint8`, `single`, and the like. However, before mathematical operations can be performed on such types, they must be converted to double precision using the `double` function

    · `double`: Represents double-precision floating-point numbers with 64 bits of precision.

Listing 2.1 Creating a double-precision variable

```
x = 3.14159; % x is a double-precision floating-
    point number
```

    · `single`: Represents single-precision floating-point numbers with 32 bits of precision.

Listing 2.2 Creating a single-precision variable

```
y = single(1.23456); % y is a single-precision
    floating-point number
```

    · `int8`, `int16`, `int32`, `int64`: Represent signed integers with different bit sizes (8, 16, 32, 64).

Listing 2.3 Creating signed integer variables

```
a = int8(-128); % a is an 8-bit signed integer
b = int16(32767); % b is a 16-bit signed integer
c = int32(-2147483648); % c is a 32-bit signed
    integer
d = int64(9223372036854775807); % d is a 64-bit
    signed integer
```

    · `uint8`, `uint16`, `uint32`, `uint64`: Represent unsigned integers with different bit sizes (8, 16, 32, 64).

Listing 2.4 Creating unsigned integer variables

```
e = uint8(255); % e is an 8-bit unsigned integer
f = uint16(65535); % f is a 16-bit unsigned
    integer
g = uint32(4294967295); % g is a 32-bit unsigned
    integer
```

```
h = uint64(18446744073709551615); % h is a 64-bit
    unsigned integer
```

– **Logical Data Type**

Represents logical values, which can be either **true** or **false**. MATLAB uses logical data types for logical operations and conditional statements. Logical arrays in MATLAB can be created using relational and logical operators, and they are useful for indexing, masking, and logical operations.

Listing 2.5 Creating logical variables

```
x = true; % x is a logical value (true)
y = false; % y is a logical value (false)
z = (5 > 3); % z is a logical value (true)

% Creating a logical array
A = [1 2 3; 4 5 6];
B = (A > 3); % B is a logical array
% B = [false false false
% true true true]
```

Logical arrays can be used for indexing and masking:

Listing 2.6 Using logical arrays for indexing and masking

```
A = [1 2 3; 4 5 6];
B = (A > 3); % B is a logical array
C = A(B); % C contains elements of A where B is true
% C = [4 5 6]
```

– **String Data Type**

Strings in MATLAB are arrays of characters that provide advanced string manipulation capabilities. Strings can be created using double quotes or the `string` function, and they support various operations such as concatenation, substring extraction, and regular expressions.

Listing 2.7 Creating and manipulating string variables

```
str1 = "Hello, World!"; % str1 is a string
str2 = string('MATLAB'); % str2 is a string

% Concatenating strings
fullStr = strcat(str1, ' ', str2); % fullStr = "
    Hello, World! MATLAB"

% Extracting substrings
subStr = str1(8:12); % subStr = "World"

% Using regular expressions
matches = regexp(str2, '[AE]', 'match'); % matches =
    {'A', 'E'}
```

- **Character Data Type**

  Represents a sequence of characters, such as letters, numbers, and symbols. MATLAB uses Unicode encoding to support a wide range of characters. Character arrays in MATLAB are row vectors of characters, and they can be created using single quotes or the `char` function.

Listing 2.8 Creating and manipulating character array variables

```
char_array = 'abcd'; % char_array is a character
    array
greek_char = '\alpha'; % greek_char is the Greek
    letter 'alpha'

% Concatenating character arrays
combined_chars = [char_array, greek_char]; %
    combined_chars = 'abcd\alpha'

% Converting to string
str = string(char_array); % str = "abcd"
```

- **Table**

  Represents tabular data with named variables (columns) and observations (rows). Tables provide a convenient way to work with structured data. Tables can be created from cell arrays, numeric arrays, or other data sources, and they support various operations such as sorting, filtering, and merging.

Listing 2.9 Creating and manipulating a table

```
names = {'John', 'Jane', 'Bob', 'Alice'};
ages = [25, 32, 41, 28];
heights = [1.75, 1.68, 1.82, 1.63];
myTable = table(names, ages, heights, ...
'VariableNames', {'Name', 'Age', 'Height'});

% Sorting the table by age
sortedTable = sortrows(myTable, 'Age');

% Filtering the table by height
filteredTable = myTable(myTable.Height > 1.7, :);
```

- **Cell**

  Cell arrays in MATLAB can store different data types, including numeric arrays, character arrays, strings, structures, and other cell arrays. They provide a flexible way to store and manipulate heterogeneous data.

Listing 2.10 Creating and manipulating a cell array

```
c = {1, 'hello', true, [1 2; 3 4]}; % c is a cell
    array

% Accessing elements of a cell array
num = c{1}; % num = 1
str = c{2}; % str = 'hello'
```

```
logical_val = c{3}; % logical_val = true
matrix = c{4}; % matrix = [1 2; 3 4]

% Adding elements to a cell array
c{end+1} = struct('name', 'John', 'age', 35); % adds
    a structure to the end of the cell array
```

– **Structures**

Represents a collection of related data fields grouped together under a single variable. Each field can hold data of different types and sizes. Structures provide a way to organize and manipulate data in a structured manner.

Listing 2.11 Creating and manipulating a structure

```
person.Name = 'John Doe';
person.Age = 35;
person.Height = 1.78;
person.IsStudent = false;

% Accessing structure fields
name = person.Name; % name = 'John Doe'
age = person.Age; % age = 35

% Adding a new field
person.Email = 'john.doe@example.com';

% Creating an array of structures
employees(1) = person; % initialize the first
    employee
employees(2) = struct('Name', 'Jane Smith', 'Age',
    28, 'Height', 1.65, 'IsStudent', true);
```

These examples demonstrate the usage and manipulation of various data types in MATLAB, including logical arrays, strings, character arrays, tables, cell arrays, and structures. They cover creating, accessing, and modifying these data types, as well as performing common operations on them.

With the exceptions of function handles and tables, this matrix or array is a minimum of 0-by-0 in size and can grow to an n-dimensional array of any size. A function handle is always scalar (1-by-1). A table always has m rows and n variables, where m $\geq$ 0 and n $\geq$ 0.

- **Scalar (1-by-1)**

  – **Function Handles**

    Represents a reference to a function. Function handles allow passing functions as arguments, storing them in variables, and calling them dynamically.

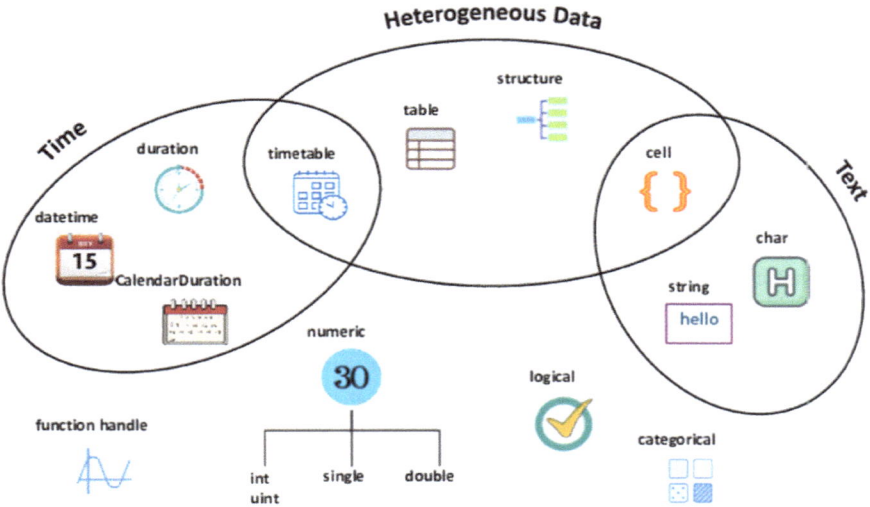

Fig. 2.2 MATLAB data types in Mindmap

Listing 2.12 Creating a function handle

```
f = @sin; % f is a function handle to the sine
    function
x = 0:pi/4:2*pi; % Create a vector of x values
y = f(x); % Evaluate the sine function at x values
```

2.2 Working with Data Types

In the exploration of MATLAB's data types, as shown in Fig. 2.2, one learns that MATLAB offers a comprehensive set of built-in types that meet the needs of both numerical and text processing. The variety and flexibility of these data types allow users to perform complex computations and data manipulation with ease. This section will delve into the instantiation and application of these types, illuminating the significance of each specific type within MATLAB's ecosystem.

The fundamental MATLAB classes are shown in the following Table 2.1.

2.2.1 Creating Variables

In MATLAB, **variables** are used to store data of various types, such as numbers, text, logical values, and more, which are used to store and manipulate data in MATLAB.

Table 2.1 Fundamental MATLAB classes

No.	Data type	Features
1	double, single	• Floating-point numbers • Required for fractional numeric data • Double- and single-precision • **double** is the default numeric type • Two-dimensional arrays can be sparse • Use realmin and realmax to show range of values
2	int8, uint8, int16, uint16, int32, uint32, int64, uint64	• Integers • Signed and unsigned whole numbers • More efficient use of memory • Choose from 4 sizes (8, 16, 32, and 64 bits) • Use intmin and intmax to show range of values
3	string, char	• Data types for text • Both data types store characters as Unicode characters • Support conversions to and from numeric representations • Use either data type with regular expressions • **string** arrays recommended for storing multiple strings • To search for and match text in strings, use pattern objects. (since R2020b)
4	logical	• Used in relational conditions or to test state • Can have one of two values: true or false • Also useful in array indexing • Two-dimensional arrays can be sparse
5	cell	• Cells store arrays of varying classes and sizes • Allows freedom to package data as you want • Manipulation of elements is similar to numeric or logical arrays • Method of passing function arguments • Use in comma-separated lists • More memory required for overhead
6	table, timetable	• Tables and timetables are rectangular containers for mixed-type, column-oriented data • Tables have row and variable names that identify contents • Timetables also provide storage for time series data in a table with rows labeled by timestamps. Timetable functions can synchronize, resample, or aggregate timestamped data • Use properties to store metadata such as variable units • Manipulation of elements similar to numeric or logical arrays • Access data by numeric or named index • Can select a subset of data and preserve the table container or can extract the data from a table

(continued)

Table 2.1 (continued)

No.	Data type	Features
7	struct	• Fields store arrays of varying classes and sizes
		• Access one or all fields/indices in single operation
		• Field names identify contents
		• Method of passing function arguments
		• Use in comma-separated lists
		• More memory required for overhead
8	function_handle	• Pointer to a function
		• Enables passing a function to another function
		• Can also call functions outside usual scope
		• Use to specify graphics callback functions
		• Save to MAT-file and restore later

They can be assigned values of different data types, such as **numeric**, **character**, **logical**, and **cell arrays**. Variables are created by assigning a value to a valid variable name, which follows certain rules. To create a variable, one simply needs to assign a value to it using the assignment operator (=). The **variable name** must start with a letter and can contain letters, digits, and underscores. MATLAB is case-sensitive, so variables with the same name but different cases are treated as different variables.

Here is an example of creating a variable in MATLAB:

Listing 2.13 Creating a numeric variable.

```
myNumber = 5; % Assigns the value 5 to the variable '
    myNumber'
x = 5; % Numeric variable
y = 'Hello'; % Character vector
z = true; % Logical variable
c = {1, 2, 3}; % Cell array
```

In the above example, a variable named myNumber is created and assigned the value of 5. MATLAB automatically determines the data type of the variable based on the assigned value.

Another example showcasing the creation of a string variable:

Listing 2.14 Creating a string variable.

```
myString = "Hello, MATLAB!"; % Assigns a string value to
    the variable 'myString'
```

Here, the variable myString is created and assigned the string value "Hello, MATLAB!". String values in MATLAB are enclosed in double quotes.

It is important to note that variables in MATLAB do not need to be explicitly declared or defined before assigning a value to them. MATLAB dynamically allocates memory for variables as needed.

● **Numeric**

Numeric classes in MATLAB include signed and unsigned integers, and single- and double-precision floating-point numbers. By default, MATLAB stores all numeric values as double-precision floating point. You can choose to store any number, or array of numbers, as integers or as single-precision. Integer and single-precision arrays offer more memory-efficient storage than double-precision

All numeric types support basic array operations, such as subscripting, reshaping, and mathematical operations. Two-dimensional double and logical matrices can be created using one of two storage formats: **full** or **sparse**. For matrices with mostly zero-valued elements, a sparse matrix requires a fraction of the storage space required for an equivalent full matrix. Sparse matrices invoke methods especially tailored to solve sparse problems

Here are two MATLAB examples demonstrating the use of numeric data types:

Listing 2.15 Example of double and single precision floating-point numbers.

```
% Double precision (default)
a = 3.14159;
disp(a);

% Single precision
b = single(3.14159);
disp(b);
```

Listing 2.16 Example of sparse matrix.

```
% Create a sparse matrix
S = sparse([1 1 2 3], [1 3 2 3], [5 7 6 8], 3, 3);
disp(S);

% Perform operations on the sparse matrix
X = S * 2; % Scalar multiplication
Y = S + S; % Addition
```

● **Character and String** MATLAB provides two data types for text: **string** and **char**. Both data types store characters as Unicode characters. They support conversions to and from numeric representations and can be used with regular expressions.

To store multiple strings, it is recommended to use **string** arrays rather than cell arrays of character vectors. However, cell arrays can still be used for this purpose. Here are two MATLAB examples demonstrating the use of character and string data types:

Listing 2.17 Example of string concatenation

```
% Create a string array
str = ["Hello", "World"];

% Concatenate strings
greeting = str(1) + " " + str(2);
disp(greeting);
```

Listing 2.18 Example of regular expression matching

```
% Create a character vector
text = 'The quick brown fox jumps over the lazy dog.';

% Find matches using a regular expression
pattern = 'the';
matches = regexp(text, pattern, 'match');
disp(matches);
```

• **Logical** The **logical** class is used in relational conditions or to test state. A logical value can have one of two values: true or false. Logical arrays are also useful in array indexing. Two-dimensional logical arrays can be sparse.

   Here are two MATLAB examples demonstrating the use of the logical data type:

Listing 2.19 Example of logical indexing

```
% Create a numeric array
A = [1 2 3; 4 5 6; 7 8 9];

% Use logical indexing to select elements
mask = A > 5;
B = A(mask);
disp(B);
```

Listing 2.20 Example of logical operations

```
% Logical operations
a = true;
b = false;

% Logical AND
c = a && b;
disp(c);

% Logical OR
d = a || b;
disp(d);

% Logical NOT
e = ~a;
disp(e);
```

In the first example, we create a numeric array A and then use a logical mask mask = A > 5 to select elements from A that are greater than 5. The selected elements are stored in the new array B.

The second example demonstrates various logical operations:

– c = a && b performs a logical AND operation between a and b.
– d = a || b performs a logical OR operation between a and b.
– e = ~a performs a logical NOT operation on a.

Table 2.2 Truth table for logical AND operation

A	B	A AND B
True	True	True
True	False	False
False	True	False
False	False	False

Table 2.3 Truth table for logical OR operation

A	B	A OR B
True	True	True
True	False	True
False	True	True
False	False	False

Table 2.4 Truth table for logical NOT operation

A	NOT A
True	False
False	True

These logical operations can be useful in various contexts, such as conditional statements, array indexing, and data filtering.

Here are the truth tables for the logical AND, OR, and NOT operations in MATLAB presented (Tables 2.2, 2.3 and 2.4).

The truth tables clearly illustrate the behavior of the logical AND, OR, and NOT operations in MATLAB:

– The logical AND operation (&&) returns true only when both operands are true.
– The logical OR operation (‖) returns true when at least one of the operands is true.
– The logical NOT operation (∼) returns the opposite truth value of the operand.

These truth tables are fundamental in understanding and working with logical operations in MATLAB, which are essential for various programming tasks such as conditional statements, data filtering, and logical indexing.

• **Table and Timetable Tables** and **timetables** are rectangular containers for mixed-type, column-oriented data. Tables have row and variable names that identify their contents. Timetables also provide storage for time series data in a table with rows labeled by timestamps. Timetable functions can synchronize, resample, or aggregate timestamped data.

The properties of a table or timetable can be used to store metadata such as variable units. Manipulation of elements is similar to numeric or logical arrays, and data can be accessed by numeric or named index.

Here are two MATLAB examples demonstrating the use of tables and timetables:

Listing 2.21 Example of creating and manipulating a table

```
% Create a table
T = table([1 2 3]', [4 5 6]', 'VariableNames', {'A', '
   B'}, 'RowNames', {'Row1', 'Row2', 'Row3'});

% Access data by variable name
disp(T.A);

% Add a new variable
T.C = [7 8 9]';
```

Listing 2.22 Example of creating and resampling a timetable

```
% Create a timetable
times = datetime(2023, 4, 1) + calmonths(0:2:6);
data = [1 2 3 4]';
TT = timetable(times, data, 'VariableNames', {'Values'
   });

% Resample the timetable
resampledTT = retime(TT, 'monthly');
disp(resampledTT);
```

- **Structure and Cell Array** **Structures** and **cell arrays** are containers that can store arrays of varying classes and sizes.
  In a structure, fields store the arrays, and field names identify the contents. Structures can access one or all fields/indices in a single operation, and they are often used as a method of passing function arguments or in comma-separated lists. However, structures require more memory overhead than other data types.
  Cell arrays allow freedom to package data as you want, and the manipulation of elements is similar to numeric or logical arrays. Like structures, cell arrays are commonly used as a method of passing function arguments or in comma-separated lists, but they also require more memory overhead than other data types.
  Here are two MATLAB examples demonstrating the use of structures and cell arrays:

Listing 2.23 Example of creating and accessing a structure

```
% Create a structure
student.Name = 'John Doe';
student.Age = 25;
student.GPA = 3.8;

% Access structure fields
disp(['Name: ' student.Name]);
disp(['Age: ' num2str(student.Age)]);
```

Listing 2.24 Example of creating and manipulating a cell array

```
% Create a cell array
C = {[1 2 3], 'hello', true};

% Access cell elements
disp(C{1}); % Display the numeric array
disp(C{3}); % Display the logical value

% Add a new cell element
C{4} = magic(3); % Add a 3x3 magic square
```

- **Function Handle** A **function handle** is a pointer to a function. It enables passing a function to another function and can also call functions outside the usual scope. Function handles are commonly used to specify graphics callback functions and can be saved to a MAT-file and restored later.

Here are two MATLAB examples demonstrating the use of function handles:

Listing 2.25 Example of a function handle as a callback

```
% Define a callback function
function handleClick(src, event)
disp('Button clicked!');
end

% Create a button with the callback function
fig = uifigure;
btn = uibutton(fig, 'Text', 'Click Me', 'Position',
    [50 50 100 30]);
btn.ButtonPushedFcn = @handleClick;
```

Listing 2.26 Example of passing a function handle as an argument

```
% Define a function that takes a function handle as an
    argument
function result = applyFunction(func, x)
result = func(x);
end

% Pass an anonymous function as a function handle
squareRoot = @(x) sqrt(x);
result = applyFunction(squareRoot, 25);
disp(result);
```

2.2.2 Accessing and Modifying Data

Once data is stored in **variables** or **arrays** in MATLAB, it is often necessary to access and modify specific elements or portions of that data. MATLAB provides several methods for accessing and modifying data using **indexing**, or extract subsets of data using **slicing**. Data can be modified by **assigning** new values or using built-in functions. Indexing and slicing are powerful techniques in MATLAB for accessing and manipulating specific elements or subsets of data within arrays, matrices, and other data structures. **Indexing** allows you to access individual elements or subarrays using numerical indices, while **slicing** enables you to extract specific rows, columns, or submatrices using colon notation or logical indexing.

Here's an example:

Listing 2.27 Accessing and modifying data in MATLAB

```
A = [1 2 3; 4 5 6]; % Create a matrix
A(2, 3) = 10; % Modify an element
B = A(1, :); % Extract a row
```

- **Assigning** Assignment is the process of storing or modifying data in a variable or array using the assignment operator (=). In addition to simple assignment, MATLAB supports various forms of assignment, such as array expansion and structure field assignment.

Listing 2.28 Modifying elements in an array using assignment

```
A = zeros(3,3); % Create a 3x3 array of zeros
A(1,:) = [1 2 3]; % Assign values to the first row
A(:,2) = [4; 5; 6]; % Assign values to the second
    column
```

- **Indexing**

  Accessing elements in MATLAB arrays is achieved through the use of indices, which can be either scalar or vectorised. The modification of array elements is similarly realisable by assigning new values to specific indices by specifying the row and column indices of the desired element(s).

  For instance, consider the following MATLAB syntax to access and modify an element in an array:

Listing 2.29 Accessing and modifying an array element

```
% Given an array 'A'
A = [1, 2, 3; 4, 5, 6; 7, 8, 9];

% Accessing the element at the second row, third
    column
element = A(2, 3);

% Modifying the element at the second row, third
    column to 10
A(2, 3) = 10;

A = [1 2 3; 4 5 6; 7 8 9]; % Create a 3x3 matrix

A(2, 3) % Access element at row 2, column 3
```

This capacity to pinpoint a specific datum is pivotal. Additionally, let us illustrate altering a section of an array through logical indexing:

Listing 2.30 Modifying elements using logical indexing

```
% Modifying elements greater than 5 to be equal to 5
A(A > 5) = 5;
```

- **Slicing**

  The **slicing** is also called **subscripting**, which are essentially the same concept. They both refer to the process of extracting a subset or a part of an array or matrix using indices or logical expressions.

  The term "subscripting" is more commonly used in MATLAB documentation and literature, but "slicing" is also sometimes used, especially when referring to extracting a contiguous subset of elements from a vector or a row/column from a matrix.

  Here are some examples that illustrate the equivalence of subscripting and slicing in MATLAB:

Listing 2.31 Subscripting/slicing examples

```
% Create a vector
x = [1 2 3 4 5 6 7 8 9];

% Subscripting/slicing to extract a subset
x(3:6) % Returns [3 4 5 6] (slicing)

% Create a matrix
A = [1 2 3; 4 5 6; 7 8 9];

% Subscripting/slicing to extract a row
A(2, :) % Returns [4 5 6] (slicing a row)

% Subscripting/slicing to extract a column
A(:, 3) % Returns [3; 6; 9] (slicing a column)

B = A(:, 2:3) % Extract columns 2 and 3 from A
```

In the examples above, the use of colon operators (e.g., 3:6, :, 2, :, :,3) is a form of subscripting or slicing that extracts a subset of elements from the array or matrix. So, while the term "subscripting" is more prevalent in MATLAB documentation, the concepts of subscripting and slicing are effectively the same and refer to the process of extracting parts of arrays or matrices using indices or logical expressions.

2.2.3 Type Conversion

In MATLAB, **data type conversion** refers to the process of transforming values from one data type to another, allowing for compatibility and flexibility in data manipulation. MATLAB provides various functions to facilitate conversions between different data types, such as numeric arrays, strings, character arrays, dates and times, categorical arrays, tables, and timetables. Common conversions include **converting numbers to text, text to numeric values, hexadecimal and binary representations**, and **converting between text and datetime or duration values**.

The **cast** function is a powerful tool for type conversion in MATLAB. It converts the data type of an input array or variable to the specified data type. The syntax for using the **cast** function is:

Listing 2.32 Syntax for the cast function

```
new_data = cast(original_data, 'data_type');
```

Here, `original_data` is the input array or variable, and `'data_type'` is a string specifying the desired data type for the output array `new_data`.

For example, to convert a double-precision array to a single-precision array:

Listing 2.33 Converting from double to single precision

```
x = [1.2345 6.7890]; % Double-precision array
y = cast(x, 'single'); % Convert to single precision
```

MATLAB also provides Specialised functions for type conversion, such as `double`, `single`, `int8`, `int16`, `int32`, `int64`, `uint8`, `uint16`, `uint32`, `uint64`, `logical`, and `char`. These functions convert the input data to the corresponding data type [4].

Listing 2.34 Converting a character array to a double-precision array

```
str = '3.14159'; % Character array
x = double(str); % Convert to double-precision array
```

Type conversion can also be performed using operators such as + and − when operating on different data types. MATLAB automatically converts the operands to the highest precision data type involved in the operation.

Listing 2.35 Automatic type conversion during arithmetic operations

```
x = int8(5); % Integer 8-bit value
y = x + 2.5; % y is a double-precision value
```

In the example above, the `int8` value `x` is automatically converted to a double-precision value before adding 2.5, resulting in `y` being a double-precision value.

Proper type conversion is essential for efficient memory usage, accurate computations, and compatibility with various functions and operations in MATLAB. The documentation [4] provides a comprehensive guide on type conversion functions and techniques. Here is to list a few in Table 2.5.

- **Numbers and Text**:

  - To convert numbers to text, functions like `string`, `char`, `cellstr`, `int2str`, `mat2str`, and `num2str` can be used.

Table 2.5 Type conversion functions in MATLAB

Function	Description
Convert numbers to text	
`string`	String array
`char`	Character array
`cellstr`	Convert to cell array of character vectors
`int2str`	Convert integers to characters
`mat2str`	Convert matrix to characters
`num2str`	Convert numbers to character array
Convert text to numbers	
`str2double`	Convert strings to double precision values
`str2num`	Convert character array or string to numeric array
`native2unicode`	Convert numeric bytes to Unicode character representation
`unicode2native`	Convert Unicode character representation to numeric bytes
Hexadecimal and binary numbers	
`base2dec`	Convert text representation of base-n integer to double value
`bin2dec`	Convert text representation of binary integer to double value
`dec2base`	Convert decimal integer to its base-n representation
`dec2bin`	Convert decimal integer to its binary representation
`dec2hex`	Convert decimal integer to its hexadecimal representation
`hex2dec`	Convert text representation of hexadecimal integer to double value
`hex2num`	Convert IEEE hexadecimal format to double-precision number
`num2hex`	Convert single- and double-precision numbers to IEEE hexadecimal format
Dates and times	
`datetime`	Arrays that represent points in time
`duration`	Lengths of time in fixed-length units
`matlab.datetime.compatibility.` `convertDatenumConvert`	Convert inputs to datetime values in a backward-compatible way
Categorical arrays, tables, and timetables	
`categorical`	Array that contains values assigned to categories
`table2array`	Convert table to homogeneous array
`table2cell`	Convert table to cell array
`table2struct`	Convert table to structure array
`array2table`	Convert homogeneous array to table

(continued)

Table 2.5 (continued)

Function	Description
cell2table	Convert cell array to table
struct2table	Convert structure array to table
array2timetable	Convert homogeneous array to timetable
table2timetable	Convert table to timetable
timetable2table	Convert timetable to table
Cell arrays and structures	
cell2mat	Convert cell array to ordinary array of the underlying data type
cell2struct	Convert cell array to structure array
mat2cell	Convert array to cell array whose cells contain subarrays
num2cell	Convert array to cell array with consistently sized cells
struct2cell	Convert structure to cell array
Convert from structure to Mindmap	
Struct2MindMap	Creates a mindmap from a given MATLAB structure

- To convert text to numeric values, functions like str2double, str2num, native2unicode, and unicode2native are available.

- **Hexadecimal and Binary Numbers**:

  - Functions like base2dec, bin2dec, dec2base, dec2bin, dec2hex, hex2dec, hex2num, and num2hex allow conversions between decimal, hexadecimal, and binary representations of numbers.

- **Dates and Times**:

  - To convert values to datetime or duration objects, functions like datetime, duration, and matlab.datetime.compatibility. convertDatenumConvert can be used.
  - To convert datetime or duration objects to text, functions like string, char, and cellstr are available.

- **Categorical Arrays, Tables, and Timetables**:

  - Functions like categorical, table2array, table2cell, table2struct, array2table, cell2table, struct2table, array2timetab, letable2timetable, and timetable2table facilitate conversions between these data types.

- **Cell Arrays and Structures**:
  - Functions like `cell2mat`, `cell2struct`, `mat2cell`, `num2cell`, `struct2cell`, and `struct2cell` enable conversions between cell arrays, structures, and regular arrays.

Proper data type conversion is essential for compatibility between different operations, efficient memory usage, and accurate computations within MATLAB.

Here's an example of converting a numeric value to text:

Listing 2.36 Converting numeric value to text

```
num = 42;
str = \textcolor{green}{num2str}(num); % Convert numeric
    value to string
```

MATLAB also provides the struct2mindmap function [5] to convert a MATLAB structure to a MindMap format, which can be useful for visualizing and sharing complex data structures. Also, to import a mindmap to MATLAB.

2.2.4 Operations and Functions on Data Types

MATLAB provides a wide range of built-in functions and operators that can be applied to various data types, enabling efficient data manipulation and analysis. These include **arithmetic operations**, **logical operations**, **matrix operations**, **string operations**, and Specialised functions for specific data types like datetime and categorical arrays [6] (Table 2.6).

1. **Arithmetic operations** such as **addition**, **subtraction**, **multiplication**, and **division** can be performed on **numeric data types** like doubles, singles, and integers. MATLAB also provides built-in functions like `sin`, `cos`, `tan`, and `sqrt` for performing mathematical calculations on these data types. For example:

Listing 2.37 Arithmetic operations and functions on numeric data types

```
a = 3.5; % double
b = int32(10); % int32
c = a + b; % Addition of double and int32
d = sin(a); % Sine function on a double
```

Listing 2.38 Matrix operations on numeric data types

```
A = [1 2 3; 4 5 6]; % Double matrix
B = [3 4; 2 1; 5 6]; % Double matrix
C = A * B; % Matrix multiplication
D = \textcolor{green}{sin}(A); % Apply the sine
    function element-wise
```

Table 2.6 Specialised functions in MATLAB

Function	Description
Numeric data types	
diff, int	**Calculus**: Numerical differentiation anc integration
interp1, interp2	**Interpolation**: Interpolation of data in 1-D and 2-D
fmincon, fminsearch	**optimisation**: Constrained and unconstrained optimisation
mean, std, var	**Statistics**: Compute mean, standard deviation, variance
fft, ifft	**Fourier Transforms**: Fast Fourier Transform and inverse
String data types	
regexp, regexprep	**Regular Expressions**: Pattern matching and replacement
tokenizedDocument, removeWords	**Natural Language Processing**: Text tokenization and cleaning
unicode2native	**String Encoding**: Convert between Unicode and native encodings
Table and timetable data types	
sortrows, unstack	**Data Manipulation**: Sort rows, restructure table layout
outerjoin, innerjoin	**Merging**: Merge tables based on key columns
synchronize, rmmissing, resample	**Time Series Analysis**: Synchronize, remove missing, resample data
Other application domains	
imfilter, medfilt2	**Image Processing**: Filter and process images
filter, deconv	**Signal Processing**: Filter and deconvolve signals
tf, step	**Control Systems**: Transfer functions, step response
cpomplexp, lteOFDMModulate	**Communications**: Complex plane, LTE OFDM modulation

2. **Matrix operations**

Matrix operations are a core part of MATLAB and can be performed on **numeric data types** like doubles, singles, and integers. MATLAB provides a rich set of functions and operators for working with matrices and arrays.

Basic matrix operations include addition, subtraction, multiplication (both element-wise and matrix multiplication), division, exponentiation, and more. These can be performed using standard arithmetic operators like additicn ($+$), subtraction ($-$), multiplication ($*$), division ($/$), and exponentiation ($\hat{}$), etc. For example:

Listing 2.39 Basic matrix arithmetic operations

```
A = [1  2;  3  4];
B = [5  6;  7  8];
C = A + B; % Matrix addition
D = A - B; % Matrix subtraction
E = A * B; % Matrix multiplication
F = A .* B; % Element-wise multiplication
G = A .^ 2; % Element-wise exponentation
```

MATLAB also provides many functions for common matrix operations like transpose (`transpose`, `permute`), inverse (`inv`), determinant (`det`), trace (`trace`), eigenvalues and eigenvectors (`eig`), norms (`norm`) and much more. For example:

Listing 2.40 Matrix functions in MATLAB

```
A = [1  2  3;  4  5  6;  7  8  9];
B = inv(A); % Matrix inverse
lambda = eig(A); % Eigenvalues of A
[V, D] = eig(A); % Eigenvectors (V) and eigenvalues (
    D) of A
normA = norm(A); % Norm of matrix A
traceA = trace(A); % Trace of matrix A
```

In addition to these basic operations, MATLAB provides advanced functions for matrix decompositions like LU, QR, SVD and more through functions like `lu`, `qr`, `svd` etc.

Sparse matrices, which efficiently store matrices with mostly zero entries, can also be created and operated on using functions like `sparse`, `issparse`, `full` etc. Overall, MATLAB's ability to perform a wide variety of matrix operations on numeric data types is a core strength that makes it suitable for linear algebra, signal processing, image processing and many other technical computing applications.

3. **Logical operations** like and, or, and not can be performed on **logical data types**, which store boolean values (true or false). These operations are often used in conditional statements and indexing. For example:

Listing 2.41 Logical operations on logical data types

```
a = true; % Logical true
b = false; % Logical false
c = a && b; % Logical AND operation
d = ~c; % Logical NOT operation
```

Listing 2.42 Indexing with logical data types

```
A = [1  2  3;  4  5  6]; % Double matrix
B = A > 3; % Logical matrix based on condition
C = A(B); % Extracting elements of A where B is true
```

4. **String operations** for **string data types**, MATLAB provides a range of functions for string manipulation, such as concatenation, splitting, searching, and replacing. These functions are available through two primary data types in MATLAB: the traditional **character arrays** (char) and the modern **string arrays** (string) introduced in MATLAB R2016b [7]. The string array functionality provides enhanced capabilities for text processing, including advanced pattern matching, array operations, and Unicode support [8]. MATLAB's string handling framework offers a comprehensive suite of functions:

- **Core String Functions**: Essential operations such as strcat, strtrim, strrep, and strcmp for basic string manipulation [7].
- **Pattern Matching**: Advanced text processing through regexp, contains, startsWith, and endsWith functions.
- **String Array Operations**: Modern functions like join, split, strip, and replaceBetween that operate efficiently on string arrays [7].
- **Conversion Utilities**: Functions for converting between different text representations, including char, string, cellstr, and various numeric conversion functions [9].

For example:

Listing 2.43 String operations and functions on string data types

```
str1 = "Hello "; % String array
str2 = "World"; % String array
fullStr = str1 + str2; % String concatenation
idx = contains(fullStr, "lo"); % Finding substring
newStr = replace(fullStr, "lo", "LO"); % Replacing
    substring
```

Listing 2.44 Regular expression operations on string data types

```
str = "The quick brown fox jumps over the lazy dog.";
pattern = "(?<animal>\w+)\s+(?<color>\w+)\s+(?<
    animal2>\w+)"; % Pattern with named groups
[start, endPos, extractor, matches] = regexp(str,
    pattern, "names"); % Regular expression matching
```

5. **Specialised functions**

MATLAB offers a wide range of **specialised functions** that cater to various data types and application domains, enabling advanced data manipulation, analysis, and processing capabilities. These functions are optimised for specific tasks and provide efficient and accurate results.

These are just a few examples of the various operations and functions available in MATLAB for different data types. It is important to understand the capabilities of each data type and the corresponding operations and functions to effectively manipulate and analyze data in MATLAB.

In addition to basic operations, MATLAB provides many **Specialised functions** for working with different data types. These functions enable advanced data manipulation, analysis, and processing capabilities.

- For **numeric data types** like doubles and singles, MATLAB offers a vast library of specialised math functions. This includes functions for calculus (diff, int), interpolation (interp1, interp2), optimisation (fmincon, fminsearch), statistical computations (sum, mean, std, var), Fast Fourier Transforms (fft, ifft). Additionally, functions such as conv, filter, and resample are available for signal processing tasks [10], and more.

For example:

Listing 2.45 Specialised math functions in MATLAB

```
x = 0:0.1:2*pi;
y = sin(x);
yprime = diff(y); % First derivative of y
area = trapz(x, y); % Numerical integration to find
    area

X = membrane(5, 7); % Test problem data
options = optimoptions('fmincon','Algorithm','
    interior-point');
[x, fval] = fmincon(@(x) myFunObj(x, X), [], [], [],
    [], [], [], [], [], options);
```

Here is to explain the example step by step:

- x = 0:0.1:2*pi; generates a vector x with values ranging from 0 to 2π with a step size of 0.1.
- y = sin(x); calculates the sine of each element in x and assigns the result to y.
- yprime = diff(y); computes the first derivative of y using the diff function, which calculates the difference between consecutive elements.
- area = trapz(x, y); numerically integrates the function represented by x and y using the trapezoidal rule and assigns the result to area.
- X = membrane(5, 7); generates test problem data using the membrane function, which is likely a user-defined function or part of a specific toolbox.
- options = optimoptions('fmincon','Algorithm','interior-point'); sets the options for the fmincon optimisation function, specifying the interior-point algorithm.
- [x, fval] = fmincon(@(x) myFunObj(x, X), [], [], [], [], [], [], [], [], options); calls the fmincon optimisation function to minimise the objective function myFunObj subject to the specified options. The empty brackets [] indicate that no linear

inequality constraints, linear equality constraints, lower bounds, or upper bounds are provided.

Note: Assuming that the `membrane` and `myFunObj` functions are properly defined, and make sure to replace `myFunObj` with the actual objective function you want to minimise when using this code snippet.

- For **string data types**, MATLAB provides functions for regular expression operations (`regexp`, `regexprep`), string comparison, searching, regular expression matching, and replacement operations (`strcmp`, `strfind`, and `strrep`), natural language processing (`tokenizedDocument`, `removeWords`), string encoding (`unicode2native`), and more. These functions simplify the manipulation and analysis of textual data. For example:

Listing 2.46 String processing functions in MATLAB

```
str = "The quick brown fox jumps over the lazy dog.";
pattern = "\w+"; % Word pattern
tokens = regexp(str, pattern, 'match'); % Extract all
    words

text = "This is good. However, it is raining outside
    .";
doc = tokenizedDocument(text);
doc = removeWords(doc, ["is", "raining"], 'IgnoreCase'
    , true); % Not case-sensitive
```

- For **table and timetable data types**, MATLAB provides functions for data manipulation (`sortrows`, `unstack`), merging (`outerjoin`, `innerjoin`), time series analysis (`synchronize`, `rmmissing`, `resample`), and more. These functions facilitate data sorting, merging, aggregation, and resampling operations, making it convenient to work with tabular and time-series data.
- MATLAB also includes specialised toolboxes and functions for application domains like image processing (`imfilter`, `medfilt2`), signal processing (`filter`, `deconv`), control systems (`tf`, `step`), communications (`cpomplexp`, `lteOFDMModulate`), and many others. Such as,

  - Image Processing: Functions such as `imfilter` and `medfilt2` are used for image enhancement and filtering techniques.
  - Signal Processing: The `filter` and `deconv` functions are fundamental for designing and analysing signal processing systems.
  - Control Systems: Functions like `tf` and `step` are essential for the design and analysis of control systems.
  - Communications: Toolboxes provide functions such as `complexp` and `lteOFDMModulate` for communication system design and modulation techniques.

These Specialised functions greatly extend MATLAB's capabilities beyond basic data operations, enabling users to perform advanced analysis, modeling, and processing tasks across various domains and data types.

2.2.5 Missing Data Handling

In real-world data analysis, missing or invalid data is a common occurrence. MATLAB provides various techniques for handling missing data, such as **identifying**, **filtering**, and **imputing** missing values. The isnan function can be used to identify missing values represented by NaN (Not a Number), and functions like rmmissing can remove missing data.

Here's an example of identifying and removing missing data:

Listing 2.47 Handling missing data in MATLAB

```
data = [1 2 NaN; 4 5 6; 7 NaN 9];
missing_indices = \textcolor{green}{isnan}(data); % Find
    missing values
clean_data = \textcolor{green}{rmmissing}(data); %
    Remove missing values
```

In this table, It has been highlighted important concepts like **NaN**, **Inf**, **missing**, and **outlier** values using the command. The table is divided into three sections: Missing Data Identification, Missing Data Imputation, and Missing Data Handling in Analysis, each with relevant MATLAB functions and their descriptions (Table 2.7).

Table 2.7 Missing data handling functions in MATLAB

Function	Description
Missing data identification	
isnan	Check for **NaN** (Not a Number) values
isfinite	Check for finite (non-**Inf** and non-**NaN**) values
ismissing	Check for **missing** values in tables and timetables
Missing data imputation	
fillmissing	Replace **missing** values with specified value or method
rmmissing	Remove **missing** observations from data
filloutliers	Replace **outlier** values with specified method
Missing data handling in analysis	
nanmean, nansum, etc.	Compute statistics ignoring **NaN** values
rmmissing	Remove **missing** observations before analysis
fillmissing	Impute **missing** values before analysis

2.3 Advanced Topics

2.3.1 Custom Data Types

In addition to the built-in **data types** that MATLAB provides, users can also create their own custom data types using **classes** and **structures**. This allows for the creation of more complex data types that can encapsulate multiple properties and behaviours.

- **Classes** Classes in MATLAB are user-defined data types that can contain **properties** (data) and **methods** (functions). They provide a way to create objects that combine data and the operations that can be performed on that data. This allows for better organisation and modularity of code, as well as the ability to create more complex data structures.
  Here is a simple example of defining a class in MATLAB:

Listing 2.48 Defining a class in MATLAB

```
classdef Person
properties
Name
Age
end
methods
    function obj = Person(name, age)
        obj.Name = name;
        obj.Age = age;
    end

    function greet(obj)
        disp(['Hello, my name is ' obj.Name ' and I am
            ' num2str(obj.Age) ' years old.']);
    end
end
end
```

In this example, the Person class has two properties (Name and Age) and two methods (Person and greet). The Person method is a constructor that initialises the object with a name and age, while the greet method prints a greeting message.
Here's another example of creating a custom data type called `Circle` in MATLAB:

Listing 2.49 Custom data type example

```
classdef Circle
properties
radius
center
end
methods
    function obj = Circle(r, x, y)
        obj.radius = r;
```

```
          obj.center = [x, y];
      end

      function area = getArea(obj)
          area = pi * obj.radius^2;
      end
 end
 end
```

In this example, the `Circle` class has two properties: `radius` and `center`. The constructor method initializes these properties, and the `getArea` method calculates the area of the circle based on its radius.

- **Structures** Structures in MATLAB are another way to create custom data types. They are similar to classes, but are generally simpler and do not support methods. Structures are useful for grouping related data together in a single variable. Here is an example of creating a structure in MATLAB:

Listing 2.50 Creating a structure in MATLAB

```
person = struct('Name', 'John Doe', 'Age', 30, 'Email'
    , 'john.doe@example.com');
disp(person)
```

This creates a structure called person with three fields: Name, Age, and Email. The values of these fields can be accessed using dot notation, e.g., person.Name. Both classes and structures provide a way to create custom data types in MATLAB, with classes offering more advanced features and functionality, while structures are simpler but more limited.

2.3.2 Enumerations

Enumerations (or `enums`) are a special data type in MATLAB that allows developers to define a set of named constant values. Enumerations are useful for representing a fixed set of choices or states, making the code more readable and maintainable.

Here's an example of defining and using an enumeration in MATLAB:

Listing 2.51 Enumeration example

```
% Define the enumeration
TrafficLight = enumeration('TrafficLight', .'Red', '
    Yellow', 'Green');

% Use the enumeration
currentLight = TrafficLight.Green;
if currentLight == TrafficLight.Red
disp('Stop!');
else
disp('Go!');
end
```

In this example, the `TrafficLight` enumeration is defined with three possible values: `'Red'`, `'Yellow'`, and `'Green'`. The enumeration is then used to represent the current state of a traffic light, and a conditional statement checks the state and displays the appropriate message.

2.3.3 Data Type Validation

MATLAB provides **data type validation** functions that allow developers to check the data type of variables or arrays and ensure they conform to the expected type. This can be useful for catching errors early and maintaining data integrity.

Here's an example of using the `isa` function to validate the data type of a variable:

Listing 2.52 Data type validation example

```
x = 3.14;
if isa(x, 'double')
disp('x is a double-precision floating-point number.');
else
error('x is not a double-precision floating-point number
    .');
end
```

In this example, the `isa` function checks if the variable x is of the `'double'` data type. If the condition is true, a message is displayed; otherwise, an error is thrown.

2.3.4 Performance Considerations

The choice of data type in MATLAB can have a significant impact on **performance**, both in terms of memory usage and computational speed. In general, it is recommended to use the smallest data type that can accurately represent the data, as smaller data types require less memory and can often be processed more efficiently

Here's an example that demonstrates the performance difference between using double-precision and single-precision floating-point numbers:

Listing 2.53 Performance comparison example

```
% Create double-precision and single-precision arrays
A_double = rand(1000, 1000);
A_single = single(A_double);

% Time matrix multiplication
tic
B_double = A_double * A_double';
t_double = toc;
```

```
tic
B_single = A_single * A_single';
t_single = toc;

disp(['Double-precision time: ', num2str(t_double), '
    seconds']);
disp(['Single-precision time: ', num2str(t_single), '
    seconds']);
```

In this example, two large matrices (A_double and A_single) are created, one using double-precision and the other using single-precision floating-point numbers. The time taken to perform matrix multiplication on each matrix is measured and displayed. Generally, the single-precision operation will be faster and require less memory than the double-precision operation, but at the cost of reduced precision.

2.3.5 Memory Allocation and Management

MATLAB provides several functions and techniques for **memory allocation and management**, which can be important for optimizing memory usage and performance, particularly when working with large datasets or running computationally intensive operations.

One way to manage memory is by using the whos function to display information about the variables in the workspace, including their data types and the amount of memory they are using. Here's an example:

Listing 2.54 Using whos to check memory usage

```
A = rand(1000, 1000);
B = single(A);
whos
```

The whos command will display information about the variables A and B, including their data types and the amount of memory they are using.

Another technique for managing memory is to use the clear function to remove variables from the workspace when they are no longer needed. This can free up memory for other operations. For example:

Listing 2.55 Using clear to free memory

```
A = rand(1000, 1000);
% ... perform some operations with A
clear A;
```

In this example, the variable A is removed from the workspace after it is no longer needed, freeing up the memory it was using.

MATLAB also provides functions for managing memory allocation and preallocation, such as maxNumCompThreads, memory, and malloc, which can be useful for optimizing performance in certain scenarios.

2.3.6 Ranges, Casting and Machine Epsilon

- **Ranges** in MATLAB refer to the span of values that a data type can represent. Different data types have different ranges, which determine the minimum and maximum values they can store. For example, the range of the double data type in MATLAB is approximately $\pm 1.7 \times 10^{308}$, while the range of the int32 data type is -2,147,483,648 to 2,147,483,647. It is crucial to be aware of the range of your data type to avoid overflow or underflow errors when performing calculations.
- **Casting** in MATLAB refers to the process of converting a value from one data type to another. MATLAB provides various casting functions, such as int8, int16, int32, int64, double, single, and so on, to facilitate data type conversions. Casting can be useful when you want to ensure that your calculations are performed with the desired data type or when you need to convert data between different types.
- **Machine Epsilon** is a concept that relates to the precision of floating-point arithmetic in a computer. It represents the smallest positive number that, when added to 1, yields a value greater than 1. In MATLAB, the value of machine epsilon can be obtained using the eps function. Machine epsilon is important to consider when dealing with numerical algorithms, as it affects the accuracy of calculations and can lead to issues such as round-off errors.

Understanding **ranges**, **casting**, and **machine epsilon** in MATLAB is crucial for ensuring accurate and reliable numerical computations. By being aware of the limitations of data types and the precision of floating-point arithmetic, you can make informed decisions when designing algorithms and performing calculations in MATLAB.

2.4 Operators

2.4.1 Arithmetic Operators

Arithmetic operators are used to perform mathematical operations like addition, subtraction, multiplication, division, and exponentiation. The basic arithmetic operators in MATLAB are:

- **Addition** (+)
- **Subtraction** (-)
- **Multiplication** (*)
- **Division** (/)
- **Exponentiation** (^)

Example 1: Basic arithmetic operations

Listing 2.56 Arithmetic operations in MATLAB

```
a = 5;
b = 3;

sum = a + b; % Addition: sum = 8
diff = a - b; % Subtraction: diff = 2
prod = a * b; % Multiplication: prod = 15
div = a / b; % Division: div = 1.6667
pow = a ^ b; % Exponentiation: pow = 125
```

Example 2: Performing operations on arrays

Listing 2.57 Array operations using arithmetic operators

```
A = [1 2 3; 4 5 6]; % 2x3 matrix
B = [7 8; 9 10; 11 12]; % 3x2 matrix

C = A * B; % Matrix multiplication
D = A .^ 2; % Element-wise exponentiation
```

2.4.2   *Relational Operators*

Relational operators are used to compare two values or expressions and return a logical value (**true** or **false**). The main relational operators in MATLAB are:

- **Equal to** (==)
- **Not equal to** (=)
- **Greater than** (>)
- **Less than** (<)
- **Greater than or equal to** (>=)
- **Less than or equal to** (<=)

Example 1: Using relational operators on scalars

Listing 2.58 Relational operations on scalar values

```
a = 5;
b = 3;

equal = (a == b); % false
notEqual = (a ~= b); % true
greaterThan = (a > b); % true
lessThan = (a < b); % false
geq = (a >= b); % true
leq = (a <= b); % false
```

Example 2: Applying relational operators on arrays

Listing 2.59 Relational operations on arrays

```
A = [1 2 3; 4 5 6];
B = [1 3 5; 4 6 8];

C = (A == B); % Element-wise comparison
D = (A > 2); % Comparison with scalar
```

2.4.3 Logical Operators

Logical operators are used to combine or negate logical values (**true** or **false**). The main logical operators in MATLAB are:

- **Logical AND:** (&)
- **Logical OR:** |
- **Logical NOT:** ∼

The **logical AND** operator (&) returns **true** if both operands are **true**, and **false** otherwise. The **logical OR** operator (|) returns **true** if at least one of the operands is **true**, and **false** if both operands are **false**. The **logical NOT** operator ( ) negates the logical value of the operand, returning **true** if the operand is **false**, and **false** if the operand is **true**.

Example 1: Using logical operators on scalar values

Listing 2.60 Logical operations on scalar values

```
a = true;
b = false;

and_result = a & b; % false
or_result = a | b; % true
not_result = ~a; % false
```

Example 2: Applying logical operators on arrays

Listing 2.61 Logical operations on arrays

```
A = [1 2 3; 4 5 6];
B = [1 3 5; 4 6 8];

C = (A > 2) & (B < 6); % Element-wise logical AND
D = ~(A == B); % Element-wise logical NOT and comparison
```

Logical operators are particularly useful in conditional statements, such as if statements and switch statements, where they are used to evaluate conditions and control the flow of the program. They can also be used to perform element-wise operations on arrays of logical values.

2.4.4 Assignment Operators

Assignment operators are used to assign values to variables. The basic assignment operator in MATLAB is the equals sign (=). There are also compound assignment operators that combine an arithmetic operation with assignment.

- **Assignment** (=)
- **Addition assignment** (+=)
- **Subtraction assignment** (-=)
- **Multiplication assignment** (*=)
- **Division assignment** (/=)

  Example 1: Using the basic assignment operator

Listing 2.62 Basic assignment in MATLAB

```
x = 10; % Assign the value 10 to variable x
y = x; % Assign the value of x to variable y
```

  Example 2: Compound assignment operators

Listing 2.63 Compound assignment operators

```
a = 5;
a = a + 3; % a = 8
a += 3; % a = 11 (equivalent to a = a + 3)
a -= 2; % a = 9 (equivalent to a = a - 2)
a *= 2; % a = 18 (equivalent to a = a * 2)
a /= 3; % a = 6 (equivalent to a = a / 3)
```

2.4.5 Special Characters

In MATLAB, there are several special characters that serve specific purposes in programming. These characters are typically used in combination with other characters or words to represent certain operations or concepts. The paragraph below discusses some of the most commonly used special characters in MATLAB.

- **Percent sign** (%) is used to denote comments in MATLAB code. Any text following the percent sign on the same line is treated as a comment and is ignored by the MATLAB interpreter.
- **Semicolon** (;) is used to suppress the output of a command or expression. When a semicolon is placed at the end of a line, MATLAB will execute the command or expression but will not display the result in the Command Window.
- **Comma** (,) is used to separate elements or arguments in various contexts, such as function calls, array creation, and cell array creation.

- **Colon** (:) has multiple uses in MATLAB. It can be used to create a range of values, represent all elements in a particular dimension of an array, or specify a step size when creating a range.
- **Dot** (.) is used for several purposes, including accessing properties or methods of objects, performing element-wise operations on arrays, and concatenating strings.
- **Underscore** (_) is commonly used in variable and function names to improve readability, particularly in cases where multiple words are involved.

Here are two examples demonstrating the use of special characters in MATLAB:

Listing 2.64 Using special characters in MATLAB.

```
% This is a comment using the percent sign (%)
a = 1:5; % Creates a range from 1 to 5
b = [1, 2, 3]; % Comma separates elements
c = a .* b; % Dot performs element-wise multiplication
disp(c); % Displays the result

d = 'Hello'; e = 'World'; % Strings
f = strcat(d, '_', e); % Underscore concatenates strings
disp(f); % Output: 'Hello_World'
```

Listing 2.65 Using special characters in Indexing.

```
A = [1 2 3; 4 5 6; 7 8 9]; % Create a 3x3 matrix
B = A(:, 2:3); % Colon selects all rows and columns 2 to
    3
disp(B); % Output:
% 2 3
% 5 6
% 8 9
```

2.5 Expressions

In MATLAB, expressions are used to perform calculations and manipulate data. An expression is a combination of variables, constants, and operators that can be evaluated to produce a value. This section provides an overview of the different types of expressions and their usage within the MATLAB language.

2.5.1 Arithmetic Expressions

Arithmetic expressions in MATLAB involve mathematical operations such as addition, subtraction, multiplication, and division. These expressions are evaluated based on the precedence of operators and can include parentheses to control the order of operations.

Example: Arithmetic Expressions.

Arithmetic expressions involve mathematical operations such as addition, subtraction, multiplication, and division. These expressions follow the standard mathematical conventions and can be used to perform calculations on numeric data in MATLAB. The basic arithmetic operators in MATLAB are:

- **Addition**: The '+' operator is used to add two values together.
- **Subtraction**: The '-' operator is used to subtract one value from another.
- **Multiplication**: The '*' operator is used to multiply two values.
- **Division**: The '/' operator is used to divide one value by another.
- **Exponentiation**: The '^' operator is used to raise a value to a power.

For example, the expression $a = b + c$ adds the values of variables b and c and assigns the result to variable a. Similarly, expressions like $d = e * f$ and $g = h/i$ perform multiplication and division operations, respectively. It is important to note that MATLAB follows the order of operations (PEMDAS/BODMAS) when evaluating arithmetic expressions.

```matlab
1   % Addition
2   a = 3 + 4;
3
4   % Subtraction
5   b = 10 - 5;
6
7   % Multiplication
8   c = 2 * 6;
9
10  % Division
11  d = 15 / 3;
12
13  % Exponentiation
14  e = 2 ^ 4;
```

2.5.2 Relational Expressions

Relational expressions are used to compare values and determine the relationship between them. These expressions return logical (Boolean) values of either true or false. The relational operators available in MATLAB are:

- **Equal to**: The '==' operator checks if two values are equal.
- **Not equal to**: The ' =' operator checks if two values are not equal.
- **Greater than**: The '>' operator checks if one value is greater than another.
- **Less than**: The '<' operator checks if one value is less than another.
- **Greater than or equal to**: The '>=' operator checks if one value is greater than or equal to another.

- **Less than or equal to**: The '$<=$' operator checks if one value is less than or equal to another.

These relational operators are commonly used in conditional statements and loops to control the flow of program execution based on certain conditions.

Example: Relational Expressions

```
% Equal to
a = 5;
b = 5;
result1 = a == b;

% Not equal to
c = 3;
d = 7;
result2 = c ~= d;

% Greater than
e = 10;
f = 7;
result3 = e > f;

% Less than
g = 4;
h = 6;
result4 = g < h;

% Greater than or equal to
i = 8;
j = 8;
result5 = i >= j;

% Less than or equal to
k = 5;
l = 7;
result6 = k <= l;
```

2.5.3 Logical Expressions

Logical expressions involve logical (Boolean) operators that combine multiple relational expressions. These operators allow for the evaluation of complex conditions by combining simple conditions. The logical operators available in MATLAB are:

- **Logical AND**: The '&&' operator returns true if both conditions are true.
- **Logical OR**: The '||' operator returns true if at least one condition is true.
- **Logical NOT**: The '~' operator negates the result of a condition.

These logical operators are fundamental in constructing conditional statements and making decisions based on multiple conditions. These operators are used to combine multiple conditions and evaluate the overall truth value of an expression.

Example: Logical Expressions.

```matlab
% Logical AND
a = 5;
b = 7;
result1 = (a > 0) && (b < 10);

% Logical OR
c = 3;
d = 11;
result2 = (c < 5) || (d > 10);

% Logical NOT
e = true;
result3 = ~e;
```

2.5.4 String Expressions

In addition to numeric computations, MATLAB also supports string manipulation. String expressions involve operations on character arrays or strings, such as concatenation, substring extraction, and comparison. Strings can be enclosed in single quotes ('') or double quotes ("") in MATLAB.

String expressions in MATLAB are used to manipulate and concatenate character arrays. String concatenation is performed using the + operator, and string comparison is done using the relational operators (==, =, etc.) as with other data types.

Example: String Expressions.

```matlab
% String concatenation
str1 = 'Hello';
str2 = 'World';
result1 = [str1, ' ', str2];

% Substring extraction
str = 'MATLAB';
result2 = str(2:4);

% String comparison
str1 = 'apple';
str2 = 'banana';
result3 = strcmp(str1, str2);
```

2.5.5 Function Expressions

Function expressions (Function calls) in MATLAB involve the use of built-in functions or user-defined functions to perform specific tasks. Function expressions are formed by providing input arguments to a function, which then returns the desired output. These expressions follow the syntax of the function name followed by parentheses, which may contain input arguments if required. Function calls allow for the execution of predefined algorithms or user-defined procedures to perform specific tasks.

Example: Function Expressions.

```
% Built-in function call
result1 = sin(pi/2);

% User-defined function call
result2 = myFunction(x, y, z);
```

In addition to these types of expressions, MATLAB also supports other advanced concepts such as indexing expressions, array operations, and control flow expressions. These topics will be covered in more detail in later sections.

2.5.6 Array and Matrix Expressions

```
% Element-wise operations
a = [1, 2, 3];
b = [4, 5, 6];
result1 = a + b;

% Matrix multiplication
c = [1, 2; 3, 4];
d = [5, 6; 7, 8];
result2 = c * d;

% Matrix inversion
e = [1, 2; 3, 4];
result3 = inv(e);
```

2.6 Statement

In MATLAB, **statements** are the building blocks of programs and are used to perform various operations and manipulations on data. Statements can be **expressions** that compute values, **assignments** that store values in variables, or **control flow** statements that control the execution of other statements based on certain conditions or loops.

MATLAB supports several types of statements, including **arithmetic**, **relational**, **logical**, and **assignment** statements. These statements can be combined using **operators** and **expressions** to create more complex computations and algorithms.

- **Arithmetic Statements**

  Arithmetic statements in MATLAB involve the use of **arithmetic operators** such as addition (+), subtraction (-), multiplication (*), division (), and exponentiation (ˆ). These operators can be used to perform arithmetic operations on numeric data types, including scalars, vectors, and matrices.

  The basic syntax for arithmetic statements in MATLAB is:

Listing 2.66 Arithmetic Statement Syntax

```
result = operand1 operator operand2;
```

To demonstrate the use of arithmetic statements, consider the following example:

Listing 2.67 Arithmetic Statement Example

```
% Arithmetic operations
a = 5;
b = 3;

c = a + b; % Addition
d = a - b; % Subtraction
e = a * b; % Multiplication
f = a / b; % Division
g = a ^ b; % Exponentiation

% Display results
disp(['Sum: ' num2str(c)]);
disp(['Difference: ' num2str(d)]);
disp(['Product: ' num2str(e)]);
disp(['Quotient: ' num2str(f)]);
disp(['Power: ' num2str(g)]);
```

This example performs various arithmetic operations on the variables 'a' and 'b', and displays the results using the **disp** and **num2str** functions.

- **Relational and Logical Statements**

  Relational statements in MATLAB involve the use of **relational operators** to compare values and produce logical results (**true** or **false**). The relational operators used in MATLAB are:

  - **Less than** ($<$): Returns true if the operand on the left is less than the operand on the right.
  - **Greater than** ($>$): Returns true if the operand on the left is greater than the operand on the right.
  - **Less than or equal to** ($<=$): Returns true if the operand on the left is less than or equal to the operand on the right.

– **Greater than or equal to** ($>=$): Returns true if the operand on the left is greater than or equal to the operand on the right.
– **Equal to** ($==$): Returns true if the operands on both sides are equal.
– **Not equal to** ($=$): Returns true if the operands on both sides are not equal.

Logical statements, on the other hand, involve the use of **logical operators** to combine or manipulate logical values (**true** or **false**). The logical operators used in MATLAB are:

– **Logical AND** (&): Returns true if both operands are true.
– **Logical OR** (|): Returns true if either or both operands are true.
– **Logical NOT** (): Returns the opposite logical value of the operand.

The basic syntax for relational and logical statements in MATLAB is:

Listing 2.68 Relational and Logical Statement Syntax

```
logical_result  =  operand1  relational_operator  operand2
    ;
combined_result  =  logical_value1  logical_operator
    logical_value2;
```

To illustrate the use of relational and logical statements, consider the following example:

Listing 2.69 Relational and Logical Statement Example

```
% Relational and logical operations
a = 10;
b = 5;

c = a > b; % Relational operation (greater than)
d = a == b; % Relational operation (equal to)
e = c & ~d; % Logical operation (AND and NOT)
f = c | d; % Logical operation (OR)

% Display results
disp(['a > b: ' num2str(c)]);
disp(['a == b: ' num2str(d)]);
disp(['(a > b) AND NOT (a == b): ' num2str(e)]);
disp(['(a > b) OR (a == b): ' num2str(f)]);
```

In this example, the relational operations a > b and a == b are performed, and the results are stored in the variables c and d, respectively. Then, the logical operations c & $\tilde{d}$ (AND and NOT) and c | d (OR) are performed, and the results are stored in the variables e and f, respectively. Finally, the results of these operations are displayed using the **disp** and **num2str** functions.

Relational and logical statements are widely used in MATLAB for various purposes, such as conditional execution, loop control, and data manipulation. They are essential for implementing decision-making and flow control in MATLAB programs.

Listing 2.70 Relational and Logical Statements Example

```
% Define some variables
a = 10;
b = 5;
c = 7;
d = 3;

% Relational statements
disp('Relational Statements:')
disp(['a > b: ' num2str(a > b)]); % Returns 1 (true)
disp(['a < b: ' num2str(a < b)]); % Returns 0 (false)
disp(['a == c: ' num2str(a == c)]); % Returns 0 (false
    )
disp(['b >= d: ' num2str(b >= d)]); % Returns 1 (true)
disp(['c ~= d: ' num2str(c ~= d)]); % Returns 1 (true)

% Logical statements
disp('Logical Statements:')
disp(['(a > b) & (c > d): ' num2str((a > b) & (c > d))
    ]); % Returns 1 (true)
disp(['(a < b) | (c > d): ' num2str((a < b) | (c > d))
    ]); % Returns 1 (true)
disp(['(a == c): ' num2str((a == c))]); % Returns 1 (
    true)

% Combination of relational and logical statements
disp('Combination of Relational and Logical Statements
    :')
disp(['((a > b) & (c > d)) | ((a < b) & (c < d)): '
    ...
num2str(((a > b) & (c > d)) | ((a < b) & (c < d)))]);
    % Returns 1 (true)
```

In this example, we first define four variables a, b, c, and d using assignment statements. Then, we demonstrate the use of relational operators ($>$, $<$, ==, $>=$, =) to perform comparisons and store the logical results (0 for false, 1 for true) in the MATLAB console.

- **Assignment Statements**

Assignment statements in MATLAB are used to store values in variables. The basic syntax for an assignment statement is:

Listing 2.71 Assignment Statement Syntax

```
variable = expression;
```

Where variable is the name of the variable to be assigned, and expression is a valid MATLAB expression that evaluates to a value.

To demonstrate the use of assignment statements, consider the following example:

Listing 2.72 Assignment Statement Example

```
% Assignment statements
a = 10;
b = 3.14;
c = 'hello';
d = true;

% Display variable values
disp(['Value of a: ' num2str(a)]);
disp(['Value of b: ' num2str(b)]);
disp(['Value of c: ' c]);
disp(['Value of d: ' num2str(d)]);
```

In this example, various data types (integer, floating-point, string, and logical) are assigned to variables 'a', 'b', 'c', and 'd', respectively. The values of these variables are then displayed using the **disp** and **num2str** functions.

2.7 Laboratory

This section provides several lab works and exercises to help reinforce the concepts covered in this chapter.

1. Arithmetic Statements

   a. Write a MATLAB script that performs the following arithmetic operations:

      - Addition of two scalars
      - Subtraction of two vectors
      - Multiplication of a scalar and a matrix
      - Division of two scalars
      - Exponentiation of a scalar and a vector

   Solution:

Listing 2.73 Arithmetic Statements Lab Work

```
      % Arithmetic statements lab work

% Addition of two scalars
a = 5;
b = 3;
c = a + b;
disp(['Addition: ' num2str(c)]); % Output:
    Addition: 8

% Subtraction of two vectors
x = [10, 20, 30];
y = [5, 10, 15];
z = x - y;
```

```
disp('Subtraction: ');
disp(z); % Output: Subtraction: 5 10 15

% Multiplication of a scalar and a matrix
A = [1, 2; 3, 4];
k = 2;
B = k * A;
disp('Multiplication: ');
disp(B); % Output: Multiplication: 2 4
% 6 8

% Division of two scalars
p = 10;
q = 5;
r = p / q;
disp(['Division: ' num2str(r)]); % Output:
    Division: 2

% Exponentiation of a scalar and a vector
s = 2;
t = [1, 2, 3];
u = s .^ t; % Element-wise exponentiation
disp('Exponentiation: ');
disp(u); % Output: Exponentiation: 2 4 8
```

b. Create a MATLAB script that computes the area and circumference of a circle given its radius. Display the results with appropriate labels.
Solution:

Listing 2.74 Circle Area and Circumference

```
% Compute area and circumference of a circle
radius = 5; % Radius of the circle

% Constants
pi = 3.14159;

% Calculations
area = pi * radius^2;
circumference = 2 * pi * radius;

% Display results
disp(['Area of the circle: ' num2str(area)]);
disp(['Circumference of the circle: ' num2str(
    circumference)]);
```

2. Relational and Logical Statements

a. Write a MATLAB script that compares two matrices element-wise and displays the indices of the elements where the condition is true.

Solution:

Listing 2.75 Matrix Comparison

```
      % Compare two matrices element-wise
A = [1, 2, 3; 4, 5, 6; 7, 8, 9];
B = [2, 3, 1; 5, 6, 4; 8, 9, 7];

% Element-wise comparison
C = A > B;

% Find indices where condition is true
[row, col] = find(C);

% Display indices
disp('Indices where A > B:');
disp([row, col]);
```

b. Create a MATLAB script that checks if a given number is even or odd using a relational and logical statement.
   Solution:

Listing 2.76 Even or Odd

```
      % Check if a number is even or odd
num = 17;

% Check for even or odd
isEven = mod(num, 2) == 0;

% Display result
if isEven
disp([num2str(num) ' is an even number.']);
else
disp([num2str(num) ' is an odd number.']);
end
```

3. Assignment Statements

   a. Write a MATLAB script that creates a vector of 10 random integers between 1 and 100, sorts the vector in ascending order, and assigns the sorted vector to a new variable.
      Solution:

Listing 2.77 Random Vector Sorting

```
      % Create a vector of 10 random integers
randomVector = randi([1, 100], 1, 10);

% Sort the vector in ascending order
```

```
sortedVector = sort(randomVector);

% Display the original and sorted vectors
disp('Original vector:');
disp(randomVector);
disp('Sorted vector:');
disp(sortedVector);
```

b. Create a MATLAB script that prompts the user to enter their name and age, stores the input values in appropriate variables, and displays a greeting message with the user's name and age.
Solution:

Listing 2.78 User Input and Greeting

```
    % Prompt user for input
name = input('Enter your name: ', 's');
age = input('Enter your age: ');

% Display greeting message
disp(['Hello, ' name '! You are ' num2str(age) '
    years old.']);
```

4. MATLAB Expressions

a. Write a MATLAB script that evaluates the following expression: $f(x) = x^3 - 2x^2 + 3x - 5$ for a given value of x.
Solution:

Listing 2.79 Evaluating a Function Expression

```
    % Evaluate the function f(x) = x^3 - 2x^2 + 3x
        - 5
x = 2; % Given value of x

% Evaluate the expression
f = x^3 - 2x^2 + 3x - 5;

% Display the result
disp(['For x = ' num2str(x) ', f(x) = ' num2str(f)
    ]);
```

b. Create a MATLAB script that computes the sum of the first n natural numbers using the formula: $S_n = \frac{n(n+1)}{2}$. Prompt the user to enter the value of n.
Solution:

Listing 2.80 Sum of First n Natural Numbers

```
% Prompt user for input
n = input('Enter the value of n: ');
```

```
% Compute the sum of the first n natural numbers
sum_n = n * (n + 1) / 2;

% Display the result
disp(['The sum of the first ' num2str(n) ' natural
    numbers is: ' num2str(sum_n)]);
```

5. Debugging

a. Debug the following MATLAB script that is supposed to compute the area of a rectangle but contains an error.

Listing 2.81 Rectangle Area (with Error)

```
% Compute the area of a rectangle
length = 5;
width = 3;

area = length * width % Missing semicolon

% Display the result
disp('The area of the rectangle is:')
disp(area)
```

Solution: The error in the script is the missing semicolon at the end of the line 'area = length * width'. MATLAB interprets this line as a command to display the result of 'length * width' instead of assigning it to the variable 'area'. To fix the error, add a semicolon at the end of the line:

Listing 2.82 Rectangle Area (Corrected)

```
% Compute the area of a rectangle
length = 5;
width = 3;

area = length * width; % Semicolon added

% Display the result
disp('The area of the rectangle is:')
disp(area)
```

b. Debug the following MATLAB script that is supposed to compute the volume of a sphere but contains an error.

Listing 2.83 Sphere Volume (with Error)

```
% Compute the volume of a sphere
radius = 3;
pi = 3.14159;
```

```
volume = 4/3 * pi * radius^3 % Missing parentheses

% Display the result
disp(['The volume of the sphere is: ' num2str(
    volume)]);
```

Solution: The error in the script is the missing parentheses around the expression '

$$V = \frac{4}{3}\pi radius^3$$

'. MATLAB evaluates the expression from left to right, so it first divides 4 by 3 and then multiplies the result by 'pi' and '

$$radius^3$$

'. To fix the error, add parentheses around the expression:

Listing 2.84 Sphere Volume (Corrected)

```
% Compute the volume of a sphere
radius = 3;
pi = 3.14159;

volume = (4/3) * pi * radius^3 % Parentheses added

% Display the result
disp(['The volume of the sphere is: ' num2str(
    volume)]);
```

6. Creating a Class

Listing 2.85 Creating a class in MATLAB.

```
classdef Vehicle
properties
Make
Model
Year
end

methods
    function obj = Vehicle(make, model, year)
        obj.Make = make;
        obj.Model = model;
        obj.Year = year;
    end

    function display(obj)
        disp(['This is a ', num2str(obj.Year), ' ',
            obj.Make, ' ', obj.Model]);
    end
```

```
end
end

% Creating an instance of the Vehicle class
myCar = Vehicle('Toyota', 'Corolla', 2018);
myCar.display(); % Output: This is a 2018 Toyota
    Corolla
```

7. Creating a Structure

Listing 2.86 Creating a structure in MATLAB.

```
student = struct('Name', 'Alice', 'Age', 20, 'Major',
    'Computer Science');

% Accessing structure fields
disp(['Name: ', student.Name]);
disp(['Age: ', num2str(student.Age)]);
disp(['Major: ', student.Major]);
```

8. Defining an enumeration in MATLAB

Listing 2.87 Defining an enumeration.

```
TrafficLight = enumeration('TrafficLight', 'Red', '
    Yellow', 'Green')

TrafficLight =

1*enumeration
Red Yellow Green

TrafficLight enumeration
```

In the above example, we define an enumeration called `TrafficLight` with three members: `Red`, `Yellow`, and `Green`. The enumeration members are assigned integer values starting from 1 by default.

Enumerations can be used in various contexts, such as switch statements, comparisons, and function arguments. Here's an example of using the `Traffic Light` enumeration in a switch statement:

Listing 2.88 Using an enumeration in a switch statement.

```
currentLight = TrafficLight.Yellow;

switch currentLight
case TrafficLight.Red
disp('Stop')
case TrafficLight.Yellow
disp('Slow down')
case TrafficLight.Green
disp('Go')
end
```

This code will output:

```
Slow down
```

9. Practice Creating Variables and Type Conversion
   1. Create a variable date1, equal to the date of today 2. Create a variable num, equal to 100. 3. Create a variable date2, equal to date1+num. 4. View the size and type of the above variables.

Listing 2.89 Practice Creating Variables and Type Conversion

```
date1 = datetime('today');
num = 100;
date2 = date1 + num;
% View the size and type of the above variables
whos date1 date2 date3 num str is_same diff
```

10. Display the range of various data types.

Listing 2.90 Display the range of various data types.

```
%   Integer
[intmin("int8")  intmax("int8")]
[intmain("int16")  intmax("int16")]
[intmain("int32")  intmax("int32")]
[intmain("int64")  intmax("int64")]
[intmin("uint8")  intmax("uint8")]
[intmain("uint16")  intmax("uint16")]
[intmain("uint32")  intmax("uint32")]
[intmain("uint64")  intmax("uint64")]

% Floating number
[-realmax('single')  -realmin('single')  realmin('
    single')  realmax('single')]
[-realmax  -realmin  realmin  realmax]
```

11. Cast various data types.

   a. Cast an all-1 array to a complex array.

Listing 2.91 Cast various data types.

```
A = ones(2,3);
p = complex(1,1);
B = cast(A, 'like', p);
```

   b. Cast a hexadecimal floating-point number to a decimal floating-point number.

Listing 2.92 Cast a hexadecimal floating-point number to a decimal floating-point number.

```
a = 'b6eae18b';
c = typecast(uint32(hex2dec(a)),'single');
```

12. Display the Machine Epsilon of the floating point number.
    Determine whether the results of the three equations (same numbers, different order) are equal.

Listing 2.93 Display the Machine Epsilon of the floating point number.

```
x1 = 0.33 - 0.5 + 0.17;
x2 = 0.33 + 0.17 - 0.5;
x3 = 0.17 - 0.5 + 0.33;

% Show results
fprintf('The value of x1 is %d.\n',x1);
fprintf('The value of x2 is %d.\n',x2);
fprintf('The value of x3 is %d.\n',x3);

% Use "==" to determine
disp(x1 == x2);
disp(x1 == x3);
disp(x2 == x3);

%  Determine with the precision
disp(abs(x1-x2)<eps);
disp(abs(x1-x3)<eps);
disp(abs(x2-x3)<eps);
```

13. Practice the use of Arithmetic Operators.
    Calculate the values of the following mathematical expressions:

$$\sin\left(\frac{|15 - 91 + 7|^2}{8}\right) \tag{2.1}$$

$$5.8^{(2.4 \times 1.9)} \tag{2.2}$$

$$\tan\left(\sqrt{2}\right) + \log\left(\cos\left(\frac{\pi}{2}\right)\right) \tag{2.3}$$

$$\exp\left(1 + \sin(10)\right) \tag{2.4}$$

Listing 2.94 Practice the use of Arithmetic Operators.

```
sin(abs(15-91+7)^2/8)
5.8^(2.4*1.9)
tan(sqrt(2))+log(cos(pi/2))
exp(1+sin(10))
```

14. Practice the use of Relational Operators.

    Find the location of numbers greater than 60 in the vector A = [203 15 9 64 52 47 87 9 11], then extract those numbers.

Listing 2.95 Practice the use of Relational Operators.

```
A = [203 15 9 64 52 47 87 9 11];
% Use the relational operator ">".

I = find(A > 60)
A(I)
```

The running result shows that the values greater than 60 in the vector are 203, 64, 87, and their positions are 1, 4, and 7 respectively.

15. Practice the use of Logical Operators.

    Find and extract all even numbers in the vector [67 83 46 92 8 332 26 583].

Listing 2.96 Practice the use of Relational Operators.

```
A = [67 83 46 92 8 332 26 583];
B = mod(A,2);
% Use the logical operator "~".
C = ~B;
A(C)
```

The results show that the even numbers in the vector are 46, 92, 8, 332, and 26.

16. Practice the use of Assignment Operators.

    a. Practice using "=" assignment and deal function assignment.

Listing 2.97 Practice the use of Assignment Operators.

```
% Direct assignment
x1 = 88;
fprintf('The value of x1 is %d.\n',x1);

% Assign values using the deal function
[y1, y2, y3] = deal(11);
fprintf('The value of y1 is %d.\n',y1);
fprintf('The value of y2 is %d.\n',y2);
fprintf('The value of y3 is %d.\n',y3);
[z1, z2, z3] = deal(21, 22, 33);
fprintf('The value of z1 is %d.\n',z1);
fprintf('The value of z2 is %d.\n',z2);
fprintf('The value of z3 is %d.\n',z3);
```

```
Results:
The value of x1 is 88.
The value of y1 is 11.
The value of y2 is 11.
The value of y3 is 11.
```

```
The value of z1 is 21.
The value of z2 is 22.
The value of z3 is 33.
```

b. Load the file height.mat

    1. Load the file height.mat. 2. Create age by converting ageMos from months to years. 3. Create avgM and avgF by converting avgMcm and avgFcm from centimeters to feet (1 ft = 30.48 cm). 4. Plot both converted heights versus age. 5. Add a title, axis labels, and a legend. 6. Calculate and plot the height difference between genders in inches. 7. Add a black dashed line where the difference is zero.

Listing 2.98 Practice the use of Assignment Operators.

```
edit heightByAge_template.mlx
edit heightByAge.mlx
```

17. Evaluate the expression

$$\frac{5 + \cos\left(\frac{49\pi}{180}\right)}{13 + \sqrt{7} - 2i} \tag{2.5}$$

Evaluate the expression 2.5, assign the result to variable x, and then determine whether variable x is greater than 0.

Listing 2.99 Evaluate the expression

```
% Evaluate the expression in Equation
numerator = 5 + cos(49*pi/180);
denominator = 13 + sqrt(7) - 2i;
x = numerator / denominator;

% Determine if x is greater than 0
if real(x) > 0
    disp('x is greater than 0.')
else
    disp('x is not greater than 0.')
end
```

2.8 Problems

This section provides several problems to help reinforce the concepts covered in this chapter.

1. Write a MATLAB script that takes two user inputs (a and b) and computes the following expression: $y = \frac{a^2 + b^3}{a - 2b}$. The script should handle the case where the denominator is zero and display an appropriate error message.

2. Create a MATLAB script that generates a random 3×3 matrix with integer values between 1 and 10. The script should then find the maximum and minimum values in the matrix and display their indices.

3. Write a MATLAB function that takes two vectors as input and returns their dot product. The function should handle the case where the input vectors have different lengths and display an appropriate error message.

4. Create a MATLAB script that generates a random vector of length 10 with integer values between 1 and 20. The script should then count the number of occurrences of each value in the vector and display the results.

5. Write a MATLAB script that prompts the user to enter a string. The script should then count the number of vowels (a, e, i, o, u) and consonants in the string and display the results.

2.9 Summary

This chapter covers the fundamental concepts of data types, expressions, and statements in MATLAB. It provides a comprehensive introduction to the different types of data that can be represented and manipulated in MATLAB, including scalars, vectors, matrices, and strings. The chapter also explores arithmetic, relational, and logical expressions, as well as assignment statements, which are essential for performing calculations and assigning values to variables.

- **Data Types**: MATLAB supports various data types, including numeric (double, single, integer), logical, character, and cell arrays. Understanding the different data types and their properties is crucial for effective data manipulation and analysis.
- **Arithmetic Expressions**: Arithmetic expressions involve mathematical operations such as addition, subtraction, multiplication, division, and exponentiation. MATLAB provides a range of arithmetic operators and functions to perform these operations on scalars, vectors, and matrices.
- **Relational and Logical Expressions**: Relational expressions compare values using operators like greater than, less than, equal to, and not equal to. Logical expressions combine relational expressions using logical operators (AND, OR, NOT) to form more complex conditions.
- **Assignment Statements**: Assignment statements are used to assign values to variables in MATLAB. These statements can involve scalars, vectors, matrices, or expressions, and they allow for efficient data manipulation and storage.
- **Expressions and Statements**: Expressions are combinations of variables, operators, and functions that produce a result. Statements, on the other hand, are instructions that perform actions or operations in MATLAB, such as assigning values, displaying results, or controlling program flow.
- **Debugging**: Debugging is an essential part of the programming process. This chapter provides examples of common errors and techniques for identifying and resolving them, such as adding missing semicolons or parentheses.

For undergraduate students, this chapter lays the foundation for understanding and working with data in MATLAB. It introduces the fundamental concepts and syntax necessary for performing basic calculations, manipulating data, and controlling program flow. By mastering these concepts, students will be better equipped to tackle more advanced topics and applications in subsequent chapters.

For postgraduate students and researchers, this chapter serves as a refresher and reinforcement of the core principles of MATLAB programming. While the concepts covered may seem basic, they are crucial for writing efficient, error-free, and maintainable code. Additionally, the examples and exercises provided can help strengthen problem-solving skills and prepare researchers for more complex data analysis and modeling tasks.

For professional engineers and practitioners, this chapter can serve as a reference for the essential building blocks of MATLAB programming. It provides a concise overview of data types, expressions, and statements, which are fundamental to many engineering applications, such as signal processing, control systems, and data analysis. By solidifying their understanding of these concepts, professionals can enhance their productivity and efficiency when working with MATLAB in various industries and domains.

Overall, this chapter serves as a comprehensive introduction to the fundamental concepts of data types, expressions, and statements in MATLAB. It provides a strong foundation for individuals at all levels, from undergraduate students to professional engineers, to effectively utilize MATLAB for a wide range of applications and domains.

References

1. MathWorks, "Data Types," [Online]. Available: https://www.mathworks.com/help/matlab/data-types.html, accessed on Feb. 17, 2024
2. MathWorks, "Fundamental MATLAB Classes," [Online]. Available: https://www.mathworks.com/help/matlab/matlab_prog/fundamental-matlab-classes.html, accessed on Feb. 17, 2024
3. MathWorks, "Data Type Identification," [Online]. Available: https://www.mathworks.com/help/matlab/data-type-identification.html, accessed on Feb. 17, 2024
4. MathWorks, "Data Type Conversion," [Online]. Available: https://www.mathworks.com/help/matlab/data-type-conversion.html, accessed on Feb. 17, 2024
5. MathWorks, "Convert from Structure to Mindmap," [Online]. Available: https://www.mathworks.com/matlabcentral/fileexchange/43654-struct2mindmap-a-structure-to-mindmap-converter, accessed on Feb. 17, 2024
6. MathWorks, "Operators and Special Characters," [Online]. Available: https://www.mathworks.com/help/matlab/matlab_prog/matlab-operators-and-special-characters.html, accessed on Feb. 17, 2024
7. MathWorks, "String Functions", 2023. [Online]. Available: https://www.mathworks.com/help/matlab/characters-and-strings.html
8. R. Pratap, "Getting Started with MATLAB: A Quick Introduction for Scientists and Engineers", 7th ed. Oxford: Oxford University Press, 2016
9. S. Attaway, "MATLAB: A Practical Introduction to Programming and Problem Solving", 5th ed. Oxford: Butterworth-Heinemann, 2018
10. MathWorks, "Signal Processing Toolbox" [Online]. Available: https://uk.mathworks.com/help/signal/index.htm, accessed on Feb. 17, 2024

Chapter 3
Vectors, Arrays, Matrices, and Data Structures

Chapter Learning Outcomes

- Understand the concepts of **vectors**, **arrays**, and **matrices** in MATLAB, and their differences.
- Create and manipulate **vectors** and **matrices** using various techniques.
- Perform **array operations** such as addition, subtraction, multiplication, and element-wise operations.
- Utilise **indexing** and **slicing** to access and modify specific elements or subsets of arrays.
- Work with **multi-dimensional arrays** and understand their applications.
- Explore **cell arrays** and **structures** as data structures for storing heterogeneous data.
- Apply **logical indexing** and **masking** techniques to extract and manipulate data based on conditions.

Chapter Key Words

- **Vector**: A one-dimensional array of elements, represented as a row or column in MATLAB. Vectors can store numeric, logical, or character data.
  Explanation: Vectors are fundamental data structures in MATLAB. They are used to represent ordered collections of scalar values, such as a list of numbers or characters. Vectors can be created using square brackets or colon notation, and they support various operations like arithmetic, indexing, and concatenation.
- **Matrix**: A two-dimensional array of elements, arranged in rows and columns. Matrices are widely used for representing and manipulating data in MATLAB.
  Explanation: Matrices are essential for numerical computations and data analysis in MATLAB. They can be created using square brackets or specific functions like 'zeros', 'ones', and 'eye'. Matrices support a wide range of operations, including arithmetic, matrix multiplication, transposition, and various specialized functions for linear algebra and signal processing.

© The Author(s) 2025
Y. Chen and L. Huang, *MATLAB Roadmap to Applications*,
https://doi.org/10.1007/978-981-97-8788-3_3

- **Array Operations**: Mathematical operations performed on arrays, such as addition, subtraction, multiplication, and element-wise operations (like squaring or taking the square root of each element).
  Explanation: Array operations are fundamental in MATLAB for performing computations on vectors and matrices. These operations can be performed element-wise (applying the operation to each element individually) or using matrix operations like multiplication and transposition, depending on the specific operation and the dimensions of the arrays involved.
- **Indexing and Slicing**: Techniques for accessing and modifying specific elements or subsets of arrays using indices or ranges of indices.
  Explanation: Indexing and slicing are powerful tools for accessing and manipulating data in MATLAB arrays. Indexing allows you to retrieve or modify individual elements or specific rows/columns of an array, while slicing enables you to extract or modify subsets of array elements based on specified ranges of indices.

3.1   Vector

MATLAB is a high-level language and interactive environment for numerical computation, visualisation, and programming, uses various types of data structures to store and manage data. Among these, **vectors**, **arrays**, **matrices**, and more general **data structures** are fundamental.

Vectors are **one-dimensional arrays** that can hold a sequence of numerical or categorical data. They are the fundamental building blocks for more complex data structures in MATLAB. MATLAB distinguishes between row vectors and column vectors, where a row vector is represented by a single row of elements and a column vector by a single column.

- A row vector is created by enclosing the elements within square brackets, separated by space or commas.

```
rowVec = [1 2 3 4]; % Creates a row vector
```

- A column vector is created by separating the elements with semicolons or by using the transpose operator on a row vector.

```
colVec = [1; 2; 3; 4]; % Creates a column vector
```

Listing 3.1 Vector Example

```
% Row vector
row_vector = [1, 2, 3]; % Creates a 1-by-3 row vector
% Column vector
column_vector = [1; 2; 3]; % Creates a 3-by-1 column vector
```

Vectors are crucial for performing mathematical operations and are commonly used in linear algebra and calculus, which can be created using various MATLAB functions and operators, such as:

- Using the [...] operator: x = [1, 2, 3, 4, 5];
- Using Constant values with the **zeros**(n) function: x = zeros(5, 1);
- Using Constant values with the **ones**(n) function: x = ones(5, 1);
- Using the Linear spacing function '**linspace**(a, b, n)', e.g. : x = linspace(0, 10, 5);
- Using the random numbers with '**rand**(n)'.

MATLAB provides various operations that can be performed on vectors, including:

- Addition: **x** = [1, 2, 3] +[4, 5, 6];
- Subtraction: **x** = [1, 2, 3] −[4, 5, 6];
- Multiplication: **x** = [1, 2, 3] ×[4, 5, 6];
- Division: **x** = [1, 2, 3] /[4, 5, 6];
- Modulus: **x** = [1, 2, 3]   mod [4, 5, 6];
- Dot Product: **x** = [1, 2, 3] ·[4, 5, 6];
- Magnitude: **x** = ‖[1, 2, 3]‖ or abs([1, 2, 3]);

Vectors can also be created using the colon notation, which generates a sequence of values between a specified start and end value. For instance:

Listing 3.2 Creating a vector using colon notation.

```
x = 1:5 % Creates the vector [1 2 3 4 5]
y = 10:-2:2 % Creates the vector [10 8 6 4 2]
```

Once created, vectors support various operations such as arithmetic operations, indexing, and concatenation. Here's an example of performing element-wise multiplication on two vectors:

Listing 3.3 Element-wise vector multiplication.

```
a = [1 2 3];
b = [4 5 6];
c = a .* b % Element-wise multiplication, c = [4 10 18]
```

Indexing allows accessing and modifying specific elements of a vector. For instance:

Listing 3.4 Vector indexing.

```
v = [10 20 30 40 50];
v(3) % Returns the third element, 30
v(2:4) = [25 35 45] % Modifies elements 2 to 4
```

Vectors play a crucial role in various mathematical operations, data analysis, and Visualisation tasks in MATLAB.

3.2 Arrays

Arrays are fundamental data structures in MATLAB that allow you to store and manipulate collections of values of the same data type. MATLAB arrays can be **multidimensional**, meaning they can have more than two dimensions, and can store various data types, including **numeric**, **logical**, **character**, and **cell** data.

Here is an example of creating and manipulating a numeric array in MATLAB:

Listing 3.5 Creating and manipulating a numeric array.

```
% Creating a numeric array
A = [1 2 3; 4 5 6; 7 8 9];

% Accessing elements of the array
element = A(2, 3); % Returns 6

% Modifying elements of the array
A(1, 2) = 10;

% Reshaping the array
B = reshape(A, 1, 9); % Converts A to a row vector

% Performing arithmetic operations
C = A + 2; % Adds 2 to each element of A
```

MATLAB also supports **logical arrays**, which are arrays containing only the logical values `true` and `false`. Logical arrays are commonly used for indexing, conditional operations, and boolean operations. Here's an example:

Listing 3.6 Working with logical arrays.

```
% Creating a logical array
A = [1 2 3; 4 5 6; 7 8 9];
B = A > 5; % B is a logical array

% Using logical arrays for indexing
C = A(B); % C contains only the elements of A greater than 5

% Performing logical operations
D = B & (A < 9); % D is a logical array with elements that satisfy
    both conditions
```

These examples demonstrate the versatility and power of arrays in MATLAB, which is a core feature of the language and essential for many numerical and data analysis tasks.

MATLAB provides a wide range of functions and operations for creating, manipulating, and analyzing arrays. These include functions for creating arrays, accessing and modifying array elements, reshaping arrays, performing arithmetic operations, and more.

One of the key features of MATLAB arrays is their **vectorization**, which allows element-wise operations to be performed on entire arrays simultaneously, without the need for explicit looping constructs. This vectorization capability contributes to the efficiency and conciseness of MATLAB code, making it a powerful tool for numerical computing and data analysis.

The term **array** in MATLAB refers to both one-dimensional (vectors) and multi-dimensional collections of elements. Thus, a vector is a special case of an array. An **array** in MATLAB, on the other hand, can be multi-dimensional, allowing for more complex data organisation, can contain elements of the same or different data types.

The simplest form of an array is a two-dimensional matrix.

Listing 3.7 Array Example

```
% 2D array
two_d_array = ones(2, 3); % Creates a 2-by-3 array of ones
% Multi-dimensional array
multi_d_array = zeros(2, 2, 2); % Creates a 2-by-2-by-2 array of
    zeros
```

Arrays can have multiple dimensions, making them suitable for a wide range of applications, including machine learning and data analysis.

- **Key Differences**: While a vector is a one-dimensional array, the term array encompasses vectors, matrices, and other multi-dimensional structures. The main differences between them include:
- **Dimensionality**: Vectors are one-dimensional, whereas arrays can be two or more dimensions.
- **Functionality**: Vectors are often used for mathematical operations, while arrays are used for complex data representation.
  - Vectors are typically used for operations that require a single series of elements, such as in physics for force vectors.
  - Arrays, especially multi-dimensional ones, are suited for more complex data representations, like images (which can be 2D) or time-series data across multiple variables (which can be 3D or higher).
- **Flexibility**: Arrays provide a more flexible data structure to accommodate higher-dimensional data.
  Arrays can be created and manipulated using MATLAB's array operations, such as:
  - Concatenation with the '**[ ]**' operator.
  - Element-wise operations with functions like '**sum()**', '**mean()**', and '**std()**'.
- **Usage in MATLAB Functions** Many MATLAB functions are designed to operate on both vectors and arrays. However, the behaviour of these functions may differ depending on whether they are applied to a vector or an array.

Listing 3.8 Function behaviour on vectors vs. arrays

```
% Applying the 'sum' function on a vector
sum_vector = sum(row_vector); % Sums the elements of the vector
```

```
% Applying the 'sum' function on a 2D array
sum_array = sum(matrix);  % Sums the elements of each laikocolumn
```

In conclusion, understanding the difference between vectors and arrays is pivotal in MATLAB programming, as it influences data structure organisation and the application of built-in functions.

3.3 Matrix

A matrix is a fundamental data type in **MATLAB**, representing a rectangular array of numerical values. Matrices are the building blocks for many operations and computations in MATLAB. They can store and manipulate data of various numeric types, including **double**, **single**, **integer**, and **logical** values. Usually, matrices are denoted by capital letters, such as A, B, M, N, and their elements can be accessed using row and column indices.

The matrix data type is versatile and powerful, enabling efficient numerical computations, linear algebra operations, and data analysis tasks. It supports a wide range of operations such as arithmetic calculations, matrix multiplication, transposition, and more. MATLAB provides numerous built-in functions and operators specifically designed for working with matrices, making it a highly efficient and convenient tool for scientific and engineering applications.

Here are two examples demonstrating the usage of matrices in MATLAB:

Listing 3.9 Creating and manipulating matrices

```
% Create a 3x3 matrix
A = [1 2 3; 4 5 6; 7 8 9]

% Access a specific element
element = A(2, 3) % Returns 6

% Perform matrix multiplication
B = [1 0; 2 1];
C = A * B

% Find the determinant of a matrix
det_A = det(A)
```

Listing 3.10 Working with sparse matrices

```
% Create a sparse matrix
S = sparse([1 1 2 3], [1 3 2 1], [5 7 6 8], 3, 4)

% Perform operations on sparse matrices
D = S * S' % Matrix multiplication
nnz(D) % Count non-zero elements
```

Matrices are a fundamental concept in linear algebra and are widely used in various scientific and engineering domains, such as signal processing, image analysis, control systems, and machine learning. MATLAB's extensive support for matrix operations makes it a powerful tool for developing and implementing algorithms that involve matrix computations.

- **Defining Matrices**
  A **matrix** is a two-dimensional array, which is particularly useful in linear algebra. In MATLAB, matrices are fundamental and are used for representing systems of linear equations, transformations, and more.

Listing 3.11 Matrix example

```
% Matrix
matrix = [1, 2, 3; 4, 5, 6; 7, 8, 9]; % Creates a 3-by-3 matrix
```

A matrix is a **two-dimensional array** of numbers, which is a common data structure used in linear algebra and numerical computations.

- **Matrix Operations**
  Matrix operations in MATLAB include:

  - Addition and subtraction.
  - Multiplication, including element-wise (.*) and matrix (*) multiplication.
  - Transpose with the '"' operator.
  - Sorting: MATLAB provides built-in functions for sorting data in various ways, including sorting vectors or matrices in ascending or descending order based on values or specific criteria.

3.4 Data Structures

In MATLAB, *vectors* and *arrays* are both data structures capable of storing a collection of elements. Despite their similarities, they are utilised for different purposes and possess distinct characteristics.

MATLAB also provides more advanced **data structures** like structures (struct arrays), cell arrays, and tables, which are used to organise related data of varied types and sizes.

In MATLAB, there are several data structures available to organise and manipulate data efficiently. These data structures serve different purposes and offer various functionalities. Here is a list of some commonly used data structures in MATLAB:

- **Vector**: A **vector** is a one-dimensional array that can store a sequence of elements. It is often used to represent a set of values, such as time series data or a list of coordinates. The basic syntax is:

Listing 3.12 Creating a vector

```
v = [1, 2, 3, 4]; % Row vector
v = [1; 2; 3; 4]; % Column vector
```

- **Array**: An **array** is a multi-dimensional data structure that can store elements of the same data type. It can have two or more dimensions, making it suitable for representing matrices or higher-dimensional data. The basic syntax is:

Listing 3.13 Creating a 2D array

```
A = [1, 2, 3; 4, 5, 6; 7, 8, 9];
```

- **Matrix**: A **matrix** is a two-dimensional array with rows and columns. It is extensively used in linear algebra operations, such as matrix multiplication, solving linear equations, and eigenvalue calculations. The basic syntax is:

Listing 3.14 Creating a matrix

```
M = [1, 2; 3, 4]; % 2x2 matrix
```

- **Cell**: A **cell** is a container that can hold elements of different data types. It is commonly used to store heterogeneous data, such as different-sized arrays or a combination of numbers, strings, and structures. The basic syntax is:

Listing 3.15 Creating a cell array

```
C = {1, 'text', [3, 4, 5]; 6, {7, 8}, 'string'};
```

- **Structures**: **Structures** allow grouping related data together under a single variable name. Each field within a structure can store different types of data, providing a convenient way to organise and access complex data. The basic syntax is:

Listing 3.16 Creating a structure

```
S.name = 'John';
S.age = 32;
S.scores = [85, 90, 88];
```

- **String**: A **string** is a sequence of characters enclosed in single quotes (' ') or double quotes (" "). MATLAB provides various string manipulation functions and operations to work with textual data efficiently. The basic syntax is:

Listing 3.17 Creating a string

```
str1 = 'Hello, world!';
str2 = "Hello, MATLAB!";
```

- **Table**: A **table** is a two-dimensional data structure that can store heterogeneous data in a tabular format. It is similar to a spreadsheet or a database table and offers easy indexing, sorting, and filtering capabilities. The basic syntax is:

Listing 3.18 Creating a table

```
T = table([1; 2; 3], {'A'; 'B'; 'C'}, [true; false; true],
    ...
    'VariableNames', {'ID', 'Label', 'Flag'});
```

It is important to note that the choice of data structure depends on the specific requirements of the problem at hand. Each data structure has its own advantages and trade-offs in terms of storage efficiency, access speed, and ease of manipulation. Understanding the characteristics and appropriate usage of these data structures is essential for efficient data organisation and manipulation in MATLAB.

Each of the data structures discussed has its own set of practical applications. Here, we provide a few examples to illustrate how these data structures can be used in real-world scenarios.

- **Vectors**

  - Example in Signal Processing

Listing 3.19 Example of using vectors in signal processing

```
% Create a time vector
t = 0:0.01:1;

% Generate a sine wave signal
signal = sin(2 * pi * 10 * t);

% Plot the signal
plot(t, signal);
xlabel('Time (s)');
ylabel('Amplitude');
title('Sine Wave Signal');
```

- **Arrays and Matrices**

  - Example in Image Processing

Listing 3.20 Example of using arrays in image processing

```
% Read an image into an array
img = imread('example.jpg');

% Convert the image to grayscale
gray_img = rgb2gray(img);

% Display the original and grayscale images
subplot(1, 2, 1);
imshow(img);
title('Original Image');

subplot(1, 2, 2);
imshow(gray_img);
title('Grayscale Image');
```

- **Cell Arrays and Structures**

  – Example of Structures for Experimental Data

Listing 3.21 Example of using structures for experimental data

```matlab
% Create a structure for experimental data
experiment.subject = 'Subject A';
experiment.date = '2024-05-16';
experiment.data = {rand(1, 10), rand(1, 10)};

% Access and display experiment data
disp(['Subject: ', experiment.subject]);
disp(['Date: ', experiment.date]);
disp('Data:');
disp(experiment.data);
```

- **Strings**

  – Example in Text Analysis

Listing 3.22 Example of using strings for text analysis

```matlab
% Define a string
text = "MATLAB is a powerful tool for numerical computation and
    visualization.";

% Find the position of a substring
pos = strfind(text, 'powerful');

% Extract a substring
substring = extractBetween(text, pos, pos + 7);

% Display the results
disp(['The word "powerful" starts at position: ', num2str(pos)]);
disp(['Extracted substring: ', substring]);
```

- **Tables**

  – Example in Data Analysis

Listing 3.23 Example of using tables for data analysis

```matlab
% Create a table with survey data
SurveyData = table([1; 2; 3], {'Male'; 'Female'; 'Female'}, [25;
    30; 35], ...
    'VariableNames', {'ID', 'Gender', 'Age'});

% Calculate the mean age
mean_age = mean(SurveyData.Age);\xm{\para}

% Display the results
```

```
disp(['The mean age of the survey participants is: ', num2str(
    mean_age)]);
```

3.5 Advanced Data Structures

In MATLAB, several advanced data structures are utilised to handle various computational tasks efficiently:

- **Sets**: Sets are not inherently provided as a built-in data structure in MATLAB. However, sets can be represented and manipulated using arrays or logical indexing operations in MATLAB.
- **Queues**: Similar to sets, queues are not directly provided as a built-in data structure in MATLAB. Nevertheless, queue-like behaviour can be implemented using MATLAB's arrays or cell arrays, leveraging appropriate indexing and operations.
- **Stacks**: Stacks are also not inherently provided as a built-in data structure in MATLAB. Like queues, you can simulate stack behaviour using MATLAB's arrays or cell arrays, employing suitable indexing and operations.
- **Graphs**: MATLAB offers a built-in data structure known as the **graph object**, specifically designed for representing and manipulating graphs consisting of nodes (vertices) and edges. The graph object in MATLAB provides various functions for creating, modifying, and analysing graphs.
- **Trees**: Trees are not natively provided as a built-in data structure in MATLAB. However, you can represent and work with trees using MATLAB's arrays or cell arrays, utilizing appropriate indexing and operations to navigate and manipulate the tree structure.
- **Tall Arrays**: **Tall arrays** are a specialised data structure in MATLAB designed for the efficient handling of large-scale data that exceeds the available memory. Tall arrays enable processing of data stored outside of memory, such as in a database or on disk, using lazy evaluation and parallel computing. MATLAB's tall arrays are specifically designed for efficient processing of big data sets .

While sets, queues, and stacks are not directly provided as built-in data structures in MATLAB, they can be implemented using arrays or cell arrays along with suitable indexing and operations. On the other hand, MATLAB provides native data structures for graphs and tall arrays, offering specialised functionality for working with graph-based data and large-scale data processing, respectively.

The basic syntax for creating and manipulating these data structures is as follows:

Listing 3.24 Basic syntax for data structures in MATLAB

```
% Example of Stack implementation using arrays
stack = [];
stack = [stack, new_element];  % Push operation
stack(end) = [];  % Pop operation
```

```
% Example of Queue implementation using arrays
queue = [];
queue = [queue, new_element];  % Enqueue operation
queue(1) = [];  % Dequeue operation
```

3.5.1 MATLAB Examples

Example 1: Stack Implementation

Listing 3.25 Stack implementation in MATLAB

```
% MATLAB code for stack operations
stack = [];

% Push operation
stack = [stack, 10];
stack = [stack, 20];
disp('Stack after push operations:');
disp(stack);

% Pop operation
stack(end) = [];
disp('Stack after pop operation:');
disp(stack);
```

Example 2: Queue Implementation

Listing 3.26 Queue implementation in MATLAB

```
% MATLAB code for queue operations
queue = [];

% Enqueue operation
queue = [queue, 10];
queue = [queue, 20];
disp('Queue after enqueue operations:');
disp(queue);

% Dequeue operation
queue(1) = [];
disp('Queue after dequeue operation:');
disp(queue);
```

3.5.2 Graph Data Structure

The **graph object** in MATLAB allows for creating, modifying, and analysing graphs. Below is an example demonstrating its use:

Listing 3.27 Graph object in MATLAB

```
% MATLAB code for creating and visualising a graph
nodes = {'A', 'B', 'C', 'D'};
edges = [1 2; 1 3; 2 4; 3 4];
G = graph(edges(:,1), edges(:,2), [], nodes);
plot(G);
title('Graph Representation');
```

3.5.3 Tree Data Structure

MATLAB does not natively support tree data structures, but they can be implemented using cell arrays or structures. Here is an example of a simple binary tree
implementation:

Listing 3.28 Binary tree implementation in MATLAB

```
% MATLAB code for binary tree operations
tree = struct('value', 10, 'left', [], 'right', []);

% Insert nodes
tree.left = struct('value', 5, 'left', [], 'right', []);
tree.right = struct('value', 15, 'left', [], 'right', []);
disp('Binary Tree:');
disp(tree);
```

3.5.4 Tall Arrays

Tall arrays are essential for handling large-scale data that cannot fit into memory.
They allow for out-of-memory computations using parallel processing. Below is an
example of how to create and use tall arrays:

Listing 3.29 Tall arrays in MATLAB

```
% MATLAB code for using tall arrays
ds = datastore('largeDataset.csv');
tallArray = tall(ds);

% Perform operations on tall arrays
meanValue = mean(tallArray.Var1);
gather(meanValue); % Collect result
```

Example 1: Creating Tall Arrays

Listing 3.30 Creating tall arrays in MATLAB

```
% MATLAB code for creating tall arrays from datastore
ds = datastore('data.csv', 'TreatAsMissing', 'NA', 'MissingValue',
    0);
ta = tall(ds);

% Display first few rows of the tall array
preview(ta);
```

Example 2: Processing Tall Arrays

Listing 3.31 Processing tall arrays in MATLAB

```
% MATLAB code for processing tall arrays
ds = datastore('data.csv');
ta = tall(ds);

% Calculate mean of a tall array column
meanValue = mean(ta.Var1);
result = gather(meanValue); % Gather result into memory
disp('Mean value:');
disp(result);
```

3.6 Laboratory

1. Working with Vectors

   a. Practice the use of vector.
      Given f(x) and g(x),

$$f(x) = 3x^5 - 5x^4 - x^3 + 9x + 6 \tag{3.1}$$

$$g(x) = 7x^4 + 5x - 3 \tag{3.2}$$

to solve:

$$f(x) + g(x) = (3x^5 - 5x^4 - x^3 + 9x + 6) + (7x^4 + 5x - 3) \tag{3.3}$$

$$f(x) - g(x) = (3x^5 - 5x^4 - x^3 + 9x + 6) - (7x^4 + 5x - 3) \tag{3.4}$$

$$f(x) \times g(x) = (3x^5 - 5x^4 - x^3 + 9x + 6) \times (7x^4 + 5x - 3) \tag{3.5}$$

$$f(x) \div g(x) = \frac{3x^5 - 5x^4 - x^3 + 9x + 6}{7x^4 + 5x - 3} \tag{3.6}$$

Solution:

Listing 3.32 Practice the use of vector

```
f = [3 -5 -1 0 9 6];
g = [7 0 0 5 -3];

result1 = f + [0 g]              % addition
result2 = f - [0 g]              % subtract
result3 = conv(f,g)         % multiply
[result4 r] = deconv(f,g)   % divide
```

Results

$$result1 = \begin{bmatrix} 3 \ 2 \ -1 \ 0 \ 14 \ 3 \end{bmatrix} \qquad (3.7)$$

$$result2 = \begin{bmatrix} 3 \ -12 \ -1 \ 0 \ 4 \ 9 \end{bmatrix} \qquad (3.8)$$

$$result3 = \begin{bmatrix} 21 \ -35 \ -7 \ 15 \ 29 \ 52 \ 3 \ 45 \ 3 \ -18 \end{bmatrix} \qquad (3.9)$$

$$result4 = \begin{bmatrix} 0.4286 \ -0.7143 \end{bmatrix} \qquad (3.10)$$

$$r = \begin{bmatrix} 0 \ 0 \ -1.0000 \ -2.1429 \ 13.8571 \ 3.8571 \end{bmatrix} \qquad (3.11)$$

b. Sort the elements

   i Create a row vector with elements from 1 to 10.

   ii Create a column vector with the square roots of the elements in the row vector.

   iii Calculate the dot product of the two vectors.

   vi Find the length (norm) of the column vector.

   iv Sort the elements of the row vector in descending order.

Solution:

Listing 3.33 Working with vectors

```
% Create a row vector with elements from 1 to 10
row_vector = 1:10

% Create a column vector with the square roots of the elements in
    the row vector
col_vector = sqrt(row_vector)'

% Calculate the dot product of the two vectors
dot_product = row_vector * col_vector

% Find the length (norm) of the column vector
vector_length = norm(col_vector)

% Sort the elements of the row vector in descending order
sorted_vector = sort(row_vector, 'descend')
```

2. Creating and Manipulating Matrices

   a. Create a 3x4 matrix

       i Create a 3x4 matrix with random integer values between 1 and 10.

      ii Find the maximum and minimum values in the matrix.

     iii Calculate the sum of all elements in the matrix.

     vi Extract the second row of the matrix.

      v Transpose the matrix.

Solution:

Listing 3.34 Creating and manipulating matrices

```
% Create a 3x4 matrix with random integer values between 1 and 10
A = randi(10, 3, 4)

% Find the maximum and minimum values in the matrix
max_value = max(A(:))
min_value = min(A(:))

% Calculate the sum of all elements in the matrix
sum_elements = sum(A, 'all')

% Extract the second row of the matrix
second_row = A(2, :)

% Transpose the matrix
A_transpose = A'
```

   b. Find All Elements Find all elements in the array A whose absolute value is greater than 5.

$$A = \begin{pmatrix} -7 & -1 & 5 \\ -5 & 1 & 7 \\ -3 & 3 & 9 \end{pmatrix} \tag{3.12}$$

Solution:

Listing 3.35 Creating and manipulating matrices

```
% Create a 3x4 matrix with random integer values between 1 and 10
A = zeros(3,3);
A(:) = -7:2:9; % Use the "full element" assignment method to obtain
    A
L = abs(A)>5; % Generates a logical array with the same dimension
    as A
B = A(L); % Take the element in A corresponding to the logical
    value 1 in L
B = B';
disp(A);
disp(B);
```

c. Solve equation group

Given the equation group in 3.13,

$$\begin{aligned}
4.6x_1 - 2.31x_2 + 8.3x_3 + 29.4x_4 &= 40.34 \\
20.5x_1 + 8.7x_2 + 40.1x_3 - 11.9x_4 &= 1.15 \\
36.4x_1 + 0.92x_2 - 3.7x_3 + 64.3x_4 &= 32.4 \\
7.84x_1 + 40.01x_2 - 2.68x_3 - 7.92x_4 &= 27.55
\end{aligned} \tag{3.13}$$

Solution:

Listing 3.36 Solve equation group

```
A = [4.6 -2.31 8.3 29.4; 20.5 8.7 40.1 -11.9; 36.4 0.92 -3.7 64.3;
     7.84 40.01 -2.68 -7.92];
b = [40.34;1.15;32.4;27.55];
x = A\b;
disp(x')
```

Thus, the values are:

$$\begin{aligned}
x_1 &\approx -1.6082 \\
x_2 &\approx 1.3572 \\
x_3 &\approx 0.9872 \\
x_4 &\approx 1.4517
\end{aligned} \tag{3.14}$$

3. Practice the use of cell arrays. Store the data listed in Table 3.1 in a cell array S with the following structure:

- The cell array S should have 8 rows and 3 columns.
- The 1st column of S should contain the names of the students.
- The 2nd column of S should contain the scores of the three subjects fcr each student.
- The 3rd column of S should contain the total score for each student.

Table 3.1 Grade sheet for 8 students

Name	Chinese	Math	English
Zhao	88	95	91
Qian	92	94	99
Sun	74	64	81
Li	98	77	83
Zhou	61	41	80
Wu	52	25	33
Zheng	82	73	79
Wang	79	95	77

Table 3.2 Grade sheet for 8 students with total and grade

Name	Chinese	Math	English	Total	Grade
Zhao	88	95	91	274	A
Qian	92	94	99	285	A
Sun	74	64	81	219	B
Li	98	77	83	258	A
Zhou	61	41	80	182	B
Wu	52	25	33	110	C
Zheng	82	73	79	234	B
Wang	79	95	77	251	A

Table 3.3 Grade sheet for 8 students with average, total, and grade

Name	Chinese	Math	English	Average	Total	Grade
Zhao	88	95	91	91.33	274	A
Qian	92	94	99	95.00	285	A
Sun	74	64	81	73.00	219	B
Li	98	77	83	86.00	258	A
Zhou	61	41	80	60.67	182	B
Wu	52	25	33	36.67	110	C
Zheng	82	73	79	78.00	234	B
Wang	79	95	77	83.67	251	A

After storing the data, add a fourth column to the cell array S to indicate the grade of each student based on their total score (Tables 3.2 and 3.3):

- If the total score is greater than 240, assign the grade 'A'.
- If the total score is between 180 and 240 (inclusive), assign the grade 'B'.
- If the total score is less than 200, assign the grade 'C'.

Solution:

Listing 3.37 Grade sheet for 8 students

```
names = ["zhao", "qian", "sun" , "li", "zhou", "wu", "zheng", "wang
    "];
scores = [88, 92, 74, 98, 61, 52, 82, 79;
    95, 94, 64, 77, 41, 25, 73, 95;
    91, 99, 81, 83, 80, 33, 79, 77];
S = cell(8,3);
for i = 1:8
    name = names(i);
    score = scores(:,i);
    total_score = sum(score);
    S(i,:) = {name, score, total_score};
end
```

```
for j = 1:size(S,1)
    total_score = S{j,3};
    % Judge grades
    if total_score > 240
        grade = {'A'};
    elseif total_score >= 180
        grade = {'B'};
    else
        grade = {'C'};
    end
    S(j,4) = grade;
end
```

4. Arithmetic Operations on Matrices

   a. Create two 3x3 matrices with random values between -5 and 5.
   b. Add the two matrices element-wise.
   c. Multiply the two matrices.
   d. Calculate the determinant of the first matrix.
   e. Find the inverse of the second matrix (if it exists).

Solution:

Listing 3.38 Arithmetic operations on matrices

```
% Create two 3x3 matrices with random values between -5 and 5
A = randi([-5, 5], 3, 3)
B = randi([-5, 5], 3, 3)

% Add the two matrices element-wise
C = A + B

% Multiply the two matrices
D = A * B

% Calculate the determinant of the first matrix
det_A = det(A)

% Find the inverse of the second matrix (if it exists)
if det(B) ~= 0
inv_B = inv(B)
else
disp('Matrix B is singular and does not have an inverse.')
end
```

5. Sparse Matrices

   a. Create a 5x5 sparse matrix with random non-zero values between 1 and 10.
   b. Count the number of non-zero elements in the sparse matrix.
   c. Perform matrix multiplication with the sparse matrix and a full matrix.

    d. Convert the sparse matrix to a full matrix.

    e. Find the maximum value in the sparse matrix and its corresponding indices.

Solution:

Listing 3.39 Sparse matrices

```
% Create a 5x5 sparse matrix with random non-zero values between 1
    and 10
S = sprand(5, 5, 0.3) * 10

% Count the number of non-zero elements in the sparse matrix
nnz_count = nnz(S)

% Perform matrix multiplication with the sparse matrix and a full
    matrix
A = randi(10, 5, 5)
B = S * A

% Convert the sparse matrix to a full matrix
S_full = full(S)

% Find the maximum value in the sparse matrix and its corresponding
    indices
[max_value, max_indices] = max(S(:))
```

6. Array Reshaping and Indexing

    a. Create a 1x12 vector with consecutive integers from 1 to 12.

    b. Reshape the vector into a 3x4 matrix.

    c. Extract the sub-matrix consisting of the first two rows and the last two columns.

    d. Create a logical matrix to select elements greater than 5 from the original vector.

    e. Replace the selected elements with their squares.

Solution:

Listing 3.40 Array reshaping and indexing

```
% Create a 1x12 vector with consecutive integers from 1 to 12
vector = 1:12

% Reshape the vector into a 3x4 matrix
matrix = reshape(vector, 3, 4)

% Extract the sub-matrix consisting of the first two rows and the
    last two columns
sub_matrix = matrix(1:2, 3:4)

% Create a logical matrix to select elements greater than 5 from
    the original vector
logical_mask = vector > 5
```

```
% Replace the selected elements with their squares
vector(logical_mask) = vector(logical_mask).^2
```

7. Practice the use of structure

   Create a structure that contains information about the 8 students in Table 3.1 and calculate the total and average scores of these students.

Solution:

Listing 3.41 Array reshaping and indexing

```
students = struct('name', {'zhao', 'qian', 'sun' , 'li', 'zhou', '
    wu', 'zheng', 'wang'},...
'chinese', {88, 92, 74, 98, 61, 52, 82, 79},...
'math', {95, 94, 64, 77, 41, 25, 73, 95},...
'english', {91, 99, 81, 83, 80, 33, 79, 77});
scores = [students.chinese; students.math; students.english];
names = {students.name};
x = sum(scores);
disp('The total score of each student is:');
disp(x);
disp('The average score of each student is:');
disp(x/3);
```

8. Practice the use of string

   A fruit store manages two warehouses that store various types of fruits. To optimize their inventory and distribution, the store wants to analyze the fruits stored in each warehouse. The task is to use 'set operations' to:

   a. Identify the fruits that are present in both warehouses (common fruits).
   b. Determine the fruits that are exclusively stored in each warehouse (unique fruits).
   c. Create a comprehensive list of all the fruits available in the two warehouses combined.

   By performing these set operations, the fruit store can gain insights into their inventory distribution and make informed decisions about stock management and fruit allocation between the warehouses. Let W_1 and W_2 denote the sets of fruits stored in Warehouse 1 and Warehouse 2, respectively. The set operations can be defined as follows:

   $$\text{Common Fruits} = W_1 \cap W_2$$
   $$\text{Unique Fruits in Warehouse 1} = W_1 \setminus W_2$$
   $$\text{Unique Fruits in Warehouse 2} = W_2 \setminus W_1$$
   $$\text{All Fruits} = W_1 \cup W_2$$

Table 3.4 Fruit sets in warehouses

Set	Fruits
Common fruits	"banana", "pear"
Unique fruits in warehouse 1	"apple", "strawberry", "watermelon"
Unique fruits in warehouse 2	"litchi", "orange", "pineapple"
All fruits	"apple", "banana", "litchi", "orange", "pear", "pineapple", "strawberry", "watermelon"

Table 3.5 Stock data for day 3

Date	Open	High	Low	Close	Average	Difference
2024-03-03	35.18	35.31	34.5	35.16	34.905	0.02

The results of these set operations will provide the fruit store with the necessary information to effectively manage their inventory and optimize their fruit distribution strategies (Table 3.4).

Solution:

Listing 3.42 Practice the use of string

```
warehouse1 = ["apple" "banana" "strawberry" "pear" "watermelon"];
warehouse2 = ["banana" "orange" "pear" "pineapple" "litchi"];
x1 = intersect(warehouse1, warehouse2);
x2 = setdiff(warehouse1, warehouse2);
x3 = setdiff(warehouse2, warehouse1);
x4 = union(warehouse1, warehouse2);
disp(x1); disp(x2); disp(x3); disp(x4)
```

9. Practice the use of tables

   a. Read the stock data from the Excel table and store it in a tabular format.
   b. Calculate the average value of the highest and lowest prices for each day and add it as a new column to the table.
   c. Compute the absolute difference between the close price and the open price for each day and include it as another column in the table.
   d. Display the data corresponding to the third day from the updated table (Table 3.5).

Solution:

Listing 3.43 Practice the use of tables

```
data = readtable('data.xlsx');
data{:,'average'}=(data.high+data.low)/2; % add average value
data{:,'difference'}=abs(data.close-data.open); % add absolute
    difference value
disp(data(3,:)); % display the data of the third day
```

The results are

10. Practice the use of sort and index

    a. Calculate the total score for each student based on their scores in Chinese, Math, and English, as obtained in lab work 7.

    b. Sort the students in descending order according to their total scores.

    c. Print the names of the students in the sorted order.

Solution:

Listing 3.44 Practice the use of sort and index

```
%% (this part is same as previous lab work)
%%%
students = struct('name', {'zhao', 'qian', 'sun' , 'li', 'zhou'  '
    wu', 'zheng', 'wang'},...
'chinese', {88, 92, 74, 98, 61, 52, 82, 79},...
'math', {95, 94, 64, 77, 41, 25, 73, 95},...
'english', {91, 99, 81, 83, 80, 33, 79, 77});
scores = [students.chinese; students.math; students.english];
names = {students.name};
x = sum(scores);
disp('The total score of each student is:');
disp(x);
disp('The average score of each student is:');
disp(x/3);


%%%

x = sum(scores);
disp('The total score of each student is:');
disp(x);
[B,sy] = sort(x, 'descend');
disp('The ranking is:')
for i = 1:8
e = B(i);
d = find(x==e);
disp(names(d))
end
```

Then the result are (Table 3.6)

11. Sudoku Verification

Sudoku is a puzzle game played on a 9×9 grid, which is divided into 9 sub-grids of size 3×3, called chambers or boxes. The objective is to fill the grid with digits from 1 to 9, subject to the following constraints:

    1. Each row must contain the digits 1–9, without any repetition.

    2. Each column must contain the digits 1–9, without any repetition.

    3. Each 3×3 chamber must contain the digits 1–9, without any repetition.

Given a completed Sudoku grid, as shown in Fig. 3.1, the task is to verify whether it satisfies all three requirements mentioned above.

Table 3.6 Ranking of students' total scores

Student	Total score
Zhao	285
Qian	274
Wang	258
Wu	251
Zhou	234
Li	219
Zheng	182
Sun	110

Fig. 3.1 Sudoku grid

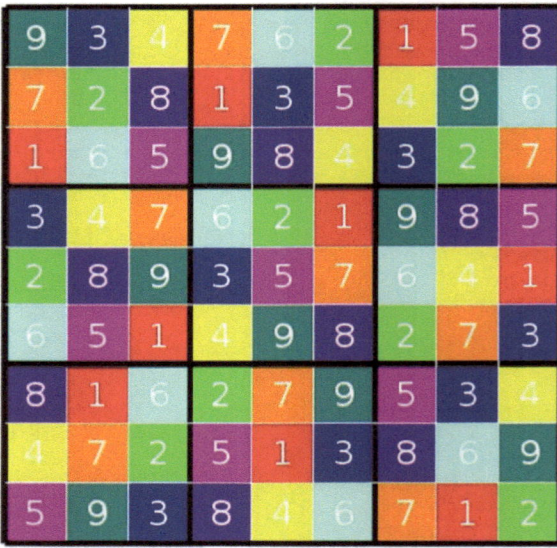

Let the Sudoku grid be represented by a 9×9 matrix S, where $S_{i,j}$ denotes the digit in the i-th row and j-th column, and $1 \leq S_{i,j} \leq 9$.

To verify the Sudoku grid, we need to check the following conditions:

  a. Row Check: For each row i, the set $S_{i,1}, S_{i,2}, \ldots, S_{i,9}$ must contain all digits from 1 to 9 without repetition.
  b. Column Check: For each column j, the set $S_{1,j}, S_{2,j}, \ldots, S_{9,j}$ must contain all digits from 1 to 9 without repetition.
  c. Chamber Check: For each 3×3 chamber, the set of digits within that chamber must contain all digits from 1 to 9 without repetition.

If all three conditions are satisfied, the Sudoku grid is considered valid. Otherwise, if any of the conditions are violated, the Sudoku grid is invalid.

The verification process can be implemented using set operations or by iterating over the rows, columns, and chambers to check for duplicates and missing digits.
Solution:

Listing 3.45 Sudoku verification

```matlab
% Define the Sudoku grid
sudoku_grid = [
    9 3 4 7 6 2 1 5 8;
    7 2 8 1 3 5 4 9 6;
    1 6 5 9 8 4 3 2 7;
    3 4 7 6 2 1 9 8 5;
    2 8 9 3 5 7 6 4 1;
    6 5 1 4 9 8 2 7 3;
    8 1 6 2 7 9 5 3 4;
    4 7 2 5 1 3 8 6 9;
    5 9 3 8 4 6 7 1 2
];

% Function to check if a set contains all digits from 1 to 9
    without repetition
function result = check_set(set)
    result = all(ismember(1:9, set)) && length(unique(set)) ==
        9;
end

% Step 1: Row Check
row_check = all(arrayfun(@check_set, sudoku_grid));

% Step 2: Column Check
column_check = all(arrayfun(@check_set, sudoku_grid.'));

% Step 3: Chamber Check
chamber_check = true;
for i = 1:3:7
    for j = 1:3:7
        chamber = sudoku_grid(i:i+2, j:j+2);
        chamber_check = chamber_check & check_set(chamber(:));
    end
end

% Final Verification
if row_check & column_check & chamber_check
    disp('The Sudoku grid is valid.');
else
    disp('The Sudoku grid is invalid.');
end
```

The result from https://matlab.mathworks.com/, the Sudoku grid is invalid.

3.7 Problems

1. Given a matrix A, create a new matrix B by replacing all negative elements in A with their absolute values.
2. Given a matrix A, find the row and column indices of the maximum element in the matrix.
3. Given a matrix A, create a new matrix B by swapping the elements along the main diagonal with the elements along the secondary diagonal.
4. Given a matrix A, create a new matrix B by shifting each element in A one position to the right, wrapping around to the beginning of the row when reaching the end.
5. Given a matrix A, create a new matrix B by extracting the elements along the diagonals parallel to the main diagonal.

3.8 Summary

In this chapter, we covered the following key concepts related to Arrays and Matrices in MATLAB:

- **Array Creation**: We learned how to create arrays in MATLAB using different methods, such as direct assignment, colon notation, and functions like `zeros`, `ones`, `eye`, and `rand`.
- **Array Indexing and Slicing**: We explored how to access and manipulate elements of arrays using various indexing techniques, including linear indexing, subscripted indexing, and slicing with colon notation.
- **Array Operations**: We studied various array operations, such as arithmetic operations (element-wise and matrix operations), relational operations, logical operations, and array concatenation.
- **Matrix Operations**: We covered important matrix operations, including matrix multiplication, transpose, inverse, determinant, and solving linear systems of equations using techniques like Gaussian elimination and matrix decompositions.
- **Special Matrix Types**: We discussed special matrix types like **diagonal, triangular**, **sparse**, and **banded** matrices, and learned how to create and manipulate them in MATLAB.
- **Array Reshaping and Manipulation**: We explored functions like `reshape`, `repmat`, `permute`, and `circshift` for reshaping, replicating, and rearranging arrays in various ways.
- **Array and Matrix Visualisation**: We learned how to visualize arrays and matrices using different plotting techniques, such as image plots, surface plots, and contour plots.

For undergraduate students, this chapter provides a solid foundation in working with arrays and matrices, which are fundamental data structures in MATLAB. Understanding these concepts is crucial for various applications in engineering, science, and data analysis.

For postgraduate students and professional researchers or engineers, this chapter serves as a refresher and introduces more advanced topics like special matrix types, matrix decompositions, and linear system solvers. These concepts are essential for numerical analysis, optimization, signal processing, and many other domains that heavily rely on matrix computations.

Overall, this chapter aims to equip readers with a comprehensive understanding of arrays and matrices in MATLAB, enabling them to efficiently manipulate and analyze data, perform numerical computations, and solve real-world problems across various disciplines.

Chapter 4
Conditional Statements

Chapter Learning Outcomes

- Understand the concept and importance of **conditional statements** in programming.
- Implement **if** statements to execute code based on a specific condition.
- Use **if-else** statements to handle alternative cases when a condition is not met.
- Apply **nested if** statements to handle multiple conditions within a single statement.
- Utilise **switch-case** statements for more efficient handling of multiple conditions.
- Employ **logical operators** (**&&**, ||, ~) to combine and negate conditions.
- Write well-structured and readable conditional statements following best practices.

Chapter Key Words

- **Conditional Statement**: A statement in programming that allows the execution of different code blocks based on whether a specified condition is true or false. Conditional statements are used to control the flow of a program based on certain conditions being met or not met.
- **If Statement**: The most basic form of a conditional statement. It executes a block of code if a specified condition is true. If the condition is false, the code block is skipped, and the program continues to the next statement.
- **If-Else Statement**: An extension of the if statement that allows for the execution of an alternative code block if the condition is false. It provides a way to handle both cases (true and false) of a condition.
- **Nested If Statement**: A conditional statement where one or more if statements are nested within another if statement. This allows for more complex decision-making processes by evaluating multiple conditions in a hierarchical manner.

© The Author(s) 2025
Y. Chen and L. Huang, *MATLAB Roadmap to Applications*,
https://doi.org/10.1007/978-981-97-8788-3_4

- **Switch-Case Statement**: A control flow statement that allows for the evaluation of a single expression against multiple cases. It provides a more concise and efficient way to handle multiple conditions compared to using multiple if-else statements.
- **Logical Operators**: Operators used to combine or negate conditions in conditional statements. The commonly used logical operators are AND (**&&**), OR (||), and NOT (~). These operators allow for more complex conditions to be evaluated.
- **Condition**: A boolean expression that evaluates to either true or false. Conditions are used in conditional statements to determine which code block should be executed based on the outcome of the evaluation.

4.1 Introduction to Conditional Statements

Conditional statements are a cornerstone of algorithmic logic and programming. MATLAB provides a comprehensive set of conditional constructs that are instrumental for formulating effective algorithms. They allow a programme to execute different code segments based on certain conditions at runtime. In MATLAB, conditional statements are pivotal for decision-making processes within code, enabling dynamic and responsive programming. Accordingly, this chapter will delineate the essentials of formulating conditional statements in MATLAB.

- Overview and Role of **Conditional Statements** Conditional statements are a fundamental concept in programming that allow you to control the flow of execution based on certain conditions being met or not. They enable programs to make decisions and perform different actions depending on the evaluation of logical expressions or conditions. Conditional statements are essential for implementing complex logic, handling various scenarios, and creating interactive and responsive applications.
  In MATLAB, conditional statements play a crucial role in data analysis, algorithm development, and scientific computing tasks. They provide a powerful mechanism to explore and manipulate data, implement mathematical models, and automate decision-making processes based on specific criteria or thresholds.
- Understanding **Control Flow** and **Decision-Making Control flow** refers to the order in which statements in a program are executed. **Decision-making** is the process of evaluating conditions and determining which set of statements should be executed based on the outcome of those conditions. Conditional statements are the primary constructs used for decision-making and controlling the flow of execution in a program [1].
  By incorporating conditional statements, MATLAB programs can adapt their behavior based on the state of variables, user inputs, or the results of computations. This flexibility allows for the creation of dynamic and intelligent systems that can respond to changing conditions and make informed decisions.

- Types of **Conditional Statements** in MATLAB MATLAB provides several types of conditional statements to facilitate decision-making and control flow. The main conditional statements in MATLAB are:

  - **if statement**: Executes a block of code if a specified condition is true.
  - **if-else statement**: Executes one block of code if a condition is true, and another block of code if the condition is false.
  - **if-elseif-else statement**: Evaluates multiple conditions in a hierarchical manner and executes the corresponding block of code for the first true condition.
  - **switch-case statement**: Evaluates an expression against multiple cases and executes the corresponding block of code for the matched case.

  These conditional statements, along with logical operators (e.g., **&&**, ||, ~), enable programmers to create complex decision structures and implement sophisticated algorithms in MATLAB [2].

4.2 The if Statement

Syntax and Structure of the **if Statement**:

Listing 4.1 If statement syntax.

```
if condition
statements
end
```

- Example 1: Check if a number is positive.

Listing 4.2 Check if a number is positive.

```
num = 5;
if num > 0
disp('The number is positive.')
end
```

- Example 2: Evaluate the value of a function.

Listing 4.3 Evaluate the value of a function.

```
x = -2;
if x >= 0
y = \textcolor{green}{sqrt}(x);
else
y = -\textcolor{green}{sqrt}(-x);
end
disp(['The value of y is ', num2str(y)])
```

- Evaluating **Logical Expressions** and **Conditions** The **condition** in an **if statement** is a **logical expression** that evaluates to either true (nonzero) or false (zero). The statements inside the **if block** are executed if the condition is true.
- Examples of Using the **if Statement**

  - With Different Data Types Example 1: Check if a number is positive.

Listing 4.4 Check if a number is positive.

```
num = 5;
if num > 0
disp('The number is positive.')
end
```

Example 2: Check if a string is empty.

Listing 4.5 Check if a string is empty.

```
str = '';
if \textcolor{green}{isempty}(str)
disp('The string is empty.')
else
disp(['The string is: ', str])
end
```

  - **Nested if Statements** for Complex **Conditions** Example 3: Determine the sign of a number using nested **if statements**.

Listing 4.6 Determine the sign of a number.

```
num = -3;
if num > 0
disp('The number is positive.')
else
if num < 0
disp('The number is negative.')
else
disp('The number is zero.')
end
end
```

4.3   The elseif and else Statements

The **elseif** clause is used to specify additional conditions to check if the previous conditions are false. The basic syntax is:

Listing 4.7 If-elseif statement syntax.

```
if condition1
statements1
elseif condition2
statements2
...
else
statements_else
end
```

- The **else** Clause for Alternative Cases The **else** clause is used to specify a block of statements to execute if none of the previous conditions are true.
- Examples of **if-elseif-else Statements**

  • Handling Multiple **Conditions** Example 4: Categorize a score into different grade levels.

Listing 4.8 Categorize a score into grade levels.

```
score = 85;
if score >= 90
grade = 'A';
elseif score >= 80
grade = 'B';
elseif score >= 70
grade = 'C';
elseif score >= 60
grade = 'D';
else
grade = 'F';
end
disp(['The grade is: ', grade])
```

  • Different Scenarios and Applications Example 5: Determine the state of water based on temperature.

Listing 4.9 Determine the state of water.

```
temperature = 25;
if temperature < 0
state = 'solid';
elseif temperature >= 0 && temperature < 100
state = 'liquid';
else
state = 'gas';
end
disp(['At ', num2str(temperature), ' degrees Celsius,
    water is in the ', state, ' state.'])
```

4.4   The switch Statement

The **switch statement** is an alternative to the **if-elseif-else** construct for evaluating multiple conditions. It is often used when there are multiple cases to consider based on the value of a single expression.

The basic syntax for the **switch statement** is:

Listing 4.10 Switch statement syntax.

```
switch expression
case case1
statements1
case case2
statements2
...
otherwise
statements_otherwise
end
```

– Using **case** and **otherwise** Clauses The **case** clauses specify the values or conditions to match against the expression. The **otherwise** clause is optional and specifies the block of statements to execute if none of the cases match.
– Examples of **switch Statements**

  • With Different Data Types Example 6: Perform different operations based on user input.

Listing 4.11 Perform operations based on user input.

```
operation = input('Enter operation (+, -, , /): ', 's');
switch operation
case '+'
disp('You selected addition.')
case '-'
disp('You selected subtraction.')
case ''
disp('You selected multiplication.')
case '/'
disp('You selected division.')
otherwise
disp('Invalid operation.')
end
```

  • Comparison with **if-elseif-else Statements** Example 7: Determine the day of the week based on a numerical input.

Listing 4.12 Determine the day of the week.

```
% Using a switch statement
day_num = 3;
switch day_num
case 1
day = 'Monday';
case

ant = b^2 - 4ac;
if discriminant > 0
% Two real roots
root1 = (-b + sqrt(discriminant)) / (2a);
root2 = (-b - sqrt(discriminant)) / (2a);
roots = [root1, root2];
elseif discriminant == 0
% One real root
root = -b / (2*a);
roots = root;
else
% Complex roots
roots = [];
end
end
```

4.5 The end Keyword

The **end** keyword in MATLAB is crucial for delimiting the scope of various programming constructs, such as loops, conditional statements, and function definitions. It
serves as a terminator that marks the end of a block of code, ensuring proper execution
and readability of the program.

In the context of conditional statements, the **end** keyword is used to terminate the
block of code associated with an **if**, **elseif**, or **switch** statement. It ensures that the
program flow follows the intended path based on the evaluated conditions. Failing
to include the **end** keyword or misplacing it can lead to syntax errors or unintended
behavior.

The basic syntax for using **end** in conditional statements is as follows:

Listing 4.13 If statement syntax with end.

```
if condition
% Statements to execute if condition is true
end

if condition
% Statements to execute if condition is true
else
% Statements to execute if condition is false
end
```

```
if condition1
% Statements to execute if condition1 is true
elseif condition2
% Statements to execute if condition2 is true
else
% Statements to execute if both conditions are false
end
```

– Common Mistakes and Best Practices Related to **end** One common mistake when using the **end** keyword is forgetting to include it or mismatching it with the corresponding control structure. This can lead to syntax errors or logical errors in the program. To avoid such issues, it is recommended to follow best practices, such as:

- Indent code blocks properly to maintain a clear visual structure.
- Use comments to clarify the purpose of each code block and its associated **end** keyword.
- Consider using an integrated development environment (IDE) or text editor with syntax highlighting and code folding features, which can help identify missing or misplaced **end** keywords.

– Examples and Applications

1. Practical Examples in Various Domains Conditional statements, including the use of the **end** keyword, are widely used across various domains and applications in MATLAB. Here are a few examples:

    a. **Signal Processing**: Applying conditional statements to filter or manipulate signals based on specific criteria, such as noise levels or frequency ranges.
    b. **Image Processing**: Implementing image segmentation, edge detection, or object recognition algorithms using conditional statements to classify pixels or regions based on their properties.
    c. **Control Systems**: Designing control algorithms with conditional statements to adjust system parameters or apply different control strategies based on the system's state or environmental conditions.
    d. **Data Analysis**: Applying conditional statements to handle missing data, detect outliers, or apply different data processing techniques based on the characteristics of the dataset.

2. Solving Real-World Problems with **Conditional Statements** Conditional statements are essential for solving real-world problems that involve decision-making or branching logic. Here are two examples that demonstrate their usage:

Listing 4.14 Temperature converter.

```
function convertedTemp = temperatureConverter(temp,
    fromScale)
```

```
if strcmp(fromScale, 'C')
% Convert from Celsius to Fahrenheit
convertedTemp = (temp * 9/5) + 32;
elseif strcmp(fromScale, 'F')
% Convert from Fahrenheit to Celsius
convertedTemp = (temp - 32) * 5/9;
else
error('Invalid temperature scale. Use "C" for Celsius or
    "F" for Fahrenheit.')
end
end
```

Listing 4.15 Grade classification.

```
function gradeCategory = classifyGrade(grade)
if grade >= 90
gradeCategory = 'A';
elseif grade >= 80
gradeCategory = 'B';
elseif grade >= 70
gradeCategory = 'C';
elseif grade >= 60
gradeCategory = 'D';
else
gradeCategory = 'F';
end
end
```

3. Integration with Other Programming Concepts

- **Loops** and **Conditional Statements** Conditional statements are often used in conjunction with loops to create more complex program flows. For example, a conditional statement inside a loop can be used to skip or perform specific operations based on certain conditions, allowing for more efficient and targeted code execution.

Listing 4.16 Loop with conditional statement.

```
% Calculate the sum of positive numbers in an array
numbers = [5, -2, 8, 0, -7, 3];
sum_positive = 0;

for i = 1:length(numbers)
if numbers(i) > 0
sum_positive = sum_positive + numbers(i);
end
end

disp(['The sum of positive numbers is: ', num2str(
    sum_positive)]);
```

- **Functions** and **Conditional Statements** Conditional statements are also commonly used within functions to handle different input scenarios or perform specific computations based on certain conditions. This allows for more modular and reusable code.

Listing 4.17 Function with conditional statement.

```
function result = computeValue(x, operation)
if strcmp(operation, 'square')
result = x^2;
elseif strcmp(operation, 'cube')
result = x^3;
else
error('Invalid operation. Use "square" or "cube".')
end
end
```

4.6 Laboratory

1. The 'If-Else' selection statement
   For any input natural number, determine whether it is odd or even.

Solution:

Listing 4.18 'If-Else' selection statement

```
x=input('Please input a natural number');

if mod(x,2)==1
        disp('It is a odd number.')
else
        disp('It is a even number.')
end
```

2. The 'elseif' selection statement
   Reassign any input value, assigning 1 if greater than 0, -1 if less than 0, and not replacing if equal to 0.

Solution:

Listing 4.19 The 'elseif' selection statement

```
x=input('Please input the value for x:');

If    x>0
        x=1;
elseif x<0
```

```
        x = -1;
else
        x = 0;
end
disp(x)
```

3. The 'nested elseif' selection statement

Please assign a grade based on the test score using the following criteria:

- A score of 90 or above receives an **A**.
- A score between 60 and 89 (inclusive) receives a **B**.
- A score below 60 receives a **C**.

Solution:

Listing 4.20 The 'nested elseif' selection statement

```
% Prompt the user to enter the score
score = input('Please enter the test score: ');

% Determine the grade based on the score
if score >= 90
    grade = 'A';
elseif score >= 60
    grade = 'B';
else
    grade = 'C';
end

% Display the grade
disp(grade);
```

4. The switch selection statement

Determine the value of season based on the entered month.

Solution:

Listing 4.21 The switch selection statement

```
% Prompt the user to enter the month as a number (1-12)
month = input('Please enter the month as a number (1-12)
    : ');

% Determine the season based on the month
if month == 12 || month == 1 || month == 2
    season = 'Winter';
elseif month >= 3 && month <= 5
    season = 'Spring';
elseif month >= 6 && month <= 8
    season = 'Summer';
elseif month >= 9 && month <= 11
    season = ''Autumn';
```

```
else
    season = 'Invalid month. Please enter a number
        between 1 and 12.';
end

% Display the season
disp(season);
```

5. Supermarket Discounting

A supermarket is conducting a significant reward activity and is discounting the goods sold. The discount criteria are as follows:

- **price < 500**: 2% discount
- **500** ≤ price < **700**: 3% discount
- **700** ≤ price < **1100**: 5% discount
- **1100** ≤ price < **1500**: 7% discount
- **1500** ≤ price < **2500**: 11% discount
- **price ≥ 2500**: 15% discount

Please use MATLAB to implement this. When entering the price, output the discounted price.

Solution:

Listing 4.22 Supermarket Discounting

```
% Prompt the user to enter the price
price = input('Please enter the price: ');

% Determine the discount based on the price
if price < 500
    discount = 0.02;
elseif price >= 500 && price < 700
    discount = 0.03;
elseif price >= 700 && price < 1100
    discount = 0.05;
elseif price >= 1100 && price < 1500
    discount = 0.07;
elseif price >= 1500 && price < 2500
    discount = 0.11;
else
    discount = 0.15;
end

% Calculate the discounted price
discounted_price = price * (1 - discount);

% Display the discounted price
disp(['The discounted price is: ', num2str(
    discounted_price)]);
```

6. Temperature Converter Write a MATLAB function that takes a temperature value and a character ('C' or 'F') representing the input scale (Celsius or Fahrenheit). The function should convert the temperature to the other scale and return the converted value.

Solution:

Listing 4.23 Temperature Converter Function.

```
function converted_temp = convertTemperature(temp,
    input_scale)
if upper(input_scale) == 'C'
% Convert from Celsius to Fahrenheit
converted_temp = (temp * 9/5) + 32;
elseif upper(input_scale) == 'F'
% Convert from Fahrenheit to Celsius
converted_temp = (temp - 32) * 5/9;
else
error('Invalid input scale. Use ''C'' for Celsius or ''F
    '' for Fahrenheit.')
end
end
```

4.6.1 Digit Counter

Write a MATLAB function that takes a positive integer as input and returns the count of each digit (0-9) present in the number.

Solution:

Listing 4.24 Digit Counter Function.

```
function digit_counts = countDigits(num)
digit_counts = zeros(1, 10);
num_str = num2str(num);
for i = 1:length(num_str)
digit = str2double(num_str(i));
digit_counts(digit + 1) = digit_counts(digit + 1) + 1;
end
end
```

7. Vowel Counter Write a MATLAB function that takes a string as input and returns the count of vowels (a, e, i, o, u) present in the string.

Solution:

Listing 4.25 Vowel Counter Function.

```matlab
function vowel_count = countVowels(str)
vowel_count = 0;
vowels = 'aeiou';
for i = 1:length(str)
char = lower(str(i));
if any(char == vowels)
vowel_count = vowel_count + 1;
end
end
end
```

8. Palindrome Checker Write a MATLAB function that takes a string as input and determines if it is a palindrome (reads the same forward and backward) or not.

Solution:

Listing 4.26 Palindrome Checker Function.

```matlab
function isPalindrome = checkPalindrome(str)
% Remove non-alphanumeric characters and convert to
    lowercase
clean_str = regexprep(lower(str), '[^a-z0-9]', '');

% Check if the string is equal to its reverse
isPalindrome = strcmp(clean_str, flip(clean_str));

end
```

9. Rock-Paper-Scissors Game Write a MATLAB script that implements the classic Rock-Paper-Scissors game. The script should prompt the user to enter their choice (rock, paper, or scissors), randomly generate the computer's choice, and determine the winner based on the rules of the game.

Solution:

Listing 4.27 Rock-Paper-Scissors Game.

```matlab
% Define the choices
choices = {'rock', 'paper', 'scissors'};

% Get user's choice
user_choice = input('Enter your choice (rock, paper, or
    scissors): ', 's');

% Generate computer's choice randomly
computer_choice = choices{randi(3)};

% Determine the winner
if strcmp(user_choice, computer_choice)
```

```
disp ('It''s a tie!')
elseif strcmp (user_choice, 'rock') && strcmp (
    computer_choice, 'scissors') || ...
strcmp (user_choice, 'paper') && strcmp (computer_choice,
    'rock') || ...
strcmp (user_choice, 'scissors') && strcmp (
    computer_choice, 'paper')
disp ('You win!')
else
disp ('Computer wins!')
end
```

4.6.2   Grade Statistics

Write a MATLAB script that prompts the user to enter a series of grades (integers between 0 and 100) and calculates the following statistics:

Minimum grade Maximum grade Mean grade Number of passing grades ($>= 60$) Number of failing grades (< 60).

Solution:

Listing 4.28 Grade Statistics Script.

```
grades = [];
while true
grade = input ('Enter a grade (0-100, or -1 to stop): ');
if grade == -1
break;
elseif grade >= 0 && grade <= 100
grades = [grades, grade];
else
disp ('Invalid grade. Please enter a value between 0 and
    100.');
end
end

if isempty (grades)
disp ('No grades entered.');
return;
end

min_grade = min (grades);
max_grade = max (grades);
mean_grade = mean (grades);
num_passing = sum (grades >= 60);
num_failing = sum (grades < 60);

disp (['Minimum grade: ', num2str (min_grade)]);
disp (['Maximum grade: ', num2str (max_grade)]);
```

```
disp(['Mean grade: ', num2str(mean_grade)]);
disp(['Number of passing grades: ', num2str(num_passing)
    ]);
disp(['Number of failing grades: ', num2str(num_failing)
    ]);
```

4.6.3 Menu-Driven Calculator

Write a MATLAB script that implements a simple menu-driven calculator. The script should display a menu with options for addition, subtraction, multiplication, and division. Based on the user's choice, the script should prompt for two numbers and perform the selected operation, displaying the result.

Solution:

Listing 4.29 Menu-Driven Calculator Script.

```
while true
% Display menu
disp('Menu:');
disp('1. Addition');
disp('2. Subtraction');
disp('3. Multiplication');
disp('4. Division');
disp('5. Exit');
% Get user choice
choice = input('Enter your choice (1-5): ');

% Exit condition
if choice == 5
    break;
end

% Get operands
num1 = input('Enter the first number: ');
num2 = input('Enter the second number: ');
% Perform operation based on choice
switch choice
    case 1
        result = num1 + num2;
        operation = 'Addition';
    case 2
        result = num1 - num2;
        operation = 'Subtraction';
    case 3
        result = num1 * num2;
        operation = 'Multiplication';
    case 4
        if num2 ~= 0
```

```
            result = num1 / num2;
            operation = 'Division';
        else
            disp('Error: Division by zero');
            continue;
        end
    otherwise
        disp('Invalid choice. Please try again.');
        continue;
end

% Display result
disp(['Result of ', operation, ': ', num2str(result)]);
end
```

These lab works cover a variety of scenarios and applications, allowing you to practice and reinforce your understanding of conditional statements in MATLAB. Feel free to modify and extend these exercises to further enhance your skills.

4.7 Problems

1. Write a MATLAB function that takes a number as input and returns the absolute value of that number using an **if** statement.
2. Implement a MATLAB script that prompts the user to enter a character and determines whether it is a vowel or a consonant using a **switch** statement.
3. Create a MATLAB function that takes three numbers as input and returns the maximum value among them using nested **if** statements.
4. Write a MATLAB script that generates a random number between 1 and 10, and based on the value, displays a corresponding message using a **switch** statement.
5. Implement a MATLAB function that takes a year as input and determines whether it is a leap year or not using an **if-elseif-else** statement.
6. Create a MATLAB script that prompts the user to enter their age and displays a message indicating their age category (e.g., child, teenager, adult) using an **if-elseif-else** statement.
7. Write a MATLAB function that takes a character as input and determines whether it is a digit, an uppercase letter, a lowercase letter, or a special character using nested **if** statements.
8. Implement a MATLAB script that prompts the user to enter a number and displays whether it is positive, negative, or zero using an **if-elseif-else** statement.
9. Create a MATLAB function that takes two numbers as input and returns their sum if both numbers are positive, their difference if one number is positive and the other is negative, or zero if both numbers are negative, using nested **if** statements.
10. Write a MATLAB script that generates two random numbers between 1 and 6 (representing dice rolls) and displays a message indicating the outcome (e.g., "You rolled a double," "You rolled a high number," etc.) using a **switch** statement.

4.8 Summary

In this chapter, we have explored the fundamental concept of **conditional statements** in MATLAB, which are essential for controlling the flow of program execution based on certain conditions. The key points covered in this chapter are summarized as follows:

- **Conditional statements** allow you to execute different blocks of code based on the evaluation of logical expressions or conditions.
- The **if** statement is the most basic form of conditional statement, which executes a block of code if a specified condition is true.
- The **if-else** statement provides an alternative block of code to execute if the condition in the **if** statement is false.
- The **if-elseif-else** statement allows you to specify multiple conditions and execute different blocks of code based on the first condition that evaluates to true.
- The **switch** statement is an alternative to nested **if-elseif** statements and is used to execute different blocks of code based on different possible values of a single expression.
- The **end** keyword is crucial for delimiting the scope of conditional statements and ensuring proper execution and readability of the program.
- Conditional statements are widely used in various domains and applications, such as signal processing, image processing, control systems, and data analysis, to solve real-world problems involving decision-making or branching logic.
- Conditional statements can be integrated with other programming concepts, such as loops and functions, to create more complex program flows and modular code.
- Best practices for writing readable and maintainable conditional statements include using descriptive variable and function names, consistent indentation, and adding comments to explain complex logic.
- Optimizing code efficiency and performance can be achieved by avoiding unnecessary computations, using vectorized operations, precomputing or caching frequently used values, and utilizing MATLAB's built-in functions and optimized algorithms.
- Debugging techniques for conditional statements include using MATLAB's built-in debugger, adding print statements, simplifying complex conditions, employing logging or tracing mechanisms, and writing unit tests or test cases.
- Guidelines for choosing the appropriate conditional statement (e.g., **if-elseif-else** or **switch**) depend on factors such as the complexity of the conditions, the need for logical operations or string comparisons, and code readability and maintainability considerations.

For Undergraduate (UG) Students:

Conditional statements are a fundamental concept in programming that allows you to control the flow of your code based on specific conditions. This chapter provides a comprehensive introduction to conditional statements in MATLAB, including their syntax, usage, and applications. By understanding and mastering conditional statements, you will be able to write more efficient and effective programs that can handle a wide range of scenarios and make intelligent decisions based on specific criteria.

For Postgraduate (PG) Students and Professional Researchers or Engineers:

Conditional statements are a powerful tool for implementing decision-making logic and branching in your MATLAB programs. This chapter delves into the nuances of conditional statements, including best practices for writing readable and maintainable code, optimizing performance, and debugging techniques. Additionally, it explores the integration of conditional statements with other programming concepts, such as loops and functions, enabling you to create more sophisticated and modular solutions. Whether you are working on research projects, developing algorithms, or building applications, a solid understanding of conditional statements will equip you with the skills necessary to tackle complex problems effectively.

For Postgraduate (PG) Students and Professional Researchers or Engineers:

Mastering conditional statements is crucial for developing efficient and robust algorithms, implementing decision-making logic in simulations and models, and building intelligent systems that can adapt to varying conditions and inputs. This chapter equips you with the necessary knowledge and techniques to leverage the power of conditional statements in MATLAB, enabling you to tackle complex research problems, optimize performance, and write maintainable and scalable code. Additionally, the integration of conditional statements with other programming concepts opens up a wide range of possibilities for creating sophisticated solutions and exploring new avenues in your field of study or professional practice.

By thoroughly understanding and applying the concepts covered in this chapter, you will be well-prepared to develop advanced applications, conduct rigorous research, and contribute to the advancement of your discipline through the effective use of conditional statements in MATLAB.

Here are some additional points:

- **Nested conditional statements** allow you to combine multiple conditions and execute different blocks of code based on their evaluation, providing greater control and flexibility in program flow.
- **Short-circuit evaluation** of logical expressions in conditional statements can improve performance by avoiding unnecessary computations when the final result can be determined early.
- **Conditional statements** are essential for implementing **control flow** in programs, enabling them to make decisions and respond accordingly based on various conditions and inputs.

- The **ternary operator** in MATLAB provides a concise way to write simple conditional expressions, making the code more compact and readable in certain situations.
- **Logical operators** (**&&**, ||, and ~) can be used to combine multiple conditions in conditional statements, allowing for more complex decision-making logic.
- **Relational operators** (<, >, <=, >=, ==, and =) are used to compare values and evaluate conditions in conditional statements.
- **Conditional statements** can be used in conjunction with other programming concepts, such as **arrays**, **matrices**, and **data structures**, to perform operations or manipulations based on specific conditions.
- **Code refactoring** and **modularisation** can improve the readability and maintainability of code involving conditional statements, especially in larger and more complex programs.

References

1. MathWorks, "MATLAB Fundamentals," [Online]. Available: https://www.mathworks.com/help/matlab/, accessed on Feb. 17, 2024
2. MathWorks, "Conditional Statements," [Online]. Available: https://www.mathworks.com/help/matlab/matlab_prog/conditional-statements.html, accessed on Feb. 17, 2024

Chapter 5
Loop Statements

Chapter Learning Outcomes

- Understand the concept of iteration and when **loop constructs** are necessary in a programme.
- Learn to implement and differentiate between **for loops** and **while loops**.
- Develop the ability to translate repetitive tasks into efficient **loop statements**.
- Recognise the importance of **loop control structures** such as **break** and **continue**.
- Write **nested loops** for problems requiring multiple levels of iteration.
- Analyse loop performance and identify cases of **infinite loops**.
- Apply best practices to ensure readability and maintainability of loop constructs.

Chapter Key Words

- **for loop**: A **for loop** is a control flow statement that allows code to be executed repeatedly based on a given Boolean condition. In MATLAB, the for loop is used to iterate over a range or collection of values, executing a block of code a specific number of times. This is particularly useful for repetitive tasks that have a predictable number of iterations.
- **while loop**: A **while loop** in MATLAB is another looping construct that executes a block of code as long as a specified condition is true. Unlike the for loop, while loops are ideal when the number of iterations is not known beforehand and depends on dynamic conditions during runtime.
- **nested loop**: A **nested loop** occurs when one loop is placed inside another. MATLAB allows for loops to be nested within for or while loops. This is useful for multi-dimensional data structures or complex iterative processes that require multiple levels of iteration.
- **break statement**: The **break statement** is used within loop constructs to terminate the loop prematurely when a certain condition is met. It provides a way to escape from the loop without having to wait for the loop condition to be false.
- **continue statement**: In contrast to the break statement, the **continue statement** is used to skip the rest of the code inside the loop and proceed with the next

© The Author(s) 2025
Y. Chen and L. Huang, *MATLAB Roadmap to Applications*,
https://doi.org/10.1007/978-981-97-8788-3_5

iteration. This can be used to avoid executing certain parts of the loop when specific conditions are encountered.

- **infinite loop**: An **infinite loop** is a sequence of instructions in a computer program that repeats indefinitely. This is usually due to a loop condition that never becomes false. In MATLAB, it is crucial to avoid infinite loops as they can cause the program to become unresponsive or enter an endless execution cycle.
- **loop performance**: **Loop performance** is an important aspect of writing efficient code. In MATLAB, loop performance can often be improved by vectorisation, preallocating arrays, or using built-in functions instead of loops where applicable.

5.1   Introduction

Loop statements are control structures that enable the repetitive execution of a block of code until a specified condition is met or a certain number of iterations is reached. They play a crucial role in programming by allowing developers to automate repetitive tasks, iterate over data structures, and implement algorithms and simulations efficiently. The significance of loop statements lies in their ability to control the program flow based on predetermined conditions or criteria, thereby enhancing the flexibility and versatility of software applications.

In MATLAB, loop statements are indispensable for various tasks, such as data processing, numerical computations, simulations, and algorithm implementation. They enable developers to perform operations on large datasets, iterate over matrices and arrays, and implement iterative algorithms for solving complex problems. By leveraging loop statements, programmers can create dynamic and responsive programs that can adapt to different inputs, scenarios, and requirements.

Efficient and optimised loop statements are crucial for ensuring the performance and scalability of MATLAB applications. Poorly designed or inefficient loops can lead to significant performance bottlenecks, particularly when dealing with large datasets or computationally intensive operations. Consequently, understanding the different types of loop statements, their syntax, and best practices for optimisation is essential for developing high-performance and robust MATLAB programs.

- **Definition and significance of loop statements in programming**:
  - Loop statements enable the repetitive execution of a block of code spsciteMAT-LABLoops
  - Allow automation of repetitive tasks and iteration over data structures
  - Facilitate implementation of algorithms and simulations
- **Role of loop statements in controlling program flow**:
  - Provide control over program execution based on conditions or criteria
  - Enable dynamic and responsive program behavior
  - Allow programs to adapt to different inputs, scenarios, and requirements

- **Importance of efficient and optimised loop statements**:
  - Ensure performance and scalability of MATLAB applications
  - Avoid performance bottlenecks, especially with large datasets or computationally intensive operations
  - Require understanding of different loop types, syntax, and optimisation techniques

5.2 Types of Loop Statements

This chapter provides an overview of the different types of loop statements in MAT-LAB. Loop statements are essential in programming to perform repetitive tasks. The three main types of loop statements in MATLAB are for loops, while loops, and do-while loops. Each type has its own syntax, structure, and specific use cases.

5.2.1 For Loops

The **for loop** is one of the most commonly used loop statements in MATLAB. It is used to iterate over a range of values or elements in a vector or matrix. For loops are commonly used when the number of iterations is **known** or when iterating over a fixed set of values.

The basic syntax of a for loop in MATLAB is as follows:

Listing 5.1 Syntax of a for loop.

```
for index = start_value : step : end_value
% Loop block of code
end
```

In a for loop, the loop index is initialized to a start value, and the loop block is executed until the end value is reached, with each iteration incrementing the index by the specified step value. For loops are commonly used in scenarios where a task needs to be repeated a specific number of times, such as iterating over an array or performing mathematical calculations. The loop body, enclosed between the `for` and `end` keywords.

Example 1: An example of a for loop that calculates the sum of the first 10 positive integers:

Listing 5.2 Example of a for loop.

```
sum = 0;
for ind = 1:10
sum = sum + i;
end
disp(sum);
```

This code initializes a variable sum to 0 and then uses a for loop to iterate from 1 to 10, adding each value of i to the sum. The final result is displayed using the disp function.

Example 2: Iterating over a range of values.

Listing 5.3 Iterating over a range of values.

```
% Iterate from 1 to 10
for ind = 1:10
disp(ind)
end
```

This code will print the values from 1 to 10 in the MATLAB Command Window.

Example 3: Iterating over the elements of a vector.

Listing 5.4 Iterating over a vector.

```
% Create a vector
myVector = [1, 2, 3, 4, 5];

% Iterate over the vector elements
for elem = myVector
disp(elem)
end
```

This code will print each element of the vector myVector in the MATLAB Command Window.

5.2.2 While Loops

While loops are used when the number of iterations is **not known** in advance or when the loop needs to continue until a certain condition is met. It executes a block of code repeatedly as long as a specified condition is true.

The basic syntax of a while loop in MATLAB is as follows:

Listing 5.5 Syntax of a while loop.

```
while condition
% Loop block of code
end
```

In a while loop, the loop block is executed repeatedly as long as the condition specified is true. The loop condition is evaluated before each iteration, and if it is false, the loop is exited. While loops are commonly used in scenarios where the number of iterations depends on dynamic conditions, such as reading data from a file or listening for user input.

Example 1: An example of a while loop that calculates the factorial of a number.

Listing 5.6 Example of a while loop.

```
n = 5;
factorial = 1;
while n > 1
factorial = factorial * n;
n = n - 1;
end
disp(factorial);
```

This code calculates the factorial of a number n by multiplying it with all the numbers from n down to 1. The loop continues until n becomes 1, and the final result is displayed using the disp function.

Example 2: Counting up to a specific value.

Listing 5.7 Counting up to a value.

```
count = 1;
maxValue = 10;

while count <= maxValue
disp(count)
count = count + 1;
end
```

This code will print the values from 1 to 10 in the MATLAB Command Window.

Example 3: Iterating until a specific condition is met.

Listing 5.8 Iterating until a condition is met.

```
x = 1;
threshold = 1e-6;

while abs(x - exp(x)) > threshold
x = exp(x);
end

disp(['The value of x is: ', num2str(x)])
```

5.2.3 Do-While Loops

Do-while loops are similar to while loops, but with one key difference: the loop block is executed at least once before the loop condition is evaluated. The basic syntax of a do-while loop in MATLAB is as follows.

Listing 5.9 Syntax of a do-while loop.

```
do
% Loop block of code
while condition
```

In a do-while loop, the loop block is executed first, and then the loop condition is evaluated. If the condition is true, the loop block is executed again, and this process continues until the condition becomes false. Do-while loops are advantageous in situations where the loop block needs to be executed at least once, regardless of the initial condition.

Here is two examples:

- Example 01 A do-while loop that prompts the user for input until a valid number is entered:

Listing 5.10 Example of a do-while loop.

```
number = 0;
do
number = input('Enter a positive number: ');
while number <= 0
disp(number);
```

This code prompts the user to enter a positive number using the input function. The loop continues to prompt the user until a valid number greater than 0 is entered. The final result is displayed using the disp function.

- Example 02

Listing 5.11 Example 1: Repeat-until user enters a positive number.

```
repeat = true;
while repeat
    number = input('Enter a positive number: ');
    repeat = number <= 0;
end
```

Listing 5.12 Example 2: Imitating do-while to perform a task at least once.

```
repeat = true;
doSomethingImportant();
while repeat
    % Check if we need to repeat
    repeat = askToRepeat();
end

function doSomethingImportant()
    disp('Performing an important task.');
end

function repeat = askToRepeat()
    answer = input('Repeat the task? y/n: ','s');
    repeat = (lower(answer) == 'y');
end
```

5.3 Loop Optimisation Techniques

Loop optimisation techniques play a crucial role in improving the efficiency of code execution. By optimising loops, unnecessary iterations can be reduced, resulting in faster and more efficient programs. This section discusses the importance of loop optimisation, strategies for reducing unnecessary iterations, and various loop optimisation techniques such as loop unrolling, loop fusion, and loop interchange. Additionally, best practices for enhancing loop performance will be explored.

5.3.1 Importance of Loop Optimisation for Efficiency

The basic syntax for a **for** loop in MATLAB is:

Listing 5.13 For loop syntax.

```
for index = startValue:endValue
    % Statements
end
```

Loops are often the most time-consuming parts of a programme. In computational tasks, especially those involving large datasets or complex algorithms, **loop optimisation** plays a critical role in enhancing efficiency. Optimising loops can significantly reduce the execution time and resource consumption of MATLAB programmes. Loop optimisation is essential for improving the efficiency of code execution, especially when dealing with large data sets or computationally intensive tasks. By optimising loops, we can minimize the number of iterations required, reduce redundant calculations, and improve memory access patterns. This leads to faster execution times and more efficient resource utilisation.

Optimising loops can have a significant impact on the overall performance of a program. In many cases, a majority of the execution time is spent within loops, making them an important target for optimisation. By carefully analyzing and optimising loops, we can achieve substantial performance improvements in our code. MATLAB provides several built-in functions and techniques for loop optimisation.

5.3.2 Strategies for Reducing Unnecessary Iterations

One of the key strategies for loop optimisation is to reduce unnecessary iterations in loop optimisation. Unnecessary iterations occur when a loop continues to execute even when further iterations will not contribute to the final result. By identifying and eliminating these unnecessary iterations, we can improve the efficiency of our code.

There are several techniques for reducing unnecessary iterations in loops.

- One common approach is loop termination based on a condition. By setting appropriate loop termination conditions, we can avoid executing the loop when further iterations are not required. This can be achieved by using conditional statements such as **if** statements within the loop.
- Another strategy is loop Vectorisation, which involves performing operations on entire arrays or vectors instead of individual elements. Vectorised operations can eliminate the need for explicit loops and reduce the number of iterations required. This technique leverages the power of MATLAB's built-in functions, such as **sum** and **dot**, that can operate on arrays efficiently.
- Some other common strategies include pre-computing invariant expressions outside the loop and using logical indexing instead of loops where possible.

To,

1. **Preallocating arrays**: Allocate memory for arrays before the loop to avoid costly resizing operations.
2. **Vectorisation**: Replace loops with vectorised operations when possible to take advantage of MATLAB's optimised built-in functions.
3. **Loop-invariant code motion**: Move computations that produce the same result in each iteration outside the loop.

Example 1: Using preallocation:

Listing 5.14 Reducing iterations by preallocating arrays.

```
% Inefficient loop
for ind = 1:n
result(ind) = computeValue(ind);
end

% optimised loop with preallocation
result = zeros(1, n);
for ind = 1:n
result(ind) = computeValue(ind);
end
```

Example 2: Pre-computing invariant expressions:

Listing 5.15 Pre-computing invariant expressions.

```
% Inefficient way
for i = 1:1000
    y(i) = i * cos(0.5 * pi);
end

% Efficient way
constant = cos(0.5 * pi);
for i = 1:1000
    y(i) = i * constant;
end
```

Example 3: Using logical indexing:

Listing 5.16 Using logical indexing.

```
% Inefficient way
for i = 1:length(data)
    if data(i) > threshold
        result(i) = data(i);
    else
        result(i) = 0;
    end
end

% Efficient way
result = zeros(size(data));
result(data > threshold) = data(data > threshold);
```

5.3.3   Loop Unrolling, Loop Fusion, and Loop Interchange Techniques

Advanced loop optimisation techniques such as **loop unrolling**, **loop fusion**, and **loop interchange** can further improve performance by reducing overhead and enhancing data locality.

- Loop unrolling involves replicating loop iterations to reduce the overhead of loop control. By executing multiple loop iterations in a single iteration, we can reduce the number of loop control instructions and improve the overall performance. However, loop unrolling can increase code size and may not be beneficial for all types of loops.

Listing 5.17 Loop unrolling example.

```
% Original loop
for ind = 1:n
result(ind) = computeValue(ind);
end

% Unrolled loop
for ind = 1:4:n
result(ind) = computeValue(ind);
if i+1 <= n, result(i+1) = computeValue(i+1); end
if i+2 <= n, result(i+2) = computeValue(i+2); end
if i+3 <= n, result(i+3) = computeValue(i+3); end
end
```

- Loop fusion combines multiple loops that perform similar operations into a single loop. By eliminating redundant loop iterations and merging loop bodies, we can reduce the overhead of loop control and improve cache utilisation. This technique can be particularly effective when dealing with nested loops.

Here is to combine two loops that iterate over the same range can reduce loop overhead:

Listing 5.18 Loop Fusion.

```
% Before fusion
for i = 1:N
    A(i) = A(i) + 1;
end
for i = 1:N
    B(i) = B(i) * 2;
end

% After fusion
for i = 1:N
    A(i) = A(i) + 1;
    B(i) = B(i) * 2;
end
```

- Loop interchange involves swapping the order of nested loops to improve memory access patterns. By changing the loop order, we can optimise the memory access patterns and enhance data locality. This can result in improved cache utilisation and reduced memory access times.

  This example is to change the order of nested loops, which can improve cache performance.

Listing 5.19 Loop interchange

```
% Before interchange
for i = 1:M
    for j = 1:N
        A(i, j) = A(i, j) + 1;
    end
end

% After interchange
for j = 1:N
    for i = 1:M
        A(i, j) = A(i, j) + 1;
    end
end
```

5.3.4 Best Practices for Enhancing Loop Performance

In addition to specific optimisation techniques, there are some general best practices that can enhance loop performance. These practices include:

- Minimising the use of conditional statements within loops, as they can introduce branching and impact performance.

- Minimising memory accesses within loops by reducing the frequency of array indexing operations.
- Utilising appropriate data types and avoiding unnecessary type conversions within loops.
- Taking advantage of MATLAB's built-in functions, such as **max** and **min**, for optimised computations.
- Considering parallelisation techniques, such as MATLAB's Parallel Computing Toolbox, to exploit multi-core processors and accelerate loop execution.

By following these best practices and applying appropriate loop optimisation techniques, we can significantly improve the performance of our code. Specifically,

5.3.4.1 Minimise the Use of Nested Loops

Deeply nested loops can lead to significant performance degradation. Whenever possible, try to minimise the levels of nesting by restructuring your code or using vectorisation techniques.

Example of reducing nested loops:

Listing 5.20 Reducing nested loops

```
% Deeply nested loop
for i = 1:N
    for j = 1:M
        A(i, j) = i + j;
    end
end

% Using matrix operations
[i, j] = meshgrid(1:N, 1:M);
A = i + j;
```

5.3.4.2 Avoid Growing Arrays in Loops

As mentioned earlier, dynamically growing arrays inside loops can severely impact performance. Always preallocate arrays to their maximum required size before entering the loop.

Example of avoiding array growth:

Listing 5.21 Avoiding array growth in loops

```
% Inefficient way
A = [];
for i = 1:1000
    A(i) = i^2;
end
```

```
% Efficient way
A = zeros(1, 1000);
for i = 1:1000
    A(i) = i^2;
end
```

5.3.4.3 Use MATLAB Built-in Functions

MATLAB's built-in functions are highly optimised and can often perform operations faster than custom code. Whenever possible, use these functions instead of writing your own.

Example of using built-in functions:

Listing 5.22 Using built-in functions

```
% Custom implementation
result = 0;
for i = 1:length(A)
    result = result + A(i);
end

% Using built-in function
result = sum(A);
```

5.3.4.4 Utilise Logical Indexing

Logical indexing can replace loops for certain operations, leading to more concise and faster code.

Example of using logical indexing:

Listing 5.23 Using logical indexing

```
% Using loop to find elements
result = zeros(size(A));
for i = 1:length(A)
    if A(i) > threshold
        result(i) = A(i);
    end
end

% Using logical indexing
result = A .* (A > threshold);
```

5.3.4.5 Profile and Benchmark Your Code

Regularly profile and benchmark your code to identify slow sections. Use MATLAB's profiling tools to gather performance data and make informed decisions about where optimisations are needed.

Example of profiling code:

Listing 5.24 Profiling code

```
% Start profiling
profile on;

% Code to be profiled
for i = 1:1000
    A(i) = i^2;
end

% Stop profiling
profile off;

% View profiling results
profile viewer;
```

5.3.4.6 Using tic and toc to Measure Performance

The tic and toc functions in MATLAB are simple yet effective tools for measuring the elapsed time of code execution. They are useful for quickly benchmarking sections of your code to understand where optimisations might be needed.

Example of using tic and toc:

Listing 5.25 Using tic and toc for performance measurement

```
% Start timer
tic;

% Code to be timed
for i = 1:1000
    A(i) = i^2;
end

% Stop timer and display elapsed time
elapsedTime = toc;
fprintf('Elapsed time: %.2f seconds\n', elapsedTime);
```

By using tic and toc, you can quickly measure the performance of different sections of your code, making it easier to identify areas that require optimisation.

5.3.4.7 Avoid Unnecessary Computations

Eliminate redundant calculations within loops. If a value does not change within the loop, compute it once before the loop starts.

Example of avoiding unnecessary computations:

Listing 5.26 Avoiding unnecessary computations

```matlab
% Inefficient way
for i = 1:1000
    A(i) = sqrt(i) + cos(i);
end

% Efficient way
cosValues = cos(1:1000);
for i = 1:1000
    A(i) = sqrt(i) + cosValues(i);
end
```

5.3.4.8 Choose the Right Data Types

Using appropriate data types can have a significant impact on performance. For example, using single precision instead of double precision can reduce memory usage and improve speed if the precision is sufficient for your requirements.

Example of choosing the right data types:

Listing 5.27 Choosing appropriate data types

```matlab
% Using double precision
A = zeros(1, 1000);

% Using single precision
A = zeros(1, 1000, 'single');
```

5.3.4.9 Utilise Parallel Computing

For computationally intensive tasks, consider using MATLAB's parallel computing capabilities. Functions like `parfor` can distribute loop iterations across multiple processors or cores, significantly speeding up execution.

Example of using parallel computing:

Listing 5.28 Using parallel computing with parfor

```matlab
% Normal for loop
for i = 1:1000
    A(i) = i^2;
end
```

```
% Parallel for loop
parfor i = 1:1000
    A(i) = i^2;
end
```

By adhering to these best practices, you can write MATLAB code that is not only faster but also easier to understand and maintain. Optimised loops contribute to overall program efficiency, which is particularly important in data-intensive and high-performance computing applications.

In this section, we discussed the importance of loop optimisation for efficiency and explored strategies for reducing unnecessary iterations. We also introduced advanced loop optimisation techniques such as loop unrolling, loop fusion, and loop interchange. Additionally, we provided best practices for enhancing loop performance. By applying these techniques and following the recommended practices, programmers can achieve significant performance improvements in their code.

5.4 Applications of Loop Statements

Loop statements in MATLAB are versatile tools that can be applied to solve a wide range of computational and real-world problems. This section explores the various applications of loop statements, highlighting their usefulness in solving computational tasks and real-world problem-solving scenarios.

5.4.1 Solving Computational Tasks

Loop statements can be used in various mathematical calculations and simulations, making them indispensable tools for scientific and engineering applications. By utilising loops, iterative algorithms and numerical methods can be implemented efficiently. These algorithms are particularly useful when solving complex mathematical problems that require repetitive calculations or approximations.

- Leveraging loops for **mathematical calculations** and simulations that require repeated operations, such as iterative solutions to equations or Monte Carlo simulations.
- Application of loops in **iterative algorithms** and numerical methods including fixed-point iteration, Newton-Raphson method, and gradient descent.
- Utilisation of loop statements in various scientific and engineering domains such as computational fluid dynamics, finite element analysis, and environmental modelling.
- Employing loops for **timing** tasks and benchmarks to measure performance of algorithms across different computing architectures spsciteMoler2011.
- Using loops for mathematical calculations and simulations: Loop statements allow for the repeated execution of a set of mathematical calculations or simulations.

This is beneficial when dealing with tasks such as solving equations, generating random numbers, or simulating dynamic systems.

- Iterative algorithms and numerical methods: Many computational tasks require iterative algorithms or numerical methods to reach an accurate solution. Loop statements provide a convenient way to iterate through a set of calculations until a desired level of accuracy is achieved.
- Examples from various scientific and engineering domains: Loop statements find applications in diverse scientific and engineering domains. For instance, they can be used in physics simulations, optimisation problems, financial modeling, and many other fields.
- Loops for Timing: In certain scenarios, it is necessary to measure the execution time of a code segment. Loop statements can be used to repeat the execution of a code segment multiple times, allowing for an accurate measurement of the time taken. This is especially important when optimising code performance or comparing different algorithms.

To illustrate the applications of loop statements in computational tasks, consider the following MATLAB code snippet. It calculates the sum of the first 100 natural numbers using a loop:

Listing 5.29 Calculating the sum of natural numbers using a loop

```
sum = 0;
for ind = 1:100
sum = sum + i;
end
```

In the above example, the loop iterates from 1 to 100 and adds each number to the variable 'sum', resulting in the sum of the first 100 natural numbers.

5.4.2   Real-World Problem Solving

Loop statements also play a crucial role in solving real-world problems by enabling data processing and repetitive task execution. They provide a mechanism for automating tasks that involve processing large datasets or performing repetitive operations on data.

- Application of loops in processing **data sets** and performing repetitive tasks such as data cleansing and transformation in data analysis workflows.
- Implementing loop statements for **image and signal processing** tasks including filtering, segmentation, and feature extraction relevant in biomedical imaging and communication systems.
- Streamlining **automation** and batch processing operations using loop statements, essential in manufacturing processes and robotic control systems.
- Applying loops to process data and perform repetitive tasks: Loop statements are commonly used to iterate through data structures, such as arrays or matrices, and

perform operations on each element. This allows for efficient data processing and manipulation.

- Examples from data analysis, image processing, and signal processing: Real-world problems often involve analyzing and processing large amounts of data. Loop statements can be used to implement algorithms for data analysis, image processing, signal processing, and other related tasks.
- Automation and batch processing using loop statements: In scenarios where repetitive tasks need to be performed on multiple datasets or files, loop statements can automate the process. By iterating through a list of files or datasets, the same set of operations can be applied to each, saving time and effort.

To demonstrate the applications of loop statements in real-world problem solving, consider the following MATLAB code snippet. It calculates the average value of an array of numbers using a loop:

Listing 5.30 Calculating the average value of an array using a loop

```
data = [1, 2, 3, 4, 5];
sum = 0;
for ind = 1:length(data)
sum = sum + data(ind);
end
average = sum / length(data);
```

In the above example, the loop iterates through each element of the data' array and adds it to the variable sum'. The average value is then calculated by dividing the sum by the length of the array.

Listing 5.31 Jacobi method for solving linear equations

```
% Assuming 'A' is the coefficient matrix and 'b' is the right-hand
    side vector
x = zeros(size(b)); % Initial guess of the solution
maxIter = 100; % Maximum number of iterations
tolerance = 1e-6; % Convergence tolerance

for iter = 1:maxIter
    x_new = x;
    for ind = 1:length(b)
        sum = b(ind);
        for jnd = 1:length(b)
            if i ~= j
                sum = sum - A(i,j) * x(j);
            end
        end
        x_new(ind) = sum / A(i,i);
    end
    if max(abs(x_new - x)) < tolerance
        break;
    end
    x = x_new;
end
```

Overall, loop statements in MATLAB have numerous applications in both computational tasks and real-world problem solving. By utilising their power and flexibility, complex problems can be efficiently tackled, leading to enhanced productivity and improved outcomes.

5.5 Debugging and Error Handling

This section focuses on the important aspects of debugging and error handling within loop statements. It covers common errors and pitfalls that can occur in loop statements, techniques for effective debugging and error resolution, testing loop conditions and termination conditions, as well as strategies for handling exceptions and error handling in loops.

5.5.1 Common Errors and Pitfalls in Loop Statements

In MATLAB, **loop statements** such as **for** and **while** loops are fundamental constructs for iterative operations. However, they can be prone to several common errors.

- One frequent issue is the **off-by-one error**, where the loop iterates one time too many or too few.
- Another common pitfall is failing to initialise loop variables correctly, leading to unexpected results or infinite loops. Understanding these errors is crucial for efficient debugging and optimisation.

Loop statements can be tricky to work with, and programmers often encounter common errors and pitfalls. These errors can lead to incorrect or unexpected behavior of the loop. Understanding these common errors and pitfalls is crucial for efficient debugging and error resolution.

Some common errors and pitfalls include:

- Off-by-one errors: These occur when the loop iterates one too many or one too few times, often due to incorrect indexing or condition checks.
- Infinite loops: These occur when the loop condition never evaluates to false, causing the loop to run indefinitely.
- Logic errors: These errors occur when the loop's logical structure does not match the intended behavior, leading to incorrect results.
- Variable scope issues: These occur when variables used within the loop have incorrect or unexpected values due to scope-related problems.

Understanding these common errors and pitfalls will help programmers identify and resolve issues efficiently.

The basic syntax for a for loop is:

Listing 5.32 For loop syntax

```
for index = start_value:end_value
    % Loop body
end
```

Example 1: A correctly implemented for loop.

Listing 5.33 Correct for loop example

```
for ind = 1:5
    disp(ind);
end
```

Example 2: A for loop with an off-by-one error.

Listing 5.34 Off-by-one error in for loop.

```
for ind = 0:5
    disp(['Current index is: ', num2str(ind)]);
end
```

Example 3: Example of Correct Loop Initialisation.

Listing 5.35 Example of correct loop initialisation

```
% Correct loop initialisation
for i = 1:10
    disp(i)
end
```

5.5.2   Techniques for Effective Debugging and Error Resolution

Effective debugging in MATLAB involves using built-in tools such as the **Editor** and **Command Window**. Setting breakpoints allows one to inspect variables and step through code to observe the program's behaviour. The **dbstop** function can set breakpoints programmatically. Additionally, the **try-catch** construct can help identify and handle errors gracefully.

When facing errors in loop statements, effective debugging and error resolution techniques can greatly aid in identifying and resolving the issues. Here are some techniques to consider:

- Print statements: Adding print statements within the loop can help track the flow of execution and identify any unexpected values or behaviors.
- Variable inspection: Inspecting the values of variables at different points in the loop can help pinpoint where the error occurs and why.
- Step-through debugging: utilising a debugger to step through the loop line by line can provide insights into the program's execution and help identify the source of errors.

- Code review: Seeking assistance from a colleague or peer to review the code can help identify logical errors or provide fresh perspectives on the problem.

By employing these techniques, programmers can efficiently debug and resolve errors in loop statements.

The basic syntax for setting a breakpoint in MATLAB is to use the Editor.

Example 1: Using a breakpoint to investigate a loop's behaviour.

Listing 5.36 Using breakpoints for debugging

```
% Setting a breakpoint
for i = 1:10
    if i == 5
        disp('Breakpoint here')
    end
    disp(i)
end
```

Example 2: Using try-catch for Error Handling.

Listing 5.37 Using try-catch for error handling

```
% Using try-catch for error handling
try
    for i = 1:10
        disp(i)
    end
catch ME
    disp('An error occurred')
    disp(ME.message)
end
```

5.5.3 Testing Loop Conditions and Loop Termination Conditions

Testing loop conditions and ensuring proper termination is critical to avoid infinite loops and ensure that the loops perform as expected. The **while** loop requires careful condition setting to ensure it terminates correctly. Using the **break** statement can provide an emergency exit from the loop upon meeting certain conditions.

To test loop conditions and loop termination conditions effectively, consider the following:

- Test with different input values: Ensure that the loop condition is tested with various input values to account for different scenarios and edge cases.
- Use logical operators correctly: Employ logical operators such as AND (&&), OR (||), and NOT (~) appropriately in loop conditions to capture the desired behavior.
- Validate termination conditions: Verify that loop termination conditions are correctly implemented to prevent infinite loops or premature termination.

Listing 5.38 Testing while loop conditions

```
% while loop condition testing
i = 1;
while i <= 10
    disp(i)
    i = i + 1;
end
```

Listing 5.39 Using break in loops

```
% Using break statement
for i = 1:10
    if i == 5
        break
    end
    disp(i)
end
```

By thoroughly testing loop conditions and termination conditions, programmers can ensure the reliability and correctness of their loops.

The basic syntax for a while loop is:

Listing 5.40 While loop syntax

```
while condition
    % Loop body
end
```

Example 1: A while loop with a clear termination condition.

Listing 5.41 While loop with termination condition

```
ind = 1;
while i <= 5
    disp(ind);
    ind = i + 1;
end
```

Example 2: A while loop where the termination condition is never met.

Listing 5.42 Faulty while loop with no termination

```
ind = 1;
while i > 0  % Incorrect condition, creates an infinite loop
    disp(ind);
    ind = i + 1;
end
```

The validity of loop conditions is paramount for the correct execution of loop statements:

Listing 5.43 Testing loop conditions for validity

```matlab
% Example of a for loop with a valid termination condition
for ind = 1:10
disp(ind);
end

% Example of a while loop with a valid termination condition
count = 0;
while count < 5
count = count + 1;
disp(count);
end
```

5.5.4 Strategies for Handling Exceptions and Error Handling in Loops

Handling exceptions and errors within loops is essential for robust code. Using MATLAB's **try-catch** blocks within loops allows one to handle unexpected conditions gracefully without terminating the loop abruptly. This approach ensures that the loop continues to execute even if an error occurs in one iteration. Exception handling plays a vital role in ensuring the robustness of loop statements. By employing effective strategies for handling exceptions and error handling in loops, programmers can gracefully handle errors and prevent program crashes.

Some strategies for handling exceptions and error handling in loops include:

- Try-catch blocks: Utilize try-catch blocks to catch and handle exceptions that may occur within the loop. This allows for graceful error handling and recovery.
- Error logging: Implement error logging mechanisms to record and track errors that occur during loop execution. This information can aid in debugging and troubleshooting.
- Error recovery: Define fallback mechanisms or alternate strategies to recover from errors and continue loop execution whenever possible.

By incorporating these strategies, programmers can ensure that their loops handle exceptions and errors effectively, enhancing the overall reliability of their programs.

- Example 1: Try-Catch Block

```matlab
% Example of using try-catch block for error handling
x = [1, 2, 3];
try
disp(x(4));
catch
disp('Error: Index out of bounds');
end
```

- Example 2: Graceful Error Handling

```
% Example of graceful error handling in a loop
x = [1, 2, 3];
for ind = 1:4
try
disp(x(ind));
catch
disp(['Error: Index ' num2str(ind) ' out of bounds']);
end
end
```

In this section, we have explored the importance of debugging and error handling in loop statements. We discussed common errors and pitfalls that programmers may encounter and provided techniques for effective debugging and error resolution. Additionally, we delved into testing loop conditions and termination conditions to ensure correct loop behavior. Lastly, we presented strategies for handling exceptions and error handling within loops to ensure robust and reliable code execution.

5.6 Advanced Topics in Loop Statements

In this section, we will explore some advanced concepts and techniques related to loop statements in MATLAB. We will discuss nested loops, loop control using **break** and **continue** statements, and the integration of loops with arrays and data manipulation.

5.6.1 Conditional Loops

- In MATLAB, it is possible to combine loop statements with conditional statements (**if**, **elseif**, and **else**) to create more complex control flow structures. This combination allows for selective execution of code blocks based on specified conditions within the loop iterations.
  The basic syntax for conditional loops is as follows:

Listing 5.44 Conditional loop syntax

```
for variable = expression
% Conditional statements
if condition1
% Statements to be executed
elseif condition2
% Statements to be executed
else
% Statements to be executed
end
end
```

- Alternatively, conditional statements can be used within a **while** loop:

Listing 5.45 Conditional while loop syntax

```
while condition
% Conditional statements
if condition1
% Statements to be executed
elseif condition2
% Statements to be executed
else
% Statements to be executed
end
end
```

This structure allows for the combination of a **while** loop with conditional statements, enabling more complex decision-making and control flow within the loop iterations. The **while** loop continues to execute as long as the main `condition` is true, and within each iteration, the conditional statements are evaluated to determine which block of code should be executed.

Example 1: Computing the sum of positive values in a vector.

Listing 5.46 Computing the sum of positive values

```
% Create a vector
myVector = [-2, 5, 0, 3, -1, 7];
positiveSum = 0;
ind = 1;

% Iterate over the vector elements
while i <= length(myVector)
% Check if the element is positive
if myVector(ind) > 0
positiveSum = positiveSum + myVector(ind);
end
ind = i + 1;
end

disp(['The sum of positive values is: ', num2str(positiveSum)])
```

In this example, a **while** loop is used to iterate over the elements of the vector `myVector`. Within the loop, an **if** statement checks if the current element is positive. If it is, the element is added to the `positiveSum` variable. The loop continues until all elements have been processed, and the final sum of positive values is printed.

Example 2: Finding the first positive value in a matrix.

Listing 5.47 Finding the first positive value in a matrix

```
% Create a matrix
myMatrix = [-2, 0, 5; -1, 3, -4; 2, 6, 1];
found = false;
ind = 1;
```

```
jnd = 1;

% Iterate over the matrix elements
while ~found && i <= size(myMatrix, 1) && j <= size(myMatrix, 2)
% Check if the element is positive
if myMatrix(ind, jnd) > 0
found = true;
disp(['The first positive value is: ', num2str(myMatrix(ind, jnd)
    )])
else
jnd = j + 1;
if j > size(myMatrix, 2)
jnd = 1;
ind = i + 1;
end
end
end

% If no positive value is found
if ~found
disp('No positive value found in the matrix.')
end
```

This code uses a **while** loop to iterate over the elements of the matrix myMatrix. Within the loop, an **if** statement checks if the current element is positive. If a positive value is found, the loop terminates, and the value is printed. If no positive value is found after iterating over all elements, a message is displayed indicating that no positive value was found in the matrix.

These examples demonstrate how conditional statements can be combined with **while** loops to introduce more complex control flow and decision-making logic within the loop iterations.

Conditional statements within loops provide additional control over the program flow, allowing for more sophisticated decision-making and branching logic. This is particularly useful when dealing with complex data structures, filtering or processing data based on specific criteria, or implementing algorithms with multiple conditions or edge cases.

Example 1: Summing even numbers in a vector.

Listing 5.48 Summing even numbers using a conditional loop

```
% Create a vector
myVector = [1, 2, 3, 4, 5, 6, 7, 8, 9, 10];
evenSum = 0;

% Iterate over the vector elements
for num = myVector
% Check if the number is even
if mod(num, 2) == 0
evenSum = evenSum + num;
```

```
end
end

disp(['The sum of even numbers is: ', num2str(evenSum)])
```

This code iterates over the elements of the vector myVector and checks if each number is even using the mod function. If the number is even, it is added to the evenSum variable. The final sum of even numbers is then printed.

Example 2: Finding the maximum value in a matrix with conditions.

Listing 5.49 Finding the maximum value with conditions

```
% Create a matrix
myMatrix = [5, 2, 8, 1; 3, 6, 4, 9; 7, 0, 2, 5];
maxValue = -inf;

% Iterate over the matrix elements
for ind = 1:size(myMatrix, 1)
for jnd = 1:size(myMatrix, 2)
% Check if the element is positive and greater than maxValue
if myMatrix(ind, jnd) > 0 && myMatrix(ind, jnd) > maxValue
maxValue = myMatrix(ind, jnd);
end
end
end

disp(['The maximum positive value is: ', num2str(maxValue)])
```

This code uses nested **for** loops to iterate over the elements of the matrix myMatrix. For each element, it checks if the value is positive and greater than the current maxValue. If both conditions are met, the maxValue is updated with the new maximum value. Finally, the maximum positive value found in the matrix is printed.

5.6.2 Nested Loops and Loop Control

In the realm of programming, nested loops and loop control mechanisms are pivotal for managing complex iterative processes. Nested loops allow for the execution of a loop within another loop, which is essential for iterating over multi-dimensional data structures or performing tasks that require multiple levels of looping.

- Nesting loops for complex iterations and patterns Nested loops involve placing one loop inside another to handle complex iterations and generate intricate patterns. The inner loop completes all its iterations for each iteration of the outer loop. This allows for more sophisticated control over the flow of the program.

The basic syntax for nested loops is as follows:

Listing 5.50 Nested loop syntax

```
for outer_variable = start:step:end
for inner_variable = start:step:end
% Statements to be executed
end
end
```

- Controlling loop execution using **break** and **continue** statements MATLAB provides two important statements for controlling loop execution: **break** and **continue**. The **break** statement is used to prematurely exit a loop, while the **continue** statement skips the remaining statements in the current iteration and moves to the next iteration.

Here's an example that demonstrates the usage of nested loops and the **break** statement:

Listing 5.51 Nested loop with break statement

```
for ind = 1:5
for jnd = 1:5
if ind == j
break;
end
fprintf('(%d, %d) ', i, j);
end
fprintf('\n');
end
```

In this example, the outer loop iterates from 1 to 5, and the inner loop also iterates from 1 to 5. However, when the values of i and j are equal, the **break** statement is encountered, causing the inner loop to terminate prematurely and move to the next iteration of the outer loop.

- Practical examples and scenarios requiring nested loops.

5.6.3   Integration with Arrays and Data Manipulation

Loops are instrumental in handling arrays and performing data manipulation tasks. Through iteration, loops enable the access and modification of array elements, catering to operations such as data aggregation, filtering, and transformation.

- Accessing and manipulating array elements using loops
  Here's an example that demonstrates accessing and manipulating array elements using loops:

Listing 5.52 Array manipulation using loops

```
data = [10, 20, 30; 40, 50, 60; 70, 80, 90];
[rows, cols] = size(data);

for ind = 1:rows
for jnd = 1:cols
data(ind, jnd) = data(ind, jnd) * 2;
end
end

disp(data);
```

- Array iteration techniques and multidimensional arrays
- Data aggregation, filtering, and transformation using loops.

The syntax to loop through an array is as follows:

Listing 5.53 Array iteration syntax

```
for ind = 1:length(array)
statements
end
```

Here are two examples that illustrate array manipulation using loops in MATLAB:

Listing 5.54 Example of array manipulation using loop

```
% Sum of array elements
array = [1, 2, 3, 4, 5];
sum = 0;

for ind = 1:length(array)
sum = sum + array(ind);

end

fprintf('The sum of the array elements is %d\n', sum);
```

Listing 5.55 Example of data filtering using loop

```
% Filtering negative values from an array
array = [-3, 1, -2, 4, 5];
filtered_array = [];

for ind = 1:length(array)

if array(ind) > 0
filtered_array = [filtered_array, array(ind)];
end
end

disp('Filtered array:');
disp(filtered_array);
```

The examples provided are intended to demonstrate the utility of loop statements in MATLAB for sophisticated programming tasks. They are structured with clear syntax and colour schemes to enhance readability and comprehension.

5.7   Best Practices and Guidelines

- Writing clear, readable, and maintainable loop statements
- Choosing appropriate loop types for different scenarios
- Performance considerations and optimisation strategies
- Documentation and commenting practices for loop statements.

5.8   Laboratory

This section provides a set of lab works and exercises to reinforce the concepts and practical implementation of loop statements in MATLAB.

1. Practice of for loop statement
   Calculate the sum of odd numbers between 1 and 100.

Solution:

Listing 5.56 Practice of for loop statement

```
sum = 0; % Initialisation parameter

for ind = 1:2:100
sum = sum + ind;
end

sum
```

2. The for loop iterating over vectors

   a. Write a MATLAB script that creates a vector of random integers between 1 and 100 with a length of 20. Use a **for** loop to iterate over the vector and print each element to the Command Window.

Solution:

Listing 5.57 Iterating over a vector

```
% Create a vector of random integers
myVector = randi([1, 100], 1, 20);
```

```
% Iterate over the vector
for ind = 1:length(myVector)
disp(myVector(ind))
end
```

b. Modify the script from the previous exercise to calculate the sum of all elements in the vector.

Listing 5.58 Calculating the sum of vector elements

```
% Create a vector of random integers
myVector = randi([1, 100], 1, 20);
vectorSum = 0;

% Iterate over the vector and calculate the sum
for ind = 1:length(myVector)
vectorSum = vectorSum + myVector(ind);
end

disp(['The sum of vector elements is: ', num2str(vectorSum)])
```

3. Nested Loops

a. Write a MATLAB script that creates a 5x5 matrix with random values between 1 and 10. Use nested **for** loops to iterate over the matrix and print each element to the Command Window.

Solution:

Listing 5.59 Iterating over a matrix

```
% Create a 5x5 matrix with random values
myMatrix = randi([1, 10], 5, 5);

% Iterate over the matrix
for ind = 1:size(myMatrix, 1)
for jnd = 1:size(myMatrix, 2)
disp(myMatrix(ind, jnd))
end
end
```

b. Modify the script from the previous exercise to calculate the sum of all elements in the matrix.

Listing 5.60 Calculating the sum of matrix elements

```
% Create a 5x5 matrix with random values
myMatrix = randi([1, 10], 5, 5);
matrixSum = 0;

% Iterate over the matrix and calculate the sum
for ind = 1:size(myMatrix, 1)
for jnd = 1:size(myMatrix, 2)
matrixSum = matrixSum + myMatrix(ind, jnd);
end
end

disp(['The sum of matrix elements is: ', num2str(matrixSum)])
```

4. Conditional Loops

   a. Write a MATLAB script that creates a vector of random integers between -10 and 10 with a length of 15. Use a **for** loop and conditional statements to count the number of positive, negative, and zero values in the vector.

Solution:

Listing 5.61 Counting positive, negative, and zero values

```
% Create a vector of random integers
myVector = randi([-10, 10], 1, 15);
positiveCount = 0;
negativeCount = 0;
zeroCount = 0;

% Iterate over the vector and count values
for ind = 1:length(myVector)
if myVector(ind) > 0
positiveCount = positiveCount + 1;
elseif myVector(ind) < 0
negativeCount = negativeCount + 1;
else
zeroCount = zeroCount + 1;
end
end

disp(['Number of positive values: ', num2str(positiveCount)])
disp(['Number of negative values: ', num2str(negativeCount)])
disp(['Number of zero values: ', num2str(zeroCount)])
```

   b. Write a MATLAB script that creates a 4x4 matrix with random values between 1 and 20. Use nested **for** loops and conditional statements to find the maximum and minimum values in the matrix.

Solution:

Listing 5.62 Finding maximum and minimum values in a matrix

```matlab
% Create a 4x4 matrix with random values
myMatrix = randi([1, 20], 4, 4);
maxValue = -inf;
minValue = inf;

% Iterate over the matrix and find maximum and minimum values
for ind = 1:size(myMatrix, 1)
for jnd = 1:size(myMatrix, 2)
if myMatrix(ind, jnd) > maxValue
maxValue = myMatrix(ind, jnd);
end
if myMatrix(ind, jnd) < minValue
minValue = myMatrix(ind, jnd);
end
end
end

disp(['The maximum value in the matrix is: ', num2str(maxValue)])
disp(['The minimum value in the matrix is: ', num2str(minValue)])
```

5. Vectorisation and Performance

    a. Write a MATLAB script that creates two vectors, A and B, of random integers between 1 and 100 with a length of 1,000,000. Calculate the element-wise sum of the two vectors using a **for** loop and Vectorisation, and compare the execution times of both approaches.

Solution:

Listing 5.63 Comparing loop and vectorisation performance

```matlab
% Create two large vectors
A = randi([1, 100], 1, 1000000);
B = randi([1, 100], 1, 1000000);

% Calculate the sum using a for loop
tic
sumLoop = zeros(size(A));
for ind = 1:length(A)
sumLoop(ind) = A(ind) + B(ind);
end
loopTime = toc;

% Calculate the sum using Vectorisation
tic
sumVector = A + B;
vectorTime = toc;
```

```
disp(['Time taken for loop approach: ', num2str(loopTime), '
    seconds'])
disp(['Time taken for Vectorisation approach: ', num2str(vectorTime
    ), ' seconds'])
```

b. Write a MATLAB function that computes the dot product of two vectors using a **for** loop and Vectorisation. Compare the execution times of both approaches for various vector lengths (e.g., 1,000, 10,000, 100,000).

Solution:

Listing 5.64 Comparing loop and vectorisation performance for dot product

```
unction comparePerformance(length)
% Create two random vectors
A = rand(1, length);
B = rand(1, length);
% Calculate dot product using a for loop
tic
dotProductLoop = 0;
for ind = 1:length
    dotProductLoop = dotProductLoop + A(ind) * B(ind);
    end
loopTime = toc;

% Calculate dot product using Vectorisation
tic
dotProductVector = dot(A, B);
vectorTime = toc;

disp(['For vector length ', num2str(length), ':'])
disp(['Time taken for loop approach: ', num2str(loopTime), '
    seconds'])
disp(['Time taken for Vectorisation approach: ', num2str(vectorTime
    ), ' seconds'])

end

% Call the function with various vector lengths
comparePerformance(1000)
comparePerformance(10000)
comparePerformance(100000)
```

6. Loop Unrolling and Parallelisation

a. Write a MATLAB script that creates a large vector of random integers between 1 and 100 with a length of 10,000,000. Calculate the sum of the vector elements using a **for** loop and the `codegen` function to generate unrolled C code. Compare the execution times of both approaches.

Solution:

Listing 5.65 Comparing loop and loop unrolling performance

```matlab
% Create a large vector
largeVector = randi([1, 100], 1, 10000000);

% Calculate the sum using a for loop
tic
sumLoop = 0;
for ind = 1:length(largeVector)
sumLoop = sumLoop + largeVector(ind);
end
loopTime = toc;

% Calculate the sum using loop unrolling
sumUnrolled = @(x) sumUnrolledCodegen(x);
tic
sumUnrolledResult = sumUnrolled(largeVector);
unrolledTime = toc;

disp(['Time taken for loop approach: ', num2str(loopTime), '
    seconds'])
disp(['Time taken for loop unrolling approach: ', num2str(
    unrolledTime), ' seconds'])
```

b. Write a MATLAB script that creates a large matrix of random values between 1 and 100 with dimensions 10,000x10,000. Use the `parfor` construct to parallelize the computation of the sum of all elements in the matrix. Compare the execution time with the serial implementation using a **for** loop.

Solution:

Listing 5.66 Comparing serial and parallel loop performance

```matlab
% Create a large matrix
largeMatrix = randi([1, 100], 10000, 10000);

% Calculate the sum using a for loop
tic
sumSerial = 0;
for ind = 1:size(largeMatrix, 1)
for jnd = 1:size(largeMatrix, 2)
sumSerial = sumSerial + largeMatrix(ind, jnd);
end
end
serialTime = toc;

% Calculate the sum using a parallel loop
tic
sumParallel = 0;
parfor ind = 1:size(largeMatrix, 1)
for jnd = 1:size(largeMatrix, 2)
```

```
sumParallel = sumParallel + largeMatrix(ind, jnd);
end
end
parallelTime = toc;

disp(['Time taken for serial approach: ', num2str(serialTime), '
    seconds'])
disp(['Time taken for parallel approach: ', num2str(parallelTime),
    ' seconds'])
```

7. Loop through vector and for timing

   a. Step 1: Let i be a natural number ranging from 1 to 10^7. Calculate $A(i) = \sin(i) \cdot \cos(i)$ using a for loop, and record the execution time of the loop.
   b. Step 2: Implement $A(i) = \sin(i) \cdot \cos(i)$ using matrix operations, and record the computation time.

Solution:

Listing 5.67 Comparing loop and loop unrolling performance

```
clear;

% Calculation using a for loop
tic;
for ind = 1 : 10^7
    A = sin(ind) * cos(ind);
end
t1 = toc;

% Calculation using matrix operations
B = zeros(1, 10^7);
tic;
jnd = 1 : 10^7;
B(1, j) = sin(jnd) .* cos(jnd);
t2 = toc;
```

8. The 'for' and 'while '
   Calculate the sum of the exponential function 2^x for x being a natural number between 0 and 20, using both a for loop and a while loop. Compare the structural differences between the for and while implementations.

Solution:

Listing 5.68 The 'for' and 'while'

```
% Calculate the sum using a for loop
sum_for = 0;
for x = 0:20
    sum_for = sum_for + 2^x;
end
```

```
fprintf('Sum calculated using for loop: %d\n', sum_for);

% Calculate the sum using a while loop
sum_while = 0;
x = 0;
while x <= 20
    sum_while = sum_while + 2^x;
    x = x + 1;
end
fprintf('Sum calculated using while loop: %d\n', sum_while);
```

9. While loop statement

   Enter a series of numbers from the keyboard. End the input by entering 0. Calculate and display the average and the sum of these numbers.

Solution:

Listing 5.69 While loop statement

```
% Initialize the sum and counter
total_sum = 0;
count = 0;

% Prompt the user to enter a number
number = input('Enter a number (end with 0): ');

% Continue accepting numbers until 0 is entered
while number ~= 0
    total_sum = total_sum + number;
    count = count + 1;
    number = input('Enter a number (end with 0): ');
end

% If at least one number was entered, calculate and display the sum
      and mean
if count > 0
    fprintf('Sum = %f\n', total_sum);
    average = total_sum / count;
    fprintf('Average = %f\n', average);
end
```

5.9   Problems

This section provides a set of problems to further challenge and reinforce the understanding of loop statements in MATLAB.

1. Write a MATLAB script that creates a vector of random integers between 1 and 20 with a length of 10. Use a **for** loop to iterate over the vector and print all elements that are even.

2. Write a MATLAB function that takes a vector as input and returns a new vector containing only the positive elements. Use a **for** loop to iterate over the input vector.

3. Write a MATLAB script that creates a 3x3 matrix with random values between 1 and 10. Use nested **for** loops to iterate over the matrix and print the elements in reverse row order.

4. Write a MATLAB function that takes a scalar value and a vector as input. Use a **for** loop to iterate over the vector and multiply each element by the scalar value. Return the modified vector.

5. Write a MATLAB script that creates a vector of random integers between 1 and 100 with a length of 20. Use a **while** loop to iterate over the vector and print all elements that are divisible by 3 or 5.

6. Write a MATLAB function that takes a matrix as input and computes the sum of all elements in the matrix using nested **for** loops.

7. Write a MATLAB script that creates two vectors, A and B, of random integers between 1 and 10 with a length of 5. Use Vectorisation to compute the element-wise sum, difference, and product of the two vectors.

8. Write a MATLAB function that takes a vector as input and returns the maximum and minimum values in the vector using a **for** loop.

9. Write a MATLAB script that creates a vector of random integers between 1 and 20 with a length of 15. Use a **for** loop to iterate over the vector and replace all occurrences of the number 3 with the value -1.

10. Write a MATLAB function that takes a matrix as input and computes the sum of the diagonal elements using a single **for** loop.

5.10 Summary

One way to optimise loops is to minimise the number of unnecessary iterations. This can be achieved by:

- Understanding the **importance of loop optimisation** for computational efficiency, which is critical when dealing with large data sets or simulations where execution time is of the essence.
- Implementing **strategies for reducing unnecessary iterations**, which minimises the computational load and enhances the overall speed of MATLAB programs.
- Practical application of **loop unrolling**, **loop fusion**, and **loop interchange techniques** which are advanced practices for improving loop performance in scripting.
- Adhering to **best practices for enhancing loop performance** such as preallocation of memory and vectorisation, that lead to significant improvements in script execution times.

Advanced loop optimisation techniques include:

- **Loop unrolling**: Reduce loop overhead by replicating the loop body multiple times and adjusting the iteration count accordingly.

- **Loop fusion**: Combine multiple loops that iterate over the same range into a single loop to reduce overhead and improve data locality.
- **Loop interchange**: Rearrange the order of nested loops to optimise memory access patterns and improve cache utilisation.

- **For Undergraduate (UG) Students**:
  This chapter covers the fundamental concepts of **loops** in MATLAB, specifically the **for** and **while** loops. Loops are essential for iterating over vectors, matrices, and other data structures, allowing you to perform operations on individual elements or subsets of data. The **for** loop is used when the number of iterations is known in advance, while the **while** loop is useful when the number of iterations depends on a specific condition. Understanding loops is crucial for writing efficient and effective MATLAB code, enabling you to automate repetitive tasks and perform complex data manipulations.
- **For Postgraduate (PG) Students**:
  This chapter delves into the applications of **loops** in MATLAB, demonstrating their versatility in various computational tasks. Loops are invaluable tools for **data processing, numerical analysis**, and **algorithm implementation**. The examples provided showcase the use of loops for **vector and matrix operations, element-wise computations, conditional statements**, and **data manipulation**. Additionally, the chapter introduces techniques for **optimising loop performance**, such as **Vectorisation** and **preallocation**, which are essential for efficient code execution, especially in large-scale computations.
- **For Professional Researchers or Engineers**:
  The chapter on loops in MATLAB serves as a comprehensive reference for professionals working in various domains, including **scientific computing, data analysis**, and **engineering applications**. It highlights the importance of **loop control structures** in implementing complex algorithms, performing **iterative calculations**, and **automating repetitive tasks**. The examples provided demonstrate best practices for **code readability, modularity**, and **maintainability**, ensuring that your MATLAB code is robust, efficient, and scalable. Additionally, the chapter explores advanced techniques, such as **nested loops** and **loop optimisation strategies**, equipping you with the tools necessary to tackle demanding computational challenges in your respective fields.

Chapter 6
Scripts and Functions

Chapter Learning Outcomes

By the end of this chapter, you should be able to:

- Understand the difference between **scripts** and **functions** in MATLAB
- Create and execute **scripts** to automate repetitive tasks and perform complex calculations
- Define and call **functions** to modularise code and improve code reusability
- Utilise **input arguments** and **return values** in functions for flexible and efficient programming
- Implement **local** and **global variables** in scripts and functions to manage data scope and accessibility
- Gain proficiency in organizing code into modular functions for better code reuse and maintainability.
- Apply scripts and functions to solve mathematical and computational problems in MATLAB.

Chapter Key Words

- **Scripts**: Scripts are files containing a sequence of MATLAB commands that can be executed together. They are used to automate repetitive tasks, perform complex calculations, and store a series of commands for later use. Scripts do not accept input arguments or return output values, and they operate on the variables in the current workspace.

© The Author(s) 2025
Y. Chen and L. Huang, *MATLAB Roadmap to Applications*,
https://doi.org/10.1007/978-981-97-8788-3_6

- **Functions**: Functions are self-contained units of code that perform a specific task. They can accept **input arguments**, perform computations, and **return output values**. Functions are used to modularise code, improve code reusability, and make code more readable and maintainable. Functions have their own local workspace and can access variables from the calling workspace using global variables.
- **Structure**: Refers to the organization and layout of code in MATLAB. It includes the use of indentation, comments, and proper naming conventions to enhance code readability.
- **Syntax**: The set of rules and conventions that dictate how MATLAB commands and statements should be written. Syntax governs the proper use of operators, functions, and data types.
- **Input Arguments**: Input arguments are values passed to a function when it is called. They allow functions to be more flexible and reusable by enabling them to perform computations on different data sets. Input arguments are specified in the function definition and are passed to the function in the order they are defined.
- **Output Arguments**: Variables or values returned by a function after performing its operations. They represent the results or processed data generated by the function.
- **Return Values**: Return values are the output of a function. They are the results of the computations performed by the function and can be assigned to variables in the calling workspace. Functions can return multiple values by enclosing them in square brackets.
- **Local Variables**: Local variables are variables that are defined within a function and are only accessible within that function. They are used to store temporary results and perform calculations within the function. Local variables have a separate workspace from the calling workspace and are cleared from memory when the function ends.
- **Global Variables**: Global variables are variables that are accessible from any workspace, including the base workspace and all functions. They are used to share data between different parts of a program. Global variables are declared using the `global` keyword and must be declared in each function that uses them.

6.1   Scripts

In MATLAB, **scripts** are files containing a sequence of commands saved in a file with a .m extension. that can be executed together. Scripts are used to automate repetitive tasks, perform complex calculations, and store a series of commands for later use. Unlike functions, scripts do not accept input arguments nor return output values; instead, they operate on the data available in the current workspace. This section will cover the basics of creating and executing scripts, as well as managing variables within scripts.

6.1.1 Introduction to Scripts in MATLAB

Scripts are plain text files with a .m extension, making them easy to create, edit, and share. The .m extension identifies the file as a MATLAB script, distinguishing it from other file types. Script files can be created and edited using the MATLAB Editor or any text editor that can save files with a .m extension.

Scripts contain a sequence of MATLAB commands and statements that are executed sequentially. These commands and statements can include variable assignments, mathematical operations, function calls, control flow statements (e.g., loops and conditionals), and plotting commands. The commands and statements in a script are executed in the order they appear, from top to bottom. Scripts can also include comments, which are lines of text that provide explanations or descriptions of the code. Comments start with a percent sign (

Unlike functions, scripts do not have a formal mechanism for accepting input arguments or returning output values. Scripts operate on the variables in the current workspace and can modify or create new variables, but they do not explicitly return values to the caller. If a script needs to use specific input values, these values must be assigned to variables within the script or be present in the workspace before the script is executed. If a script generates results that need to be used later, those results must be assigned to variables within the script, which can then be accessed from the workspace after the script finishes execution.

When a script is executed, it has access to all the variables in the current MATLAB workspace. Scripts can read and modify the values of existing variables in the workspace, as well as create new variables. Any variables created or modified by a script will remain in the workspace after the script finishes execution, allowing those variables to be used in subsequent commands or scripts. It is important to be aware of the state of the workspace when executing a script, as the script's behavior can be influenced by the presence or absence of certain variables. To avoid unintended consequences, it is often a good practice to start with a clean workspace before running a script.

The basic syntax for a MATLAB script is:

Listing 6.1 Basic MATLAB script syntax.

```
% A simple MATLAB script
% This script calculates the square of an array of numbers
numbers = [1, 2, 3, 4, 5];
squaredNumbers = numbers.^2;
disp(squaredNumbers);
```

6.1.2 Creating and Executing Scripts

Creating scripts in MATLAB is a straightforward process that can be accomplished using the built-in MATLAB Editor. The Editor is a powerful tool that provides a user-friendly interface for writing, editing, and debugging MATLAB code. It offers features such as syntax highlighting, auto-completion, and code folding, which enhance the coding experience and help users write more efficient and error-free scripts. The Editor can be accessed by clicking the "New Script" button in the MATLAB toolbar or by using the `edit` command in the Command Window.

To create a new script, users can either click the "New Script" button in the MATLAB toolbar or use the `edit` command followed by the desired script name in the Command Window. For example, typing `edit my_script.m` in the Command Window will open a new script file named "my_script.m" in the Editor. If the specified file does not exist, MATLAB will create a new file with that name. Once the new script file is open in the Editor, users can start writing their MATLAB code, including commands, statements, and comments.

When saving a script, it is essential to use the correct file extension, which is `.m` for MATLAB scripts. The `.m` extension identifies the file as a MATLAB script and ensures that MATLAB recognizes and executes the file correctly. To save a script, users can click the "Save" button in the Editor toolbar or use the "Save" option from the "File" menu. It is good practice to give scripts descriptive names that reflect their purpose, making it easier to identify and manage them within a project.

Executing a script in MATLAB is a simple process that can be done in two ways: typing the script name in the Command Window or clicking the "Run" button in the Editor. To execute a script from the Command Window, users should type the script name without the `.m` extension and press "Enter." MATLAB will then execute the commands and statements in the script sequentially. Alternatively, users can open the script in the Editor and click the "Run" button in the toolbar. This will execute the script and display any output or results in the Command Window. If the script contains errors or issues warnings, these will be displayed in the Command Window, helping users identify and resolve any problems with their code (Fig. 6.1). The basic syntax for creating a script is:

Listing 6.2 Script syntax.

```
% Script name: example_script.m
% This is a comment describing the script

% MATLAB commands and statements
x = 1:10;
y = sin(x);
plot(x, y);
```

Example 1: Calculate the average of a set of numbers

Listing 6.3 Average calculation script.

```
    % Script name: calculate_average.m
numbers = [4, 7, 1, 9, 3, 5];
average = mean(numbers);
disp(['The average is: ', num2str(average)]);
```

Example 2: Plot a sine wave

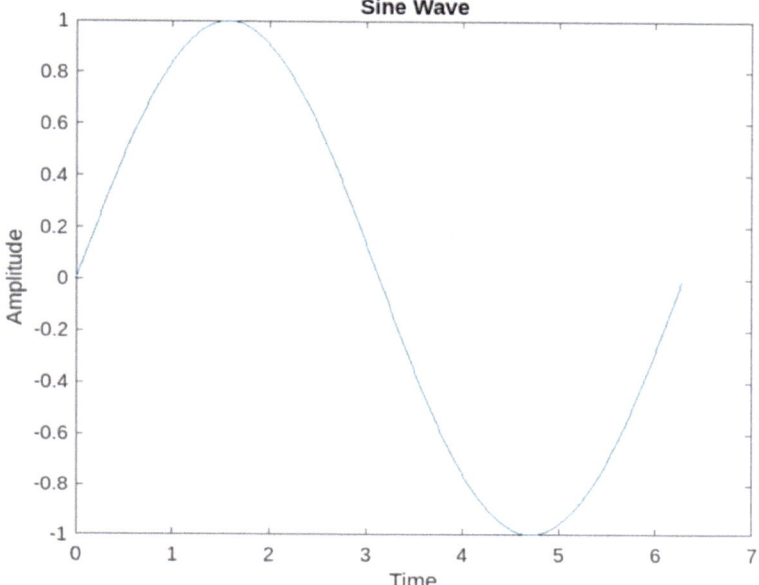

Fig. 6.1 Sine wave plot generated by the script

Listing 6.4 Sine wave plot script.

```
% Script name: plot_sine_wave.m
t = 0:0.01:2*pi;
y = sin(t);
plot(t, y);
xlabel('Time');
ylabel('Amplitude');
title('Sine Wave');
```

6.1.3 Managing Variables in Scripts

In MATLAB, variables created within a script are automatically added to the current workspace, which is a memory area where MATLAB stores all the variables, functions, and other objects currently in use. When a script assigns a value to a variable name, that variable becomes available in the workspace, allowing it to be accessed and manipulated not only within the script but also from the Command Window or other scripts and functions that are subsequently executed. This feature enables seamless integration between scripts and the broader MATLAB environment, facilitating data sharing and result propagation across different parts of a project.

Scripts have the power to access and modify variables that already exist in the workspace, providing a high degree of flexibility and interactivity. A script can read the values of variables created before its execution, either by the user in the Command Window or by previously executed scripts or functions. Moreover, scripts can modify the values of these pre-existing variables, overwriting their original values with new ones. This capability allows scripts to build upon and interact with the data and results generated by other parts of the MATLAB environment, enabling complex analyses and iterative refinements.

It is crucial to understand that variables created or modified by a script persist in the workspace even after the script has finished executing. This persistence means that any variables created within a script will remain available for use in the Command Window or in other scripts and functions executed later. While this persistence can be advantageous for maintaining data and results across different parts of a project, it can also lead to unintended consequences if the script inadvertently modifies variables used elsewhere in the program. Therefore, careful variable management is essential to ensure the integrity and reliability of MATLAB programs.

To mitigate the risk of unintended variable changes and maintain a clean and organized workspace, MATLAB provides the **clear** and **clearvars** commands. These commands allow users to remove variables from the workspace selectively. The **clear** command, when used without any arguments, removes all variables from the workspace, effectively resetting it to its initial state. Alternatively, users can specify the names of individual variables to be removed, such as `clear var1 var2`. The **clearvars** command serves a similar purpose but offers more fine-grained control over which variables are removed based on their attributes or types. By employing these commands judiciously, users can maintain a clean workspace, prevent unintended variable changes, and ensure the robustness and reliability of their MATLAB programs. Regular workspace management, including the use of clear and clearvars commands, is considered a best practice in MATLAB programming, promoting code clarity, reproducibility, and reducing the likelihood of errors stemming from variable name conflicts or unintended variable modifications.

6.2   Functions

In MATLAB, **functions** are self-contained units of code that perform a specific task. Functions can accept input arguments, perform computations or actions, and return output values. Calling functions is a fundamental aspect of programming in MATLAB, as it allows for modular and reusable code.

6.2.1 Introduction to Functions in MATLAB

In MATLAB, user-defined functions allow programmers to create their own reusable code blocks that perform specific tasks. These functions help in organising code, improving readability, and promoting code reuse. User-defined functions are stored in separate files with a .m extension and can be called from scripts or other functions.

- Why are Functions Needed?
  Functions in MATLAB are essential for several reasons:

  - **Modularity**: Functions allow the decomposition of complex problems into smaller, manageable sub-problems.
  - **Reusability**: Once defined, functions can be reused in various parts of a program or in different projects.
  - **Maintainability**: Code written in functions is easier to debug, test, and maintain.
  - **Clarity**: Functions provide a clear structure to the code, making it more readable and understandable.

- How Functions Work in MATLAB
  As given in Fig. 6.2, Functions in MATLAB operate by encapsulating a sequence of statements that perform a specific task. When a function is called, MATLAB creates a new workspace for the function, ensuring that variables within the function do not interfere with those in the global workspace.

6.2.2 Function Syntax and Structure

A MATLAB function consists of a function header, which includes the keyword function, the function name, and input and output arguments. The function body contains the code that performs the desired task. The basic syntax of a function is:

Fig. 6.2 How functions work in MATLAB

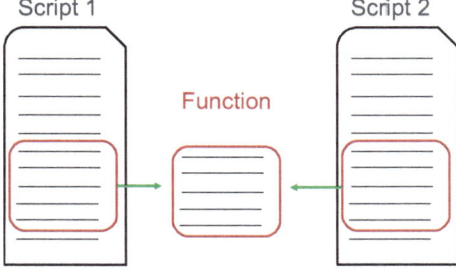

Basic Syntax:

Listing **6.5** Function definition syntax.

```
function [output1, output2, ...] = functionName(input1, input2,
    ...)
    % Function body
    statements
end
```

6.2.3 Calling Functions

To call a function in MATLAB, you simply use the function name followed by parentheses, with any required input arguments provided within the parentheses. If the function returns an output value, you can assign it to a variable. Here's the basic syntax for calling a function [1]:

Listing **6.6** Function call syntax.

```
[output] = functionName(input1, input2, ...);
```

Here are two examples demonstrating how to call functions in MATLAB:

Listing **6.7** Calling a built-in function.

```
% Call the built-in sqrt() function to compute the square root
x = 16;
sqrtResult = sqrt(x);
disp(sqrtResult);
```

Listing **6.8** Calling a user-defined function.

```
% Define a function to calculate the area of a circle
function area = circleArea(radius)
area = pi * radius^2;
end

% Call the user-defined circleArea() function
r = 5;
areaResult = circleArea(r);
disp(areaResult);
```

In the first example, the built-in sqrt() function is called with the input argument x to calculate the square root. The result is assigned to the variable sqrtResult and then displayed using the disp() function.

In the second example, a user-defined function named circleArea() is defined to calculate the area of a circle given its radius. The function is then called with the input argument r, and the result is assigned to the variable areaResult before being displayed.

   MATLAB provides a wide range of built-in functions for various tasks, such as
mathematical operations, data analysis, and visualisation. Additionally, users can
define their own functions to encapsulate specific functionality and promote code
reusability.

6.2.4   Examples of Functions in MATLAB

- **Example 1: Simple Arithmetic Function**

Listing 6.9 Addition Function.

```
function sum = addNumbers(a, b)
    % This function adds two numbers
    sum = a + b;
end
```

- **Example 2: Advanced Calculation Function**

Listing 6.10 Quadratic Equation Solver.

```
function [root1, root2] = solveQuadratic(a, b, c)
    % This function solves a quadratic equation of the form
        ax^2 + bx + c = 0
    discriminant = b^2 - 4*a*c;
    root1 = (-b + sqrt(discriminant)) / (2*a);
    root2 = (-b - sqrt(discriminant)) / (2*a);
end
```

- Figures for Examples

Listing 6.11 Plotting Quadratic Roots.

```
% MATLAB code to plot quadratic equation and its roots
a = 1;
b = -3;
c = 2;
[root1, root2] = solveQuadratic(a, b, c);
f = @(x) a*x.^2 + b*x + c;

% Plotting the quadratic function
fplot(f, [-10, 10])
hold on
plot(root1, 0, 'ro')
plot(root2, 0, 'ro')
title('Quadratic Equation and its Roots')
xlabel('x')
ylabel('f(x)')
legend('Quadratic Function', 'Roots')
grid on
hold off
```

6.2.5   Returning Values from Functions

Functions can return one or more values, or they may not return any value at all. The
number and type of output arguments are specified in the function header.

- Return One Value
  To return a single value from a function, specify the output argument in the function
  header and assign the value to the output argument within the function body.

Listing 6.12 Returning one value.

```
function result = square(x)
result = x^2;
end
```

- Return Multiple Values To return multiple values from a function, enclose the
  output arguments in square brackets in the function header and assign values to
  each output argument within the function body.

Listing 6.13 Returning multiple values.

```
function [sum, diff] = sumDiff(a, b)
sum = a + b;
diff = a - b;
end
```

- Return Nothing Functions that do not return any value are called void functions.
  These functions perform tasks without returning any output.

Listing 6.14 Void function.

```
function printMessage(message)
fprintf('%s\n', message);
end
```

6.2.6   Built-in Numerical Functions

MATLAB provides a wide range of built-in numerical functions that perform various
mathematical operations.

- eps function The eps function returns the floating-point relative accuracy of the
  machine. It is useful for determining the smallest representable difference between
  two floating-point numbers.

Listing 6.15 Using the eps function.

```
x = 1;
y = x + eps;
```

- The "Is" Functions

  MATLAB provides a set of "is" functions that test the properties of variables or arrays. Some commonly used "is" functions include `isscalar`, `isvector`, `ismatrix`, `isempty`, and `isnan`.

Listing 6.16 Using "is" functions.

```
x = [1, 2, 3];
isscalar(x) % Returns 0 (false)
isvector(x) % Returns 1 (true)
```

6.3 Variable Numbers of Arguments

MATLAB allows functions to accept a variable number of input arguments using the `varargin` keyword. This feature enables functions to handle different numbers of input arguments flexibly.

Listing 6.17 Variable number of arguments.

```
function result = sumAll(varargin)
result = sum([varargin{:}]);
end
```

6.4 Nested Functions

- Introduction to **nested functions** in MATLAB: Nested functions are functions defined within another function, known as the parent function. They are useful for encapsulating related functionality and improving code organization.
- Creating and using nested functions: Nested functions are defined within the parent function's body and can access the parent function's workspace. They are called using the nested function's name within the parent function.
- Scope and visibility of nested functions: Nested functions have access to the parent function's workspace, including local variables and input arguments. However, they are not visible outside the parent function.

Listing 6.18 Nested function example.

```
function y = parentFunction(x)
y = nestedFunction(x);

function z = nestedFunction(a)
    z = a^2;
end
end
```

6.5   Anonymous Functions and Function Handles

- Introduction to **anonymous functions** in MATLAB: Anonymous functions are small, inline functions that are defined without a specific name. They are useful for creating simple, one-time-use functions.
- Creating and using anonymous functions: Anonymous functions are created using the @ operator followed by the function's input arguments and the function body. They can be assigned to variables or passed as arguments to other functions.
- **Function handles** and their uses: Function handles are MATLAB objects that reference a function. They allow functions to be treated as data and can be used for function composition, function arrays, and more.

Listing 6.19 Anonymous function example.

```
square = @(x) x^2;
result = square(5); % Returns 25
```

6.6   Uses of Function Handles

- Passing functions as arguments to other functions: Function handles allow functions to be passed as arguments to other functions. This enables functional programming techniques and the creation of flexible, reusable code.
- Storing functions in variables or data structures: Function handles can be assigned to variables or stored in data structures like arrays or cell arrays. This allows for dynamic function selection and execution.
- Creating function arrays or cell arrays of functions: Function handles can be used to create arrays or cell arrays of functions, enabling the storage and manipulation of multiple functions as a single entity.

Listing 6.20 Function handle example.

```
functionArray = {@sin, @cos, @tan};
result = functionArray{2}(pi/4); % Calls the cos function
```

6.7   Recursive Functions

- Introduction to **recursive functions** in MATLAB: Recursive functions are functions that call themselves within their own definition. They solve problems by breaking them down into smaller subproblems and combining the results.

- Creating and using recursive functions: Recursive functions typically have a base case that terminates the recursion and a recursive case that calls the function with modified arguments. Proper termination conditions are crucial to avoid infinite recursion.
- Examples of recursive algorithms and problem-solving: Recursive functions are often used for tasks like factorial calculation, Fibonacci sequence generation, tree traversal, and divide-and-conquer algorithms.

Listing 6.21 Recursive function example.

```
function result = factorial(n)
if n == 0
result = 1;
else
result = n * factorial(n-1);
end
end
```

6.8 Live Scripts

MATLAB **Live Scripts** (.mlx files) combine executable code, formatted text, and plots in a single interactive environment, enabling users to create dynamic and engaging documents for collaboration, teaching, and learning [2]. Live Scripts support a variety of **data types**, including numeric arrays, characters and strings, tables, structures, and cell arrays [3].

One of the key features of Live Scripts is the ability to incorporate **formatted text** alongside MATLAB code. This allows users to provide explanations, context, and insights directly within the script. The text can be formatted using a range of styles, such as headings, bullet points, and equations [2].

Live Scripts also support the inclusion of **inline plots** and **interactive controls**. Inline plots enable users to visualise the results of their code directly within the script, making it easier to understand and interpret the data. Interactive controls, such as sliders and drop-down menus, allow users to dynamically adjust parameters and observe the effects on the output.

Live Scripts use the **.mlx** file extension and can be created directly from the MATLAB environment. They are particularly useful for sharing results, creating tutorials, and documenting workflows. Users can insert sections, run code interactively, and visualise outputs inline, making it a powerful tool for both teaching and research.

6.8.1 Creating Live Scripts

Creating a Live Script is straightforward. Users can start by selecting **New Live Script** from the Home tab or using the command line. Once a Live Script is created, it can be saved with the **.mlx** extension.

The basic syntax for creating a Live Script is:

Listing 6.22 Creating a Live Script

```
% Create a Live Script
filename = 'example.mlx';
edit(filename);
```

Here is an example of a simple Live Script that displays a plot:

Listing 6.23 Simple Plot in a Live Script

```
% Simple Plot Example
x = linspace(0, 2*pi, 100);
y = sin(x);
plot(x, y);
title('Sine Wave');
xlabel('x');
ylabel('sin(x)');
```

6.8.2 Adding Text, Equations, and Visualisations

Live Scripts allow the inclusion of formatted text, mathematical equations, and visualisations. This feature enables users to create comprehensive documents that explain the code and its outputs in context.

Text and equations can be added using the **Insert** tab or by typing directly into the Live Script. For example, to add a title and a description:

Listing 6.24 Adding Text and Equations

```
% Adding Text and Equations
% This is a Live Script example.

% Equation: E = mc^2
syms m c
E = m * c^2;
disp(E);
```

The inclusion of visualisations and interactive controls enhances the interactivity of Live Scripts, making them a valuable tool for exploratory data analysis and presentation.

Here are two examples demonstrating the usage of Live Scripts:

Listing 6.25 Creating a Live Script with formatted text and a plot.

```
%% Sine Wave Plot
% This section plots a sine wave with adjustable frequency and
    amplitude.

frequency = 1; % Set the frequency of the sine wave
amplitude = 2; % Set the amplitude of the sine wave

t = linspace(0, 2*pi, 100);
y = amplitude * sin(frequency * t);

plot(t, y)
xlabel('Time')
ylabel('Amplitude')
title('Sine Wave')
```

Listing 6.26 Creating a Live Script with an interactive slider.

```
%% Interactive Plot
% This section demonstrates an interactive plot using a slider.

a = 1;
x = linspace(-10, 10, 100);
y = a * x.^2;

plot(x, y)
xlabel('x')
ylabel('y')
title('Parabola')

% Create an interactive slider
a_slider = uicontrol('Style', 'slider', 'Min', -5, 'Max', 5, 'Value
    ', a, 'Position', [150 20 120 20]);
a_slider.Callback = @(src, event) updatePlot(src, event, x);

function updatePlot(src, event, x)
a = src.Value;
y = a * x.^2;
plot(x, y)
end
```

In summary, Live Scripts provide a powerful and flexible environment for creating interactive and engaging MATLAB documents. By combining executable code, formatted text, and plots, Live Scripts enable users to effectively communicate their ideas, share their results, and collaborate with others.

6.8.3 Live Code File Format (.mlx)

MATLAB stores **live scripts** and **functions** using the Live Code file format in a file with a .mlx extension [4]. The Live Code file format uses **Open Packaging**

Conventions technology, which is an extension of the zip file format. Code and formatted content are stored in an XML document separate from the output using the **Office Open XML (ECMA-376)** format [5].

- The Live Code file format offers several benefits:

  - **Interoperable Across Locales**: Live code files support storing and displaying characters across all locales, facilitating sharing files internationally.For example, a live script created with a Japanese locale setting will display correctly when opened with a Russian locale setting.
  - **Extensible**: The live code files can be extended through the ECMA-376 format, supporting a range of formatting options provided by Microsoft Word. The format also accommodates arbitrary name-value pairs, allowing further extensions beyond the standard offerings.
  - **Forward Compatible**: Future versions of live code files are compatible with previous versions of MATLAB by implementing the ECMA-376 standard's forward compatibility strategy.
  - **Backward Compatible**: Future versions of MATLAB can support live code files created by a previous version of MATLAB.
  - **Source Control** When using source control, it is essential to register the **.mlx** extension as binary. This ensures that the files are correctly managed by the version control system. MATLAB provides tools to compare live scripts or functions, such as the MATLAB Comparison Tool. For more information on registering binary files with source control systems like SVN or Git, refer to the official documentation.

To determine and display code differences between live scripts or functions, the MATLAB Comparison Tool can be used. When using source control, it is recommended to register the .mlx extension as binary.

Listing 6.27 Creating a live script.

```
% Create a new live script
livescript = mlxlive("MyLiveScript.mlx");

% Add code
livescript.Code = "x = linspace(0,10);";
livescript.Code += newline + "y = sin(x);";

% Add formatted text
para = livescript.Paragraph("This is a live script!");
para.Style = ["bold" "italic"];

% Display the live script
view(livescript);
```

Listing 6.28 Comparing live scripts.

```
% Open the comparison tool
visdiff("MyScript1.mlx", "MyScript2.mlx");
```

Note: The older .m extension used for MATLAB scripts and functions is still supported, but the .mlx format is recommended for new live scripts and functions to take advantage of the benefits outlined above.

- Office Open XML, ECMA-376

  The **Live Code file format** used by MATLAB for live scripts and functions utilises the **Office Open XML (OOXML)** format, which is an open standard (ECMA-376) developed by Microsoft [5]. This XML-based format enables the separation of code and formatted content from the output, making the files more **interoperable** across locales and **extensible** for future enhancements [4].

  The use of OOXML in MATLAB's Live Code files offers several advantages:

  - **Forward compatibility**: Future versions of live code files remain compatible with previous MATLAB versions by adhering to the ECMA-376 standard's forward compatibility strategy.
  - **Backward compatibility**: Newer MATLAB versions can support live code files created by older MATLAB versions.
  - **Extensibility**: The ECMA-376 format supports a wide range of formatting options and allows for arbitrary name-value pairs, enabling future extensions beyond the standard's current capabilities.

Listing 6.29 Creating a live script with formatted content.

```
% Create a new live script
livescript = mlxlive("MyLiveScript.mlx");

% Add formatted content
para = livescript.Paragraph("This is a bold and italic paragraph.")
    ;
para.Style = ["bold" "italic"];

% Display the live script
view(livescript);
```

Listing 6.30 Accessing live script content.

```
% Load a live script
livescript = mlxlive("MyLiveScript.mlx");

% Access the code
code = livescript.Code;

% Access formatted content
content = livescript.Paragraphs;
```

- Old Functions and Recommendations

  Over the years, MATLAB has evolved, leading to the deprecation of older functions in favour of more efficient and powerful alternatives. This section highlights some of the old functions and the recommended newer functions to use in their place.

– Old Functions

- · **findstr**–Used to find one string within another.
- · **fliplr**–Used to flip matrices left to right.
- · **fread**–Used for reading data from binary files.

– Recommended Functions

- · **strfind**–Replaces **findstr** for finding substrings within strings.
- · **flip**–Replaces **fliplr** and can be used for flipping matrices in any dimension.
- · **readmatrix**–Replaces **fread** for reading data more efficiently.

– Example: Using strfind

Listing 6.31 Using strfind to find a substring.

```matlab
% Old function findstr
oldIndex = findstr('hello world', 'world');
disp(oldIndex);

% Recommended function strfind
newIndex = strfind('hello world', 'world');
disp(newIndex);
```

– Example: Using flip

Listing 6.32 Using flip to flip a matrix.

```matlab
% Old function fliplr
oldMatrix = [1, 2; 3, 4];
flippedOld = fliplr(oldMatrix);
disp(flippedOld);

% Recommended function flip
flippedNew = flip(oldMatrix, 2); % Flip along the second
    dimension
disp(flippedNew);
```

6.9 Laboratory

1. Writing and Executing Scripts

- a. Create a new MATLAB script file called *calculate_area.m*.
- b. Inside the script, define variables **radius** and **pi** with values 5 and 3.14159, respectively.
- c. Calculate the area of a circle using the formula **area** $= \pi \times$ **radius**^{2}.
- d. Display the calculated area using the *disp()* function.
- e. Save the script and run it in the MATLAB environment.

Solution:

Listing 6.33 Script to calculate the area of a circle.

```
radius = 5;
pi = 3.14159;
area = pi * radius^2;
disp(['The area of the circle is: ', num2str(area)]);
```

Output: The area of the circle is: 78.5398

2. Defining and Using Functions

    a. Function–**calculate_volume.m**

- Create a new MATLAB function file called **calculate_volume.m**.
- Define the function to accept three input arguments: **length**, **width**, and **height**.
- Inside the function, calculate the volume of a rectangular box using the formula **volume = length * width * height**.
- Return the calculated volume from the function.
- In the MATLAB command window, call the function with sample values for length, width, and height, and display the returned volume.

Solution:

Listing 6.34 Function to calculate the volume of a rectangular box.

```
    function volume = calculate_volume(length, width, height)
volume = length * width * height;
end
```

Command window:

```
vol = calculate_volume(3, 4, 5)
vol =
60
```

    b. Function–pyramid number

Create a MATLAB function file that takes a single input argument n and returns a vector of the first n pyramid numbers. The n-th pyramid number, $P(n)$, represents the number of blocks in a pyramid made from an n-by-n square of blocks at the base, an $(n - 1)$-by-$(n - 1)$ square on top of that, and so on. It can therefore be calculated as:

$$P(n) = \sum_{k=1}^{n} k^2 \qquad (6.1)$$

Add statements to check that n is a positive integer before performing the calculations.

```
function P = pyramid_numbers(n)
    % Check that n is a positive integer
    if ~isscalar(n) || n <= 0 || n ~= floor(n)
        error('Input must be a positive integer.');
    end

    % Initialize the output vector
    P = zeros(1, n);

    % Calculate pyramid numbers
    for k = 1:n
        P(k) = sum((1:k).^2);
    end
end
```

3. Passing Arguments to Functions

    a. Create a new MATLAB function file called **calculate_hypotenuse.m**.

    b. Define the function to accept two input arguments: **side1** and **side2**.

    c. Inside the function, calculate the hypotenuse of a right triangle using the Pythagorean theorem: **hypotenuse** $= \sqrt{\text{side}1^2 + \text{side}2^2}$.

    d. Return the calculated hypotenuse from the function.

    e. In the MATLAB command window, call the function with different sets of values for side1 and side2, and display the returned hypotenuse.

Solution:

Listing 6.35 Function to calculate the hypotenuse of a right triangle.

```
function hypotenuse = calculate_hypotenuse(side1, side2)
hypotenuse = sqrt(side1^2 + side2^2);
end
```

Command window:

```
hyp1 = calculate_hypotenuse(3, 4)
hyp1 =
5
hyp2 = calculate_hypotenuse(5, 12)
hyp2 =
13
```

4. Returning Values from Functions–Statistical Measures of a Vector

    a. Create a new MATLAB function file called **calculate_stats.m**.

    b. Define the function to accept a vector of numbers as input.

    c. Inside the function, calculate the mean, median, and standard deviation of the input vector.

d. Return the calculated mean, median, and standard deviation as separate output arguments from the function.

e. In the MATLAB command window, call the function with a sample vector of numbers, and display the returned statistics.

Solution:

Listing 6.36 Function to calculate statistical measures of a vector.

```
    function [mean_val, median_val, std_val] = calculate_stats(
        data_vec)
mean_val = mean(data_vec);
median_val = median(data_vec);
std_val = std(data_vec);
end
```

Command window:

```
data = [5, 8, 2, 10, 7];
[mean_data, median_data, std_data] = calculate_stats(data)
mean_data =
6.4000
median_data =
7
std_data =
2.7639
```

5 Returning Values from Functions–Variable Numbers of Arguments Function

a. Write a MATLAB function named *who_is_bigger* that accepts two input arguments, a and b, and returns the larger of the two.

b. Test the function with the call *who_is_bigger*(2, 3).

Solution:

```
function bigger = who_is_bigger(a, b)
    % This function returns the larger of the two input arguments a
        and b.
    if a > b
        bigger = a;
    else
        bigger = b;
    end
end

% Test the function with the call who_is_bigger(2, 3)
result = who_is_bigger(2, 3);
disp(['The bigger number is: ', num2str(result)]);
```

The bigger number is: 3

6. Scope of Variables in Scripts and Functions
   a. Create a new MATLAB script file called **variable_scope.m**.
   b. Inside the script, define a variable **x** with a value of 10.
   c. Create a function called **update_x** that takes no input arguments.
   d. Inside the **update_x** function, increment the value of **x** by 5 and display the updated value.
   e. Call the **update_x** function from within the script.
   f. Display the value of **x** after calling the function.
   g. Run the script and observe the output.

Solution:

Listing 6.37 Script to demonstrate variable scope in MATLAB.

```
    x = 10; % Global variable

function update_x()
x = x + 5; % Error: Attempt to increment undefined function 'x'
disp(['Inside function, x = ', num2str(x)]);
end

update_x(); % Calling the function
disp(['Outside function, x = ', num2str(x)]);
```

Output:

```
Error: Attempt to increment undefined function 'x'
Outside function, x = 10
```

Explanation: Inside the function *update_x*, the variable **x** is treated as a local variable, which is undefined until it is assigned a value. The function cannot access the global variable **x** defined in the script. To fix this, the function should either accept **x** as an input argument or use the **global** keyword to access the global variable.

7. Anonymous Functions and Function Handles
   Use an anonymous function to compute the integral of x^2 from 0 to 1.

Solution:

```
% Define the anonymous function
f = @(x) x.^2;

% Compute the integral from 0 to 1
integral_value = integral(f, 0, 1);

% Display the result
disp(['The integral of x^2 from 0 to 1 is: ', num2str(
    integral_value)]);
```

8. Investment on stocks

   Given the prices of 20 stocks over a certain period, devise a method to select the best value stock for investment based on this data.

   One possible method is as follows:

   a. Assume that the linear trend of the stock price represents its value. The greater the slope, the higher the investment value of the stock.
   b. Calculate the slope of the price trend for each stock and select the stock with the highest slope.

   Here is the MATLAB code to achieve this:

```matlab
% Sample data: prices of 20 stocks over 10 days
prices = rand(10, 20); % Replace with actual stock price data

% Initialize variables to store slopes and best stock index
slopes = zeros(1, 20);
best_stock_index = 1;
max_slope = -inf;

% Compute the slope of the linear trend for each stock
for i = 1:20
    % Get the prices of the i-th stock
    stock_prices = prices(:, i);

    % Fit a linear model to the stock prices
    p = polyfit(1:10, stock_prices', 1);

    % Extract the slope
    slope = p(1);
    slopes(i) = slope;

    % Check if this is the highest slope
    if slope > max_slope
        max_slope = slope;
        best_stock_index = i;
    end
end

% Display the index of the best stock
disp(['The best stock for investment is stock number: ', num2str(
    best_stock_index)]);
disp(['The slope of the best stock is: ', num2str(max_slope)]);
```

6.10   Problems

1. Write a MATLAB script that prompts the user to enter their name and age, and then displays a personalised greeting.
2. Create a MATLAB function that takes two numbers as input and returns their sum, difference, product, and quotient (if applicable).

3. Write a MATLAB script that generates a random vector of 10 integers between 1 and 100, and then calculates the mean, median, and standard deviation of the vector.
4. Create a MATLAB function that takes a string as input and returns the number of vowels (a, e, i, o, u) in the string.
5. Write a MATLAB script that prompts the user to enter the coefficients of a quadratic equation (

$$ax^2 + bx + c = 0$$

) and calculates its roots using the quadratic formula.
6. Create a MATLAB function that takes a vector of numbers as input and returns a new vector containing only the unique elements.
7. Write a MATLAB script that generates a random 3x3 matrix and calculates its determinant, trace, and inverse (if applicable).
8. Create a MATLAB function that takes a positive integer as input and returns the sum of its digits.
9. Write a MATLAB script that prompts the user to enter the side lengths of a triangle and determines whether the triangle is equilateral, isosceles, or scalene.
10. Create a MATLAB function that takes a vector of numbers as input and returns a new vector containing the cumulative sum of the elements.
11. Create a MATLAB function to calculate 'the coefficient of determination or R^2'[6] with the function name 'SECF_assess_R2.m', in which, the first line is

```
R2 = SECF_assess_R2( Y_test, y_calculation)
```

12. Create two MATLAB functions to calculate 'the moving mean of the average precision (mmAP)' and 'the moving mean of standard derivation (mmSTD)'[7]. The definition of the two trend indices: mmAP and mmSTD. Both are given in equations (6.2) and (6.3), respectively.
As stated in equation (6.2), the index of mmAP is a moving average score of the mean value of vector f_j, where $i = 1, 2, \cdots, p$, p is the population of the data set, $MEAN(\cdot)$ is the average function. The index of mmSTD is a moving average score of the STD value of vector f_j, as given in equation (6.3), where $STD(\cdot)$ is the standard deviation function. Both indices are used to mitigate the short-term fluctuations by capturing the longer-term trend across the evolutionary process.

$$mmAP\left(f_j\right) = \frac{1}{p}\sum_{i=1}^{p}\left(\frac{1}{i}\sum_{j=1}^{i}MEAN\left(f_j\right)\right) \qquad (6.2)$$

$$mmSTD\left(f_j\right) = \frac{1}{p}\sum_{i=1}^{p}\left(\frac{1}{i}\sum_{j=1}^{i}STD\left(f_j\right)\right) \qquad (6.3)$$

6.11 Summary

- **User-defined functions** allow programmers to create reusable code blocks that perform specific tasks.
- Functions can return one or more values, or they may not return any value at all.
- MATLAB provides a wide range of **built-in numerical functions** for various mathematical operations.
- Functions can accept a **variable number of input arguments** using the `varargin` keyword.
- **Nested functions** are functions defined within another function and have access to the parent function's workspace.
- **Anonymous functions** are small, inline functions defined without a specific name.
- **Function handles** are MATLAB objects that reference a function and allow functions to be treated as data.
- **Recursive functions** call themselves within their own definition and solve problems by breaking them down into smaller subproblems.

For Undergraduate (UG) Students:

- This chapter serves as an excellent introduction to **MATLAB scripting** and **function development**. It covers essential concepts such as **input/output operations**, **control structures**, **arithmetic operations**, **string manipulation**, **matrix operations**, and **vector operations**.
- The step-by-step approaches, sample code, and sample outputs provided for each problem facilitate a better understanding of MATLAB's syntax and functionality.
- The extensions and variations suggested for each problem encourage students to explore further and enhance their problem-solving skills.
- Overall, this chapter lays a solid foundation for UG students to develop proficiency in MATLAB programming, which is widely used in various engineering and scientific disciplines.

For Postgraduate (PG) Students and Professional Researchers/Engineers:

- While this chapter covers fundamental MATLAB concepts, it also serves as a valuable reference for **scripting techniques** and **function development**, which are essential skills for PG students and professionals working with data analysis, simulations, and computational tasks.
- The problems presented in this chapter encompass a wide range of applications, including **string processing**, **numerical analysis**, **matrix operations**, and **data manipulation**, making it relevant for various research domains.
- The solutions provided demonstrate best practices for **code organization**, **readability**, and **modularity**, which are crucial for developing maintainable and scalable MATLAB projects.
- The extensions and variations encourage researchers and engineers to explore advanced topics and tailor the solutions to their specific requirements, fostering problem-solving skills and adaptability.

- Overall, this chapter serves as a valuable resource for PG students, researchers, and engineers, reinforcing their MATLAB proficiency and enabling them to tackle complex computational challenges effectively.

By providing a comprehensive set of problems and solutions, along with suggestions for extensions and variations, this chapter caters to learners at various levels, from UG students to professional researchers and engineers. It strikes a balance between introducing fundamental concepts and offering opportunities for advanced exploration, making it a valuable addition to any MATLAB-focused curriculum or reference material.

References

1. MathWorks, "Function," https://uk.mathworks.com/help/matlab/ref/function.html, accessed on Feb. 17, 2024
2. "Live Script Gallery," https://ww2.mathworks.cn/products/matlab/live-script-gallery.html, accessed on Feb. 17, 2024
3. MathWorks, "Data Types," [Online]. Available: https://www.mathworks.com/help/matlab/data-types.html, accessed on Feb. 17, 2024
4. MathWorks, "MATLAB Fundamentals," [Online]. Available: https://www.mathworks.com/help/matlab/, accessed on Feb. 17, 2024
5. https://learn.microsoft.com/en-us/openspecs/office_standards/ms-oe376/db9b9b72-b10b-4e7e-844c-09f88c972219, accessed on Feb. 17, 2024
6. Yi Chen, Zhang G (2013) Exchange rates determination based on genetic algorithms using Mendel's principles: investigation and estimation under uncertainty. Inf Fusion 14(3):327–333
7. Yi C, Guangfeng Z, Tongdan J, Shaomin W, Bei P (2014) Quantitative modelling of electricity consumption using computational intelligence aided design. J Clean Prod 69:143–152. https://doi.org/10.1016/j.jclepro.2014.01.058

Chapter 7
Inputs and Outputs

Chapter Learning Outcomes

- Understand the various file formats supported by MATLAB for input and output operations, including **MAT-files**, **spreadsheet files**, **binary files**, **image files**, **text files**, **audio files**, **video files**, **JSON files**, **HDF5 files**, **XML files**, and **database files**.
- Perform input and output operations using MATLAB functions specific to each file format.
- Apply appropriate techniques for reading data from files and writing data to files in MATLAB.
- Utilize lower-level file input/output functions in MATLAB for more flexible and controlled data access.
- Incorporate file input and output operations into MATLAB scripts and functions for data processing and analysis tasks.

Chapter Key Words

- **MAT-files**: MATLAB's native file format for storing and managing data, including variables, arrays, and other MATLAB objects. MAT-files provide efficient storage and retrieval of data within the MATLAB environment.
- **Spreadsheet files**: File formats such as Excel files (.xls) that are commonly used for storing and organizing tabular data. MATLAB provides functions to read from and write to spreadsheet files, enabling data exchange with external applications.
- **Binary files**: Files that store data in a binary format, which can be efficiently read and written using low-level file input/output functions in MATLAB. Binary files offer flexibility and control over data access.

© The Author(s) 2025
Y. Chen and L. Huang, *MATLAB Roadmap to Applications*,
https://doi.org/10.1007/978-981-97-8788-3_7

- **Image files**: Various file formats used for storing digital images, such as PNG, JPEG, and TIFF. MATLAB supports reading and writing image files, allowing for image processing and analysis tasks.
- **Text files**: Plain text files that store data in a human-readable format. MATLAB provides functions to read from and write to text files, enabling data exchange and processing of textual information.

7.1 Introduction

In MATLAB, inputs and outputs play a crucial role in interacting with the user and transferring data between functions and scripts. MATLAB supports a wide variety of file formats for input and output operations, extending beyond just MAT-files and spreadsheet files. These file formats include images, audio, video, JSON, XML, HDF5, binary files, and databases. The flexibility provided by MATLAB allows users to easily work with diverse data sources and seamlessly integrate with various applications. This chapter provides a comprehensive overview of the different methods and techniques for performing input and output operations in MATLAB, covering a range of file formats and their corresponding functions.

7.2 MAT-Files (.mat)

MAT-files are the native file format in MATLAB for storing and managing data. They provide an efficient means of saving and loading variables, arrays, and other MATLAB objects. MAT-files offer the advantage of preserving the data types and structures specific to MATLAB, ensuring seamless data transfer within the MATLAB environment. This section explores the functions and techniques for performing input and output operations with MAT-files, including saving variables to MAT-files using the `save` function and loading data from MAT-files using the `load` function.

The basic syntax for saving variables to a MAT-file is:

Listing 7.1 Saving variables to a MAT-file.
```
save('filename.mat', 'variable1', 'variable2', ...)
```

And the basic syntax for loading data from a MAT-file is:

Listing 7.2 Loading data from a MAT-file.
```
load('filename.mat')
```

Here are two examples demonstrating the usage of MAT-files in MATLAB:

Listing 7.3 Saving and loading a matrix to/from a MAT-file.

```
% Create a matrix
matrix = [1 2 3; 4 5 6; 7 8 9];

% Save the matrix to a MAT-file
save('matrix_data.mat', 'matrix');

% Clear the matrix variable
clear matrix;

% Load the matrix from the MAT-file
load('matrix_data.mat');
```

Listing 7.4 Saving and loading multiple variables to/from a MAT-file.

```
% Create variables
var1 = 10;
var2 = [1.5, 2.7, 3.2];
var3 = 'Hello, world!';

% Save the variables to a MAT-file
save('variables.mat', 'var1', 'var2', 'var3');

% Clear the variables
clear var1 var2 var3;

% Load the variables from the MAT-file
load('variables.mat');
```

7.3 Spreadsheet Files (.xls)

MATLAB provides functions to read and write data from spreadsheet files, such as Microsoft Excel files with the `.xls` or `.xlsx` extension. While the `xlsread()` and `xlswrite()` functions were used in earlier versions of MATLAB, it is now recommended to use the `readtable()` and `writetable()` functions for improved functionality and performance [1].

To read data from a spreadsheet file, you can use the `readtable()` function. It reads a spreadsheet file and returns a table array containing the data [1]. Here's an example:

Listing 7.5 Reading data from a spreadsheet file.

```
% Read data from a spreadsheet file
filename = 'data.xlsx';
data = readtable(filename);
```

In this example, the `readtable()` function reads the data from the file `'data.xlsx'` and stores it in the `data` variable as a table array.

To write data to a spreadsheet file, you can use the `writetable()` function. It writes a table array to a spreadsheet file [1]. Here's an example:

Listing 7.6 Writing data to a spreadsheet file.

```
% Create a table array
data = table(1:5', rand(5,1), 'VariableNames', {'ID', '
    Value'});

% Write the table to a spreadsheet file
filename = 'output.xlsx';
writetable(data, filename);
```

In this example, a table array `data` is created with two columns: `'ID'` and `'Value'`. The `writetable()` function is then used to write the table array to the file `'output.xlsx'`.

The `readtable()` and `writetable()` functions provide additional options for specifying the range of cells to read or write, the sheet name, and other parameters to customize the input and output of spreadsheet files.

By using these functions, MATLAB simplifies the process of reading and writing data from spreadsheet files, making it easier to work with data stored in popular file formats like Microsoft Excel.

7.4 Binary Files (.dat)

MATLAB provides lower-level file input/output functions that offer more flexibility and control over data access. These functions allow reading and writing data in various formats, including binary files and text files. Binary files store data in a binary format, which can be efficiently read and written using functions like `fread` and `fwrite`. This section explores the usage of lower-level file input/output functions in MATLAB for handling binary files and demonstrates their application in different scenarios.

The basic syntax for reading data from a binary file is:

Listing 7.7 Reading data from a binary file.

```
fileID = fopen('filename.dat', 'r');
data = fread(fileID, size, precision);
fclose(fileID);
```

And the basic syntax for writing data to a binary file is:

Listing 7.8 Writing data to a binary file.

```
fileID = fopen('filename.dat', 'w');
fwrite(fileID, data, precision);
fclose(fileID);
```

Here are two examples demonstrating the usage of binary files in MATLAB:

Listing 7.9 Reading binary data from a file.

```
% Open the binary file for reading
fileID = fopen('data.bin', 'r');

% Read the binary data
data = fread(fileID, 'double');

% Close the file
fclose(fileID);
```

Listing 7.10 Writing binary data to a file.

```
% Create some data
data = [1.5, 2.7, 3.2, 4.8];

% Open the binary file for writing
fileID = fopen('output.bin', 'w');

% Write the data to the file
fwrite(fileID, data, 'double');

% Close the file
fclose(fileID);
```

7.5 Image Files (.png, .jpg, .tif etc.)

MATLAB provides functions for reading and writing image files, supporting various image formats such as PNG, JPEG, and TIFF. The `imread` function is used to read image files, while the `imwrite` function is used to write image data to files. Image processing and analysis tasks can be performed on the loaded image data. This section demonstrates how to read image files, display them, and perform basic operations using MATLAB.

The basic syntax for reading an image file is:

Listing 7.11 Reading an image file.

```
imageData = imread('filename.png');
```

And the basic syntax for writing an image file is:

Listing 7.12 Writing an image file.

```
imwrite(imageData, 'filename.png');
```

Here are two examples demonstrating the usage of image files in MATLAB:

Listing 7.13 Reading and displaying an image.

```
% Read the image file
imageData = imread('image.jpg');

% Display the image
imshow(imageData);
```

Listing 7.14 Resizing and saving an image.

```
% Read the image file
imageData = imread('input.png');

% Resize the image
resizedImage = imresize(imageData, [256, 256]);

% Save the resized image
imwrite(resizedImage, 'output.png');
```

7.6 Text Files (.txt)

Text files are commonly used for storing and processing textual data. MATLAB provides functions for reading from and writing to text files, enabling data exchange and manipulation of textual information. Functions like `fileread`, `importdata`, and `textscan` can be used to read data from text files, while functions such as `fprintf` and `diary` can be used to write data to text files. This section explores the techniques for performing input and output operations with text files in MATLAB.

The basic syntax for reading data from a text file is:

Listing 7.15 Reading data from a text file.

```
fileData = fileread('filename.txt');
```

And the basic syntax for writing data to a text file is:

Listing 7.16 Writing data to a text file.

```
fileID = fopen('filename.txt', 'w');
fprintf(fileID, 'Hello, world!');
fclose(fileID);
```

Here are two examples demonstrating the usage of text files in MATLAB:

Listing 7.17 Reading and processing data from a text file.

```
% Read the text file
fileData = fileread('data.txt');

% Split the data into words
words = strsplit(fileData);
```

```
% Count the number of words
wordCount = numel(words);
```

Listing 7.18 Writing formatted data to a text file.

```
% Open the text file for writing
fileID = fopen('output.txt', 'w');

% Write formatted data to the file
fprintf(fileID, 'Name: %s\nAge: %d\nCity: %s\n', 'John',
    25, 'New York');

% Close the file
fclose(fileID);
```

7.7 Audio Files (.wav, .mp3, .flac etc.)

MATLAB supports reading and writing audio files, enabling audio processing and analysis tasks. The `audioread` function is used to read audio files, while the `audiowrite` function is used to write audio data to files. Various audio file formats, such as WAV, MP3, and FLAC, can be handled in MATLAB. This section demonstrates how to read audio files, apply basic operations, and save the modified audio using MATLAB .

The basic syntax for reading an audio file is:

Listing 7.19 Reading an audio file.

```
[audioData, sampleRate] = audioread('filename.wav');
```

And the basic syntax for writing an audio file is:

Listing 7.20 Writing an audio file.

```
audiowrite('filename.wav', audioData, sampleRate);
```

Here are two examples demonstrating the usage of audio files in MATLAB:

Listing 7.21 Reading and playing an audio file.

```
% Read the audio file
[audioData, sampleRate] = audioread('audio.wav');

% Play the audio
sound(audioData, sampleRate);
```

Listing 7.22 Modifying and saving an audio file.

```
% Read the audio file
[audioData, sampleRate] = audioread('input.mp3');
```

```
% Increase the volume by a factor of 2
modifiedAudio = audioData * 2;

% Save the modified audio
audiowrite('output.mp3', modifiedAudio, sampleRate);
```

7.8 Video Files (.avi, .mp4, .mov etc.)

MATLAB provides capabilities for reading and writing video files, enabling video processing and analysis tasks. The VideoReader class is used to read video files, while the VideoWriter class is used to write video data to files. Various video file formats, such as AVI, MP4, and MOV, can be handled in MATLAB. This section demonstrates how to read video files, extract frames, apply operations, and save the modified video using MATLAB.

The basic syntax for reading a video file is:

Listing 7.23 Reading a video file.
```
videoReader = VideoReader('filename.mp4');
```

And the basic syntax for writing a video file is:

Listing 7.24 Writing a video file.
```
videoWriter = VideoWriter('filename.avi');
open(videoWriter);
writeVideo(videoWriter, frames);
close(videoWriter);
```

Here are two examples demonstrating the usage of video files in MATLAB:

Listing 7.25 Reading and displaying video frames.
```
% Create a VideoReader object
videoReader = VideoReader('video.mp4');

% Read and display the video frames
while hasFrame(videoReader)
frame = readFrame(videoReader);
imshow(frame);
pause(1/videoReader.FrameRate);
end
```

Listing 7.26 Processing and saving video frames.
```
% Create a VideoReader object
videoReader = VideoReader('input.avi');

% Create a VideoWriter object
videoWriter = VideoWriter('output.avi');
```

```
open (videoWriter);

% Process and write the video frames
while hasFrame (videoReader)
frame = readFrame (videoReader);
grayFrame = rgb2gray (frame);
writeVideo (videoWriter, grayFrame);
end

% Close the VideoWriter object
close (videoWriter);
```

7.9 JSON Files (.json)

JSON (JavaScript Object Notation) is a lightweight data interchange format that is easy for humans to read and write and easy for machines to parse and generate. MATLAB provides functions for encoding and decoding JSON data, making it convenient to interact with web services and APIs that use JSON. The jsondecode function is used to parse JSON data into MATLAB data types, while the jsonencode function is used to convert MATLAB data types to JSON format [2].

Here's an example of parsing JSON data in MATLAB:

Listing 7.27 Parsing JSON data.

```
% JSON data
jsonData = '{"name": "John", "age": 30, "city": "New
    York"}';

% Parse JSON data into a MATLAB structure
data = jsondecode (jsonData);

% Access the parsed data
disp (data.name);
disp (data.age);
disp (data.city);
```

7.10 HDF5 Files (.hdf5)

HDF5 (Hierarchical Data Format version 5) is a file format designed for storing large and complex data sets. It provides efficient I/O performance and supports a variety of data types, including multidimensional arrays, tables, and groups. MATLAB offers functions for reading and writing HDF5 files, allowing you to work with large data sets effectively. The h5read function is used to read data from an HDF5 file, while the h5write function is used to write data to an HDF5 file [3].

Here's an example of reading data from an HDF5 file in MATLAB:

Listing 7.28 Reading data from an HDF5 file.

```
% Read data from an HDF5 file
data = h5read('data.h5', '/dataset');

% Access the read data
disp(data);
```

7.11 XML Files (.xml)

XML (eXtensible Markup Language) is a markup language that defines a set of rules for encoding documents in a format that is both human-readable and machine-readable. MATLAB provides functions for parsing XML data into MATLAB structures, making it convenient to work with data stored in XML format. The xmlread function is used to read XML data from a file or a string, while the xmlwrite function is used to write XML data to a file [4].

Here's an example of parsing XML data in MATLAB:

Listing 7.29 Parsing XML data.

```
% Read XML data from a file
xmlData = xmlread('data.xml');

% Access the parsed data
disp(xmlData.getElementsByTagName('name').item(0).
    getTextContent());
disp(xmlData.getElementsByTagName('age').item(0).
    getTextContent());
disp(xmlData.getElementsByTagName('city').item(0).
    getTextContent());
```

7.12 Database Files (.csv, .odb, etc.)

MATLAB provides functions to read and write data from various database file formats, such as comma-separated value (CSV) files and OpenDocument Database (ODB) files. In earlier versions of MATLAB, the csvread() [5] and dlmread() [6] functions were commonly used for reading CSV files. However, these functions are now considered outdated and are not recommended for use in newer versions of MATLAB [7].

Instead, it is recommended to use the readmatrix() function for reading data from CSV files and other delimited text files [8]. The readmatrix() function provides a more versatile and efficient way to read data from text files.

Here's an example of reading data from a CSV file using the `readmatrix()` function:

Listing 7.30 Reading data from a CSV file using readmatrix().

```
% Read data from a CSV file
filename = 'data.csv';
data = readmatrix(filename);
```

In this example, the `readmatrix()` function reads the data from the file `'data.csv'` and stores it in the `data` variable as a numeric matrix.

For writing data to a CSV file, you can use the `writematrix()` function [9], which is the recommended alternative to the outdated `csvwrite()` and `dlmwrite()` functions. Here's an example:

Listing 7.31 Writing data to a CSV file using writematrix().

```
% Create a numeric matrix
data = [1 2 3; 4 5 6; 7 8 9];

% Write the matrix to a CSV file
filename = 'output.csv';
writematrix(data, filename);
```

In this example, a numeric matrix `data` is created, and then the `writematrix()` function is used to write the matrix to the file `'output.csv'`.

The `readmatrix()` and `writematrix()` functions provide additional options for specifying the delimiter, numeric format, and other parameters to customize the input and output of delimited text files.

By using these recommended functions, MATLAB simplifies the process of reading and writing data from various database file formats, ensuring better performance and compatibility with newer versions of MATLAB.

7.13 Data Import and Analysis

MATLAB provides a wide range of functions for importing and analyzing data from various sources, such as text files, spreadsheets, and databases. In earlier versions of MATLAB, functions like `textread()`, `xlsread()`, and `csvread()` were commonly used for importing data [10]. However, these functions have been replaced by more versatile and efficient alternatives in newer versions of MATLAB.

- Data Import and Export

  MATLAB supports importing data from text files, spreadsheets, and other file formats. It also allows for hardware interfacing and web access. Some of the commonly used functions for data import and export include:

  - **textscan**: Read formatted data from text file or string (recommended over *textread*)

- **readtable**: Read tabular data from file (recommended over *xlsread* for Excel files)
- **writetable**: Write tabular data to file

For importing data from text files, the `readtable()`[11] and `readmatrix()` [12] functions are recommended instead of `textread()` [13]. These functions provide more flexibility in handling different file formats and data types. Here's an example of using `readtable()` to import data from a tab-delimited text file:

Listing 7.32 Importing data from a text file using readtable().

```
% Import data from a tab-delimited text file
filename = 'data.txt';
data = readtable(filename, 'Delimiter', '\t');
```

Listing 7.33 Example of importing data from a text file.

```
fileID = fopen('data.txt', 'r');
dataArray = textscan(fileID, '%f %f %f', 'Delimiter', ',
   ');
fclose(fileID);
```

For importing data from spreadsheets, the `readtable()` and `readcell()` functions are recommended instead of `xlsread()` [14]. These functions provide more options for specifying the range of data to import and handling different data types. Here's an example of using `readtable()` to import data from an Excel file:

Listing 7.34 Importing data from an Excel file using readtable().

```
% Import data from an Excel file
filename = 'data.xlsx';
data = readtable(filename, 'Sheet', 'Sheet1', 'Range', '
   A1:D10');
```

Listing 7.35 Example of reading data from an Excel file.

```
data = readtable('data.xlsx', 'Sheet', 1);
```

Once the data is imported, MATLAB provides a rich set of functions for data analysis and visualisation. The `summary()` function can be used to get a quick overview of the imported data, including the number of observations, variable names, and data types. For example:

Listing 7.36 Summarizing imported data using summary().

```
% Summarize the imported data
summary(data);
```

- Large Files and Big Data
  MATLAB provides capabilities to access and process collections of files and large data sets [15]. Some key functions include:

– **datastore**: Access collections of data files as a single entity
– **tall**: Array that stores data on disk and allows for efficient computations

Listing 7.37 Example of using datastore to read multiple files.

```
ds = datastore('*.txt', 'TextType', 'string');
data = readall(ds);
```

Listing 7.38 Example of using tall arrays for big data processing.

```
t = tall(csvDatastore('largedata.csv'));
result = gather(t.var1 + t.var2);
```

- Preprocessing Data
  MATLAB also provides functions for data preprocessing, such as handling missing values, filtering, and transforming variables. The ismissing() function can be used to identify missing values in the data, while the fillmissing() function can be used to replace missing values with specific values or interpolated values. Data preprocessing is an essential step in data analysis. MATLAB provides functions for data cleaning, smoothing, and grouping [15]. Some commonly used functions include:

  – **fillmissing**: Fill missing values in arrays
  – **smooth**: Smooth data using moving average or Savitzky-Golay filter
  – **groupsummary**: Compute summary statistics for groups of data

Listing 7.39 Example of filling missing data.

```
data = fillmissing(data, 'constant', 0);
```

Listing 7.40 Example of smoothing data using moving average.

```
smoothed_data = smooth(data, 5);
```

- Visual Exploration
  MATLAB's graphical capabilities allow for visual exploration of data through panning, zooming, rotating graphics, and modifying and saving observations [15]. Some useful functions and tools include:

  – **plot**: Create 2D line plots
  – **scatter**: Create scatter plots
  – **histogram**: Plot histograms
  – **imagesc**: Display matrix as scaled image
  – **Data Brushing**: Interactively select and highlight data points

  For data visualisation, MATLAB offers a wide range of plotting functions, including plot(), scatter(), bar(), and histogram() [16]. These functions allow you to create various types of charts and graphs to visualize and explore the imported data.

- Managing Experiments
  MATLAB provides tools to design experiments, run MATLAB code, visualize, filter, and compare results [15]. The **Experiment Manager** app allows for organizing and running experiments, while the **Parallel Computing Toolbox** enables parallel execution of experiments on multi-core processors and clusters.
- Climate Data Visualisation and Analysis
  MATLAB offers resources for climate data visualisation and analysis through the **MathWorks Teaching Resources** portal. These resources provide examples and tutorials on working with climate data, creating visualisations, and performing statistical analyses.
  By leveraging the powerful data import and analysis capabilities of MATLAB, users can efficiently load, process, and gain insights from their data, enabling them to make informed decisions and solve complex problems.

7.14   Laboratory

1. **Reading Data from a Text File** In this lab work, you will learn how to read data from a text file into MATLAB.

   a. Create a text file named "data.txt" in a directory of your choice, containing the following data:

   ```
   1.2  3.4  5.6
   7.8  9.0  2.1
   4.5  6.7  8.9
   ```

   b. In MATLAB, navigate to the directory containing the "data.txt" file using the 'cd' command.
   c. Use the 'dlmread' function to read the data from the text file into a matrix. The basic syntax is:

   Listing 7.41 Reading data from a text file.

   ```
   data = dlmread('data.txt');
   ```

   d. Print the contents of the 'data' matrix to the MATLAB command window using the 'disp' function.

   ```
   disp(data)
   ```

   Expected Output:

   ```
      1.2000  3.4000  5.6000
   7.8000  9.0000  2.1000
   4.5000  6.7000  8.9000
   ```

2. **Writing Data to a Text File**

   In this lab work, you will learn how to write data from MATLAB to a text file.

   a. Create a matrix 'data' in MATLAB with some sample data:

   ```
   data = [1.5 2.7 4.3; 6.9 8.1 3.2; 5.4 7.6 9.8];
   ```

   b. Use the 'dlmwrite' function to write the contents of the 'data' matrix to a text file named "output.txt". The basic syntax is:

   Listing 7.42 Writing data to a text file.

   ```
   dlmwrite('output.txt', data, 'delimiter', '\t');
   ```

   c. Open the "output.txt" file in a text editor to verify that the data was written correctly. The data should be tab-separated.

   Expected Output in "output.txt":

   ```
       1.5000 2.7000 4.3000
   6.9000 8.1000 3.2000
   5.4000 7.6000 9.8000
   ```

3. **Command-Line Input**

   In this lab work, you will learn how to prompt the user for input from the command line and use the input in MATLAB.

   a. Use the 'input' function to prompt the user to enter their name. The basic syntax is:

   Listing 7.43 Command-line input.

   ```
   name = input('Enter your name: ', 's');
   ```

   b. Display a greeting message using the 'fprintf' function, incorporating the user's name. The basic syntax is:

   Listing 7.44 Formatted output.

   ```
   fprintf('Hello, %s!\n', name);
   ```

   Sample Output:

   ```
       Enter your name: John
   Hello, John!
   ```

4. **Formatted Output**

   In this lab work, you will learn how to format output in MATLAB using the 'fprintf' function.

a. Define some variables in MATLAB:

```
    x  =  3.14159;
y  =  2.71828;
z  =  1.61803;
```

b. Use the 'fprintf' function to print the values of these variables with a specific number of decimal places. The basic syntax is:

Listing 7.45 Formatted output.

```
    fprintf('x  =  %.2f,  y  =  %.3f,  z  =  %.4f\n',  x,  y
        ,  z);
```

Expected Output:

```
    x  =  3.14,  y  =  2.718,  z  =  1.6180
```

5. **Data Visualisation**

In this lab work, you will learn how to create a simple plot in MATLAB using the 'plot' function.

a. Define some sample data:

```
    x  =  linspace(0,  2*pi,  100);
y  =  sin(x);
```

b. Use the 'plot' function to create a plot of the sine wave. The basic syntax is:

Listing 7.46 Plotting data.

```
    plot(x,  y);
```

c. Add a title, x-label, and y-label to the plot using the 'title', 'xlabel', and 'ylabel' functions, respectively.

```
    title('Sine  Wave');
xlabel('x');
ylabel('sin(x)');
```

d. **Reading data from an excel** In a health examination, the blood pressure data is stored in an Excel file named `bloodpressure.xls`. Please read the blood pressure data from this file. Open the file outside of MATLAB to view the data, specifically,

- Reads the blood pressure data from the Excel file `bloodpressure.xls` using the `readtable` function.
- Displays the data in the MATLAB Command Window.
- Opens the Excel file using the `winopen` function to view the data outside of MATLAB.

Here is the MATLAB code to achieve this:

```matlab
% Read the blood pressure data from the Excel file
filename = 'bloodpressure.xls';
data = readtable(filename);

% Display the data
disp(data);

% Open the file outside MATLAB to view the data
winopen(filename);
```

6. Import an image file, and output the image after being processed

- Import an image file and output the processed image.
- Read the image of a mountain from the file peak.jpg.
- Enlarge the image by a factor of 2.
- Output the enlarged image.

Here is the MATLAB code to achieve this:

```matlab
% Read the image of a mountain from the file 'peak.
    jpg'
image = imread('peak.jpg');

% Enlarge the image by a factor of 2
enlarged_image = imresize(image, 2);

% Display the original and enlarged images
figure;
subplot(1, 2, 1);
imshow(image);
title('Original Image');

subplot(1, 2, 2);
imshow(enlarged_image);
title('Enlarged Image');

% Save the enlarged image to a file
imwrite(enlarged_image, 'peak_enlarged.jpg');
```

7. Read a video file

The provided video file (visiontraffic.avi) is intended for road vehicle monitoring. Please accomplish the following tasks:

- Identify and mark the vehicles in the video with red bounding boxes.
- Develop and execute a program to count the total number of vehicles detected in the video.

The steps are listed here:

- Loads the video file visiontraffic.avi.
- Creates a video player object for displaying the video.

- Uses a foreground detector to identify moving objects (vehicles).
- Analyzes the blobs to find connected components representing vehicles.
- Draws red bounding boxes around detected vehicles.
- Counts the cumulative number of vehicles throughout the video.
- Displays the processed video and outputs the total vehicle count.

Here is the MATLAB code to achieve these tasks:

```
% Load the video file
videoReader = VideoReader('visiontraffic.avi');

% Create a video player object for displaying the
    video
videoPlayer = vision.VideoPlayer('Position', [100,
    100, 680, 520]);

% Create a foreground detector object
foregroundDetector = vision.ForegroundDetector('
    NumGaussians', 3, ...
    'NumTrainingFrames', 50);

% Create a blob analysis object to find connected
    components
blobAnalysis = vision.BlobAnalysis('
    BoundingBoxOutputPort', true, ...
    'AreaOutputPort', false, 'CentroidOutputPort',
        false, ...
    'MinimumBlobArea', 150);

% Initialize the vehicle count
vehicleCount = 0;

% Process each frame of the video
while hasFrame(videoReader)
    % Read the next frame
    frame = readFrame(videoReader);

    % Detect the foreground in the current frame
    foreground = step(foregroundDetector, frame);

    % Perform morphological operations to remove
        noise
    filteredForeground = imopen(foreground, strel('
        rectangle', [3, 3]));
    filteredForeground = imclose(filteredForeground,
        strel('rectangle', [15, 15]));
    filteredForeground = imfill(filteredForeground, '
        holes');

    % Detect connected components in the foreground
    bbox = step(blobAnalysis, filteredForeground);

    % Draw bounding boxes around detected vehicles
```

```
    result = insertShape(frame, 'Rectangle', bbox, '
        Color', 'red');

    % Update vehicle count
    vehicleCount = vehicleCount + size(bbox, 1);

    % Display the result
    step(videoPlayer, result);
end

% Release the video player object
release(videoPlayer);

% Display the total vehicle count
disp(['Total number of vehicles: ', num2str(
    vehicleCount)]);
```

These lab works and exercises cover various aspects of input/output operations in MATLAB, including reading data from files, writing data to files, command-line input, formatted output, and data visualisation. Each lab work provides step-by-step instructions, sample code snippets, and expected outputs to facilitate hands-on learning and reinforce the concepts covered in this chapter.

7.15   Problems

1. Write a MATLAB function that takes a text file as input and counts the occurrences of each unique word in the file. The function should return a struct or a cell array containing the unique words and their corresponding counts.
2. Create a MATLAB script that reads data from a CSV (Comma-Separated Values) file and performs basic data analysis tasks, such as calculating the mean, median, and standard deviation for each column of numerical data.
3. Implement a MATLAB function that takes a text file as input and removes all occurrences of a specified string from the file. The function should create a new file with the modified contents.
4. Write a MATLAB script that prompts the user to enter a series of file names and then concatenates the contents of all the specified files into a single output file.
5. Create a MATLAB function that takes a matrix as input and saves it to a binary file. The function should also include the ability to load the matrix from the binary file at a later time.
6. Develop a MATLAB script that reads data from an Excel file and creates a bar plot or a histogram to visualise the data distribution.
7. Implement a MATLAB function that takes a text file as input and performs basic text processing tasks, such as counting the number of lines, words, and characters, as well as identifying the most frequently occurring word in the file.

8. Write a MATLAB script that prompts the user to enter a directory path and then lists all the files in that directory, along with their sizes and modification dates.
9. Create a MATLAB function that takes a text file as input and replaces all occurrences of a specified word or phrase with a new word or phrase. The function should create a new file with the modified contents.
10. Develop a MATLAB script that reads data from a text file, where each line represents a data point with multiple values separated by commas or tabs. The script should plot the data using appropriate visualisation techniques (e.g., scatter plot, line plot) and allow the user to interactively explore the data.

7.16 Summary

- **File I/O operations**: The chapter covers various **file input/output** operations in MATLAB, including reading from and writing to **text files**, **Excel files**, and **binary files**.
- MATLAB supports a wide range of file formats for input and output, including **MAT-files**, **spreadsheet files**, **binary files**, **text files**, **image files**, **audio files**, **video files**, **JSON files**, **HDF5 files**, **XML files**, and **database files**.
- **MAT-files** are MATLAB-specific files that can store multiple variables, including arrays, structures, and cell arrays, in a binary format.
- **Spreadsheet files**, such as Excel files, can be read and written using functions like `xlsread` and `xlswrite`.
- **Binary files** store data in a binary format and can be efficiently read and written using functions like `fread` and `fwrite`.
- **Text files** are commonly used for storing and processing textual data, and MATLAB provides functions like `fileread`, `textscan`, and `fprintf` for reading from and writing to text files.
- **Image files** in various formats, such as PNG, JPEG, and TIFF, can be read using the `imread` function and written using the `imwrite` function.
- **Audio files**, such as WAV, MP3, and FLAC, can be read using the `audioread` function and written using the `audiowrite` function.
- **Video files**, such as AVI, MP4, and MOV, can be read using the `VideoReader` class and written using the `VideoWriter` class.
- **JSON files** are lightweight data interchange formats that can be parsed into MATLAB data types using the `jsondecode` function and encoded from MATLAB data types using the `jsonencode` function.
- **HDF5 files** are designed for storing large and complex data sets and can be read using the `h5read` function and written using the `h5write` function.
- **XML files** are markup language files that can be parsed into MATLAB structures using the `xmlread` function and written using the `xmlwrite` function.
- **Database files** such as CSV and ODB files, can be read using functions like `csvread` and `odbcread` to access data stored in databases.

- MATLAB's comprehensive file input/output capabilities enable efficient data storage, retrieval, and exchange across a wide range of file formats, facilitating data analysis, processing, and sharing of results.
- The additional file formats, including JSON, HDF5, XML, and database files, provide more options for data input and output, catering to specific data storage and exchange requirements in various domains and applications.

For Undergraduate (UG) Students:

This chapter introduces essential file handling and data processing concepts in MATLAB. It covers basic file operations like reading from and writing to text, Excel, and binary files, enabling students to work with various data formats. The chapter also demonstrates data visualisation techniques using bar plots and histograms, allowing students to visually represent and analyze data. Additionally, it explores text processing tasks such as word counting, phrase replacement, and statistical analysis, equipping students with tools for working with textual data. Lastly, the chapter discusses directory operations, teaching students how to list files and their associated metadata within a specified directory. Overall, this chapter provides a solid foundation for UG students to work with files, visualise data, process text, and manage directories in MATLAB.

For Postgraduate (PG) Students and Professional Researchers or Engineers:

Building upon the fundamental concepts covered in the UG section, this chapter offers advanced techniques and considerations for file handling, data visualisation, text processing, and directory operations in MATLAB. It delves into more complex scenarios, such as handling large files efficiently, implementing error handling and robust input validation, and exploring additional visualisation options like line plots and scatter plots. The chapter also discusses advanced text processing techniques, such as regular expression-based replacements and natural language processing integration. Furthermore, it covers strategies for optimizing performance through techniques like multithreading, parallel processing, and memory-mapped files. For researchers and engineers, the chapter highlights potential extensions and variations, such as integrating with version control systems, developing graphical user interfaces (GUIs) or command-line interfaces (CLIs), and implementing logging and error reporting mechanisms. Overall, this chapter equips PG students, researchers, and professional engineers with advanced skills and considerations for working with files, visualizing data, processing text, and managing directories in MATLAB, enabling them to tackle more complex and specialised tasks in their respective fields.

References

1. MathWorks, Read Spreadsheet Data into Table. [Online]. https://uk.mathworks.com/help/matlab/import_export/read-spreadsheet-data-into-table.html. [Accessed: Feb. 17, 2024]
2. MathWorks, JSON Processing. [Online]. https://www.mathworks.com/help/matlab/json-format.html. [Accessed: Feb. 17, 2024]

3. HDF5 Files. (n.d.). MathWorks. https://www.mathworks.com/help/matlab/hdf5-files.html

4. XML Processing. (n.d.). MathWorks. https://www.mathworks.com/help/matlab/xml-documents.html

5. MathWorks, csvread. [Online]. https://uk.mathworks.com/help/matlab/ref/csvread.html. [Accessed: Feb. 17, 2024]

6. MathWorks, dlmread. [Online]. https://uk.mathworks.com/help/matlab/ref/dlmread.html. [Accessed: Feb. 17, 2024]

7. MathWorks, csvwrite. [Online]. https://uk.mathworks.com/help/matlab/ref/csvwrite.html. [Accessed: Feb. 17, 2024]

8. MathWorks, Readmatrix. [Online]. https://uk.mathworks.com/help/matlab/ref/readmatrix.html. [Accessed: Feb. 17, 2024]

9. MathWorks, Writematrix. [Online]. https://uk.mathworks.com/help/matlab/ref/writematrix.html. [Accessed: Feb. 17, 2024]

10. MathWorks, Data Import and Export. [Online]. https://uk.mathworks.com/help/matlab/data-import-and-export.html. [Accessed: Feb. 17, 2024]

11. MathWorks, readtable. [Online]. https://uk.mathworks.com/help/matlab/ref/readtable.html. [Accessed: Feb. 17, 2024]

12. MathWorks, readmatrix. [Online]. https://uk.mathworks.com/help/matlab/ref/readmatrix.html. [Accessed: Feb. 17, 2024]

13. MathWorks, textread. [Online]. https://uk.mathworks.com/help/matlab/ref/textread.html. [Accessed: Feb. 17, 2024]

14. MathWorks, xlsread. [Online]. https://uk.mathworks.com/help/matlab/ref/xlsread.html. [Accessed: Feb. 17, 2024]

15. MathWorks, "MATLAB Fundamentals," [Online]. https://www.mathworks.com/help/matlab/. [Accessed: Feb. 17, 2024]

16. MathWorks, Data Visualization. [Online]. https://uk.mathworks.com/help/thingspeak/visualize-data.html. [Accessed: Feb. 17, 2024]

Chapter 8
Graphics and Data Visualisation

Chapter Learning Outcomes

- Understand the importance of **data visualisation** in communicating complex information effectively.
- Create various types of **2D and 3D plots** using MATLAB functions such as `plot`, `scatter`, `bar`, `contour`, `surf`, and `mesh`.
- Visualise **data distributions** using plots like **histograms**, **pie charts**, **box plots**, and **heatmaps**.
- Plot **discrete data** using **stem plots**, **stair plots**, and **Pareto charts**.
- Visualise **vector fields** and **volume data** using functions like `quiver`, `slice`, and `isosurface`.
- Display and manipulate **images** in MATLAB using functions such as `imshow`, `imadjust`, and `histeq`.
- Create **animations** using `movie` and `getframe` functions.
- Apply **formatting and annotation** techniques to enhance the clarity and aesthetics of plots.

Chapter Key Words

- **Data visualisation**: Data visualisation is the graphical representation of information and data, using visual elements like charts, graphs, and maps to provide an accessible way to understand trends, outliers, and patterns in data. It helps to communicate complex ideas and relationships clearly and efficiently, making it easier for users to analyse and derive insights from large amounts of data. MATLAB provides a wide range of powerful tools and functions for creating informative and visually appealing data visualisations.

© The Author(s) 2025
Y. Chen and L. Huang, *MATLAB Roadmap to Applications*,
https://doi.org/10.1007/978-981-97-8788-3_8

- **2D and 3D Plots**: 2D plots are graphical representations of data on a two-dimensional plane, using the x-axis and y-axis to display the relationship between two variables. 3D plots, on the other hand, introduce a third dimension (z-axis) to represent an additional variable or to create a three-dimensional surface. MATLAB offers various functions to create 2D and 3D plots, such as `plot`, `scatter`, `bar`, `contour`, `surf`, and `mesh`, enabling users to Visualise data in different ways and gain insights from multiple perspectives.
- **Data Distribution Plots**: Data distribution plots are visual representations that show how data is spread out or distributed. These plots help to identify patterns, central tendencies, and outliers in the data. Common types of data distribution plots include **histograms**, which display the frequency distribution of a dataset; **pie charts**, which show the proportional composition of categories in a dataset; **box plots**, which illustrate the distribution of data based on summary statistics; and **heatmaps**, which represent data values using color-coded matrices [1].
- **Discrete Data Plots**: Discrete data plots are used to Visualise data that takes on distinct, separate values, rather than continuous values. These plots are particularly useful for displaying data that represents counts, categories, or integers. Examples of discrete data plots in MATLAB include **stem plots**, which use vertical lines to show discrete data points; **stair plots**, which display data as a series of horizontal and vertical steps; and **Pareto charts**, which combine a bar graph and a line graph to highlight the most significant factors in a dataset [1].
- **Vector Fields and Volume visualisation**: Vector fields are used to represent the magnitude and direction of a quantity at different points in space, such as velocity or force. Volume visualisation techniques allow the display and exploration of three-dimensional scalar or vector data. MATLAB provides functions like `quiver` for plotting 2D or 3D vector fields and `slice`, `isosurface`, and `isocaps` for visualising volume data, enabling users to gain insights into complex spatial phenomena [2].
- **Plot**: A graphical representation of data, typically as a diagram, graph, or chart. Plots are used to visualise relationships between variables, identify trends, and communicate findings. MATLAB provides a wide range of plotting functions for creating various types of plots, such as line plots, scatter plots, bar charts, and more [3].
- **Figure**: In MATLAB, a figure refers to a window that contains one or more plots or other graphical objects. Figures provide a container for organising and displaying visualisations. Multiple figures can be created within a MATLAB session, allowing for the simultaneous display of different plots or graphical representations [4]. Graphics functions include 2-D and 3-D plotting functions to Visualise data and communicate results. Customise plots either interactively or programmatically [5].

8.1   Introduction

Data visualisation is the process of translating data into **graphical representations** like plots, charts, maps, and 3D visualisations, which is a powerful technique that involves translating complex data into graphical representations, such as plots, charts, maps, and three-dimensional visualisations. This process enables researchers, analysts, and decision-makers to identify patterns, trends, and outliers within the data more effectively than by examining raw numerical values. By presenting data in a visual format, data visualisation leverages the human brain's remarkable ability to perceive and process visual information, facilitating the extraction of insights and the communication of complex concepts more intuitively.

Moreover, data visualisation plays a pivotal role in converting raw data into actionable information. By transforming abstract numerical values into compelling visual representations, data visualisation empowers stakeholders to comprehend the underlying messages and relationships within the data more effectively. This enhanced understanding enables informed decision-making processes, driving strategic planning, resource allocation, and the identification of opportunities and challenges within various domains, which helps identify **patterns**, **trends**, and **outliers** in the data.

- Importance of Data visualisation
  In today's data-driven world, the ability to effectively communicate and interpret complex information is of paramount importance. Data visualisation serves as a bridge between raw data and human understanding, enabling individuals from diverse backgrounds and expertise levels to grasp intricate patterns and relationships that may be obscured in numerical form. By presenting information in a visually appealing and intuitive manner, data visualisation techniques can facilitate collaboration, enhance decision-making processes, and foster a deeper comprehension of the underlying phenomena.
- Applications of Data visualisation
  Data visualisation finds applications in a wide range of domains, spanning scientific research, business analytics, education, journalism, and beyond. In scientific research, visualisations play a crucial role in communicating findings, illustrating complex concepts, and identifying patterns that may lead to new discoveries. In the business world, data visualisation tools are employed to analyse market trends, customer behavior, and operational performance, enabling data-driven decision-making processes. Additionally, data visualisation techniques are increasingly utilized in educational settings to enhance learning experiences and facilitate the understanding of abstract concepts.
  By introducing subsections within the introduction, you can provide a more structured and organised presentation of the key aspects or topics related to the main subject. This approach allows for a more comprehensive introduction and sets the stage for the subsequent sections of the document or research paper.
- Types of Plots in MATLAB

  - **Line plots**: plot, plot3, animatedline, fplot, fplot3, loglog, semilogx, semilogy, stairs
  - **Scatter and bubble charts**: scatter, scatter3, bubblecloud, bubblechart, bubblechart3
  - **Bar graphs**: bar, bar3, bar3h, barh, Pareto
  - **Histograms**: histogram, histogram2, polarhistogram
  - **Contour plots**: contour, contour3, contourf, fcontour
  - **Surface and mesh plots**: mesh, meshc, meshz, surf, surfc, surfx, surfy, waterfall

- **Polar plots**: polarplot, polarscatter, polar, compass, ezpolar
- **Pie charts**: pie, pie3
- **Geographic plots**: geoplot, geoscatter, geobubble
- **Specialised plots**: streamline, streamtube, quiver, quiver3, stemplot, wordcloud.

This section explores various plotting options available in MATLAB, providing a comprehensive guide to visualising data effectively. MATLAB offers a wide range of plotting functions to create different types of graphs and charts, each suited to specific kinds of data and analysis needs, a few examples are as shown in Figs. 8.1, 8.2, 8.3, 8.4, 8.5, 8.6, 8.7, 8.8 and 8.9 from MATLAB gallery [6].

8.2 2D and 3D Plots

This section covers the various types of 2D and 3D plots available in MAT-LAB, which are essential for visualising and understanding data. Line plots, created using functions like `plot`, `semilogx`, `semilogy`, and `loglog`, are used to display trends and relationships between variables. Scatter and bubble charts (`scatter`, `scatter3`, `bubblechart`) are useful for visualising the distribution of data points and identifying correlations. Bar charts (`bar`, `barh`, `bar3`, `bar3h`, `pareto`) are effective for comparing categories or values. Contour plots (`contour`, `contourf`, `contour3`, `fcontour`) display the level curves of a 3D surface, while surface and mesh plots (`surf`, `mesh`, `meshc`, `meshz`, `waterfall`) create 3D visualisations of data. Geographic plots (`geoplot`, `geoscatter`, `geobubble`)

Animation

animatedline   comet   comet3

🖼 Open and explore   🖼 Open and explore   🖼 Open and explore

Contour Plots

contour   contourf   contour3   contourslice

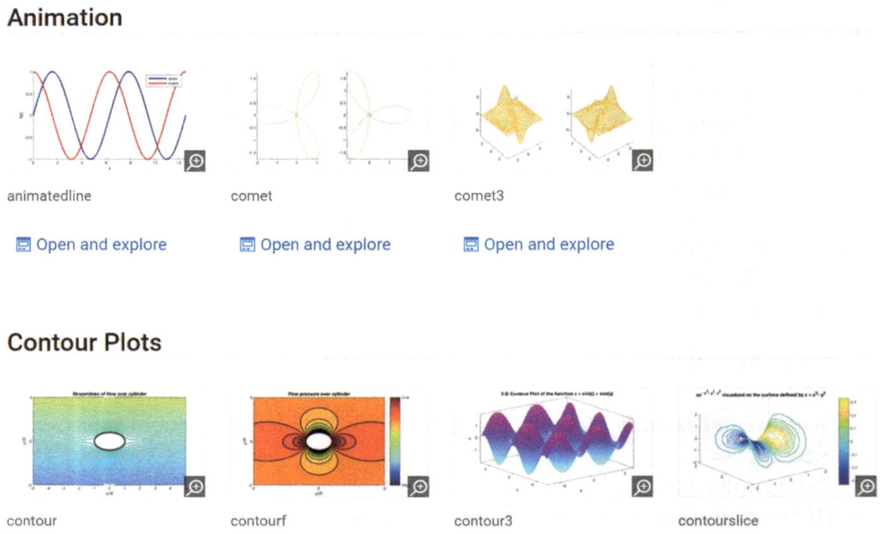

Fig. 8.1 MATLAB gallery: animation and contour plots

Data Distribution Plots

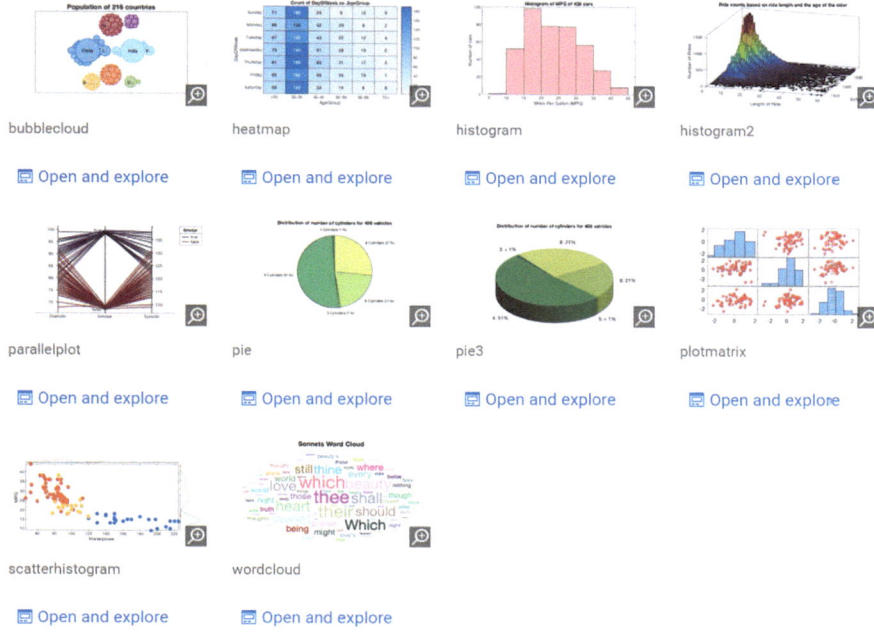

Fig. 8.2 MATLAB gallery: data distribution plots

Discrete Data Plots

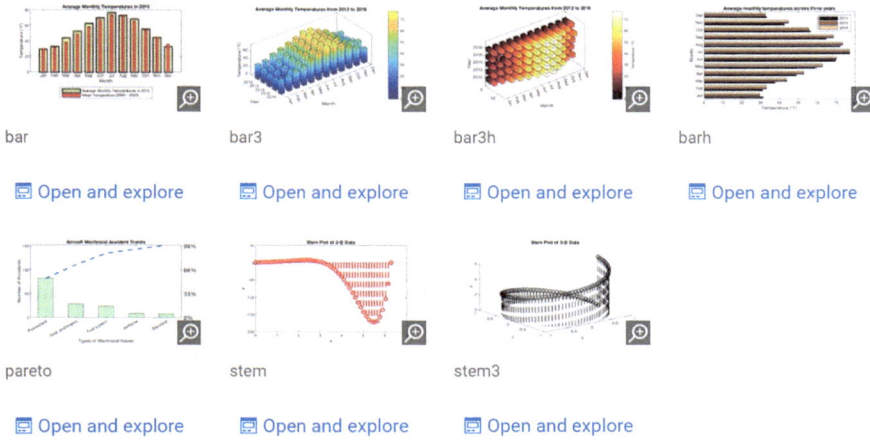

Fig. 8.3 MATLAB gallery: discrete data plots

Geographic Plots

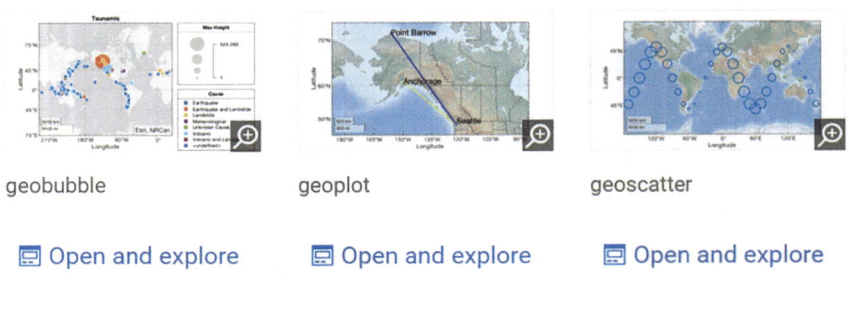

geobubble   geoplot   geoscatter

🖥 Open and explore   🖥 Open and explore   🖥 Open and explore

Images

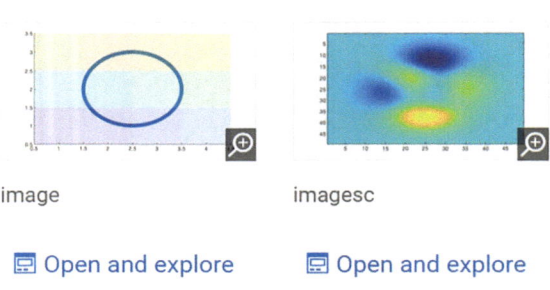

image   imagesc

🖥 Open and explore   🖥 Open and explore

Fig. 8.4 MATLAB gallery: geographic plots and images

are used to plot data on maps, and polar plots (`polarplot`, `polarscatter`, `polarhistogram`, `compass`) display data in a polar coordinate system [3].

8.2.1 Multiple Plots and Subplots

In MATLAB, it is often useful to display multiple plots or subplots within a single figure window. This allows for easy comparison and analysis of different data sets or different aspects of the same data. MATLAB provides several functions to create and manage multiple plots and subplots, such as **subplot**, **tiledlayout**, and **nexttile**.

The basic syntax for creating subplots using the **subplot** function is:

Listing 8.1 Subplot syntax.

```
subplot(m, n, p)
```

where **m** and **n** specify the number of rows and columns in the subplot grid, and **p** specifies the subplot index.

Line Plots

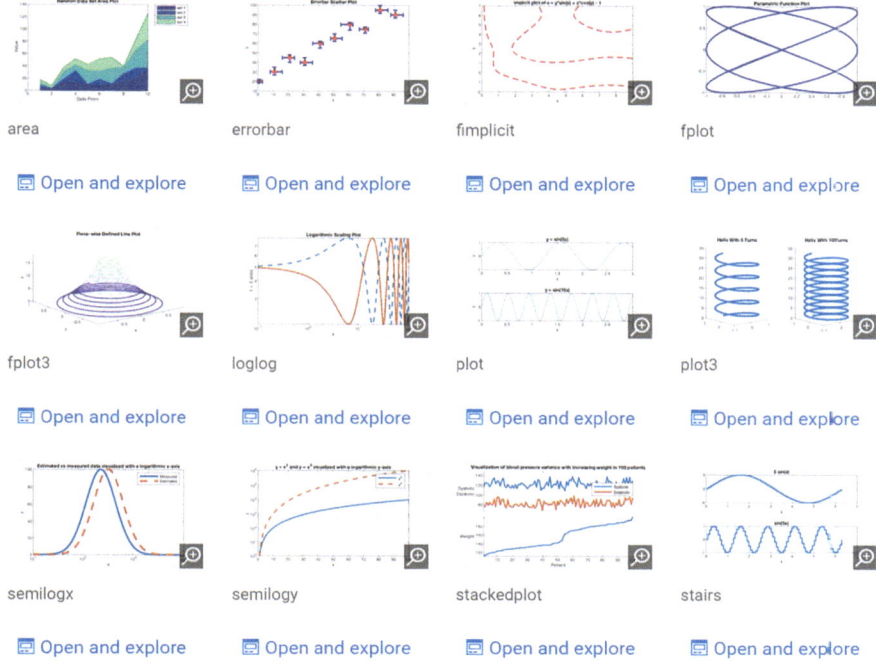

Fig. 8.5 MATLAB gallery: line plots

Polar Plots

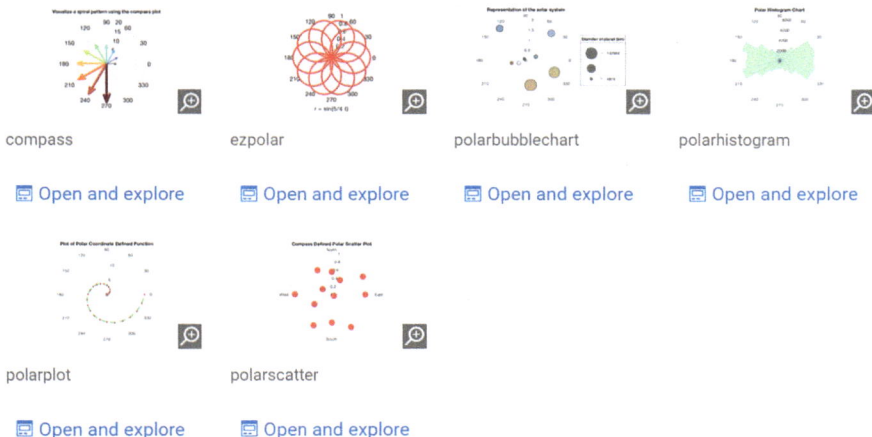

Fig. 8.6 MATLAB gallery: polar plots

Scatter and Bubble Charts

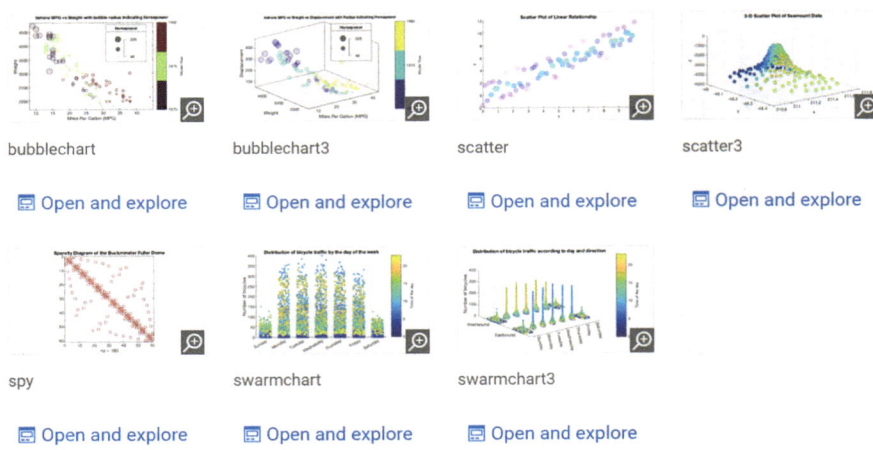

bubblechart   bubblechart3   scatter   scatter3

▣ Open and explore   ▣ Open and explore   ▣ Open and explore   ▣ Open and explore

spy   swarmchart   swarmchart3

▣ Open and explore   ▣ Open and explore   ▣ Open and explore

Fig. 8.7 MATLAB gallery: scatter and bubble charts

Surface and Mesh Plots

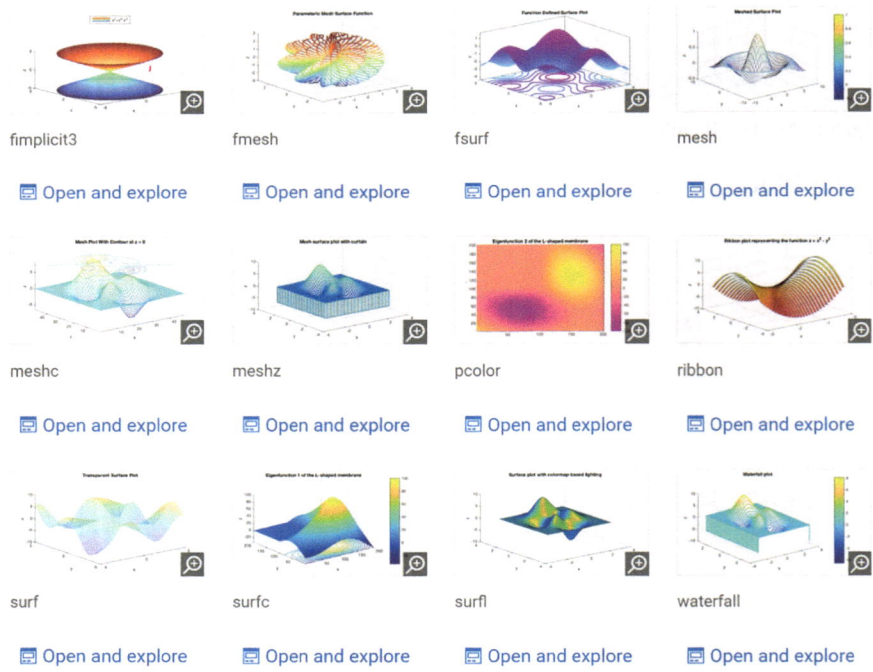

fimplicit3   fmesh   fsurf   mesh

▣ Open and explore   ▣ Open and explore   ▣ Open and explore   ▣ Open and explore

meshc   meshz   pcolor   ribbon

▣ Open and explore   ▣ Open and explore   ▣ Open and explore   ▣ Open and explore

surf   surfc   surfl   waterfall

▣ Open and explore   ▣ Open and explore   ▣ Open and explore   ▣ Open and explore

Fig. 8.8 MATLAB gallery: surface and mesh plots

Vector Fields

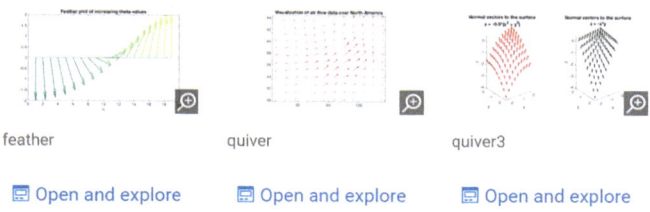

feather   quiver   quiver3

📰 Open and explore   📰 Open and explore   📰 Open and explore

Volume Visualization

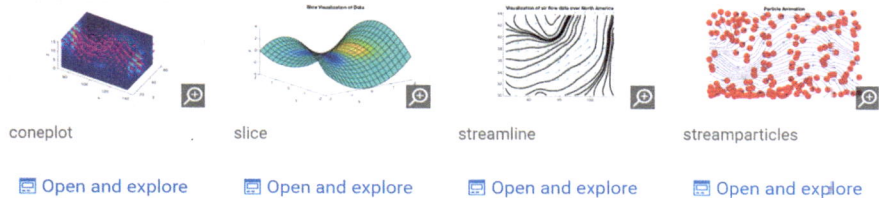

coneplot   slice   streamline   streamparticles

📰 Open and explore   📰 Open and explore   📰 Open and explore   📰 Open and explore

Fig. 8.9 MATLAB gallery: vector fields and volume visualisation

Here's an example that creates a 2x2 subplot grid and plots different functions in each subplot:

Listing 8.2 Multiple subplots example.

```
x = linspace(-pi, pi, 100);

subplot(2, 2, 1)
plot(x, sin(x))
title('Sine Function')

subplot(2, 2, 2)
plot(x, cos(x))
title('Cosine Function')

subplot(2, 2, 3)
plot(x, tan(x))
title('Tangent Function')

subplot(2, 2, 4)
plot(x, exp(x))
title('Exponential Function')
```

Another way to create subplots is using the **tiledlayout** and **nexttile** functions, which provide more flexibility in arranging the subplots [7]. Here's an example:

Listing 8.3 Tiledlayout and nexttile example.

```
x = linspace(0, 10, 100);

tiledlayout(2, 1)

nexttile
plot(x, x.^2)
title('Square Function')

nexttile
plot(x, sqrt(x))
title('Square Root Function')
```

These examples demonstrate how to create multiple plots and subplots in MAT-LAB, allowing for effective visualisation and comparison of different data sets or functions within a single figure window.

8.2.2   Customising Plot Appearance

MATLAB offers a wide range of options to customise the appearance of plots, allowing users to create visually appealing and informative graphics. These customisations include setting colours, line styles, markers, labels, titles, legends, and more. By modifying these properties, users can effectively communicate their data and enhance the readability of their plots.

The basic syntax for customising plot appearance is:

Listing 8.4 Plot customisation syntax.

```
plot(x, y, 'LineSpec')
```

where **'LineSpec'** is a character vector that specifies the line style, marker, and colour [8].

Here's an example that demonstrates customising the line style, marker, and colour:

Listing 8.5 Customising line style, marker, and colour.

```
x = linspace(0, 2*pi, 100);
y = sin(x);

plot(x, y, 'r--o', 'LineWidth', 2, 'MarkerSize', 8, '
    MarkerFaceColor', 'b')
xlabel('x')
ylabel('sin(x)')
title('Sine Function with Customised Appearance')
```

In this example, **'r–o'** specifies a red dashed line with circular markers, **'LineWidth'** sets the line width to 2, **'MarkerSize'** sets the marker size to 8, and **'MarkerFaceColor'** sets the marker face colour to blue.

Another example showcases how to customise the axes labels, title, and legend:

Listing 8.6 Customising labels, title, and legend.

```
x = 0:0.1:10;
y1 = exp(-0.2x);
y2 = exp(-0.4x);

plot(x, y1, 'b-', 'LineWidth', 2)
hold on
plot(x, y2, 'r--', 'LineWidth', 2)
xlabel('Time (s)')
ylabel('Amplitude')
title('Exponential Decay')
legend('Decay Rate 0.2', 'Decay Rate 0.4', 'Location'. '
    northeast')
```

This example plots two exponential decay functions with different decay rates, customises the line styles and colours, and adds informative labels, a title, and a legend.

By leveraging these customisation options, users can create visually stunning and informative plots that effectively communicate their data and findings.

8.2.3 Interactive Plot Features

MATLAB provides a range of interactive plot features that allow users to explore and analyze data directly within the figure window. These features include zooming, panning, data cursor, and plot rotation, which enable users to gain deeper insights into their data and make informed decisions based on their visualisations.

The basic syntax for enabling interactive plot features is:

Listing 8.7 Enabling interactive plot features.

```
figure
plot(x, y)
zoom on
pan on
datacursormode on
rotate3d on
```

Here's an example that demonstrates the use of interactive plot features:

Listing 8.8 Interactive plot features example.

```
t = linspace(0, 10, 1000);
x = sin(2pit);
y = cos(2pit);
z = t;

figure
plot3(x, y, z)
xlabel('sin(2\pit)')
ylabel('cos(2\pit)')
zlabel('t')
title('Interactive 3D Plot')

zoom on
pan on
datacursormode on
rotate3d on
```

In this example, a 3D parametric curve is plotted using **plot3**, and interactive features like zooming, panning, data cursor, and plot rotation are enabled.

Another example showcases the use of the **ginput** function for interactive data selection:

Listing 8.9 Interactive data selection using ginput.

```
x = linspace(0, 2*pi, 100);
y = sin(x);

figure
plot(x, y)
xlabel('x')
ylabel('sin(x)')
title('Interactive Data Selection')

[selected_x, selected_y] = ginput(3);
hold on
plot(selected_x, selected_y, 'ro', 'MarkerSize', 10)
```

This example plots a sine function and allows the user to select three points interactively using the **ginput** function. The selected points are then plotted as red circles on top of the original plot.

By leveraging these interactive plot features, users can explore their data in a more engaging and dynamic manner, leading to better understanding and more effective data-driven decision making.

8.2.4 Specialised Plot Types

MATLAB offers a wide range of specialised plot types that cater to specific data visualisation needs [9]. These plot types include **stem plots**, **stair plots**, **area plots**, **pie charts**, **polar plots**, and **contour plots**, among others [10]. Each of these plot types is designed to effectively communicate specific aspects of the data and provide unique insights into the underlying patterns and relationships.

The basic syntax for creating Specialised plots is:

Listing 8.10 Specialised plot syntax.

```
stem(x, y)
stairs(x, y)
area(x, y)
pie(x)
polar(theta, rho)
contour(X, Y, Z)
```

Here's an example that demonstrates the use of a stem plot:

Listing 8.11 Stem plot example.

```
x = linspace(0, 4*pi, 20);
y = cos(x);

figure
stem(x, y, 'filled')
xlabel('x')
ylabel('cos(x)')
title('Stem Plot of Cosine Function')
```

This example creates a stem plot of the cosine function, with filled markers at each data point.

Another example showcases the use of a polar plot:

Listing 8.12 Polar plot example.

```
theta = linspace(0, 2pi, 100);
rho = sin(2theta).cos(2theta);

figure
polar(theta, rho)
title('Polar Plot of sin(2\theta)cos(2\theta)')
```

This example creates a polar plot of the function $\sin(2\theta)\cos(2\theta)$, demonstrating the use of polar coordinates to visualize data.

Specialised plot types enable users to create targeted visualisations that effectively convey specific aspects of their data. By selecting the appropriate plot type for their data and analysis goals, users can gain valuable insights and communicate their findings more effectively to their audience.

Fig. 8.10 Advanced customisation techniques for scientific visualisation

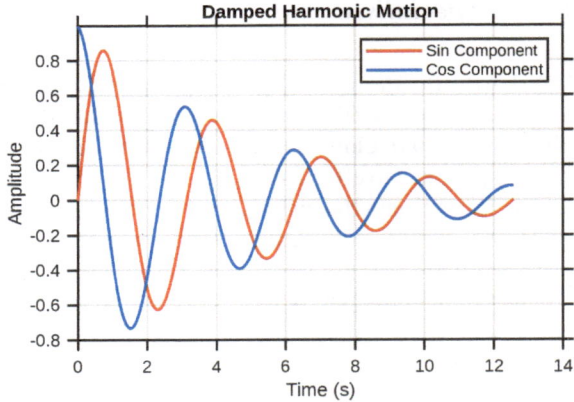

8.2.5   *Plotting Tools and Utilities*

MATLAB's comprehensive plotting tools and utilities serve as fundamental components for advanced data visualisation. These tools encompass **plot customisation**, **annotation systems**, **legend management**, **grid controls**, and **axis manipulation** capabilities [11]. Since the release of MATLAB R2024a, the **Property Inspector** has superseded the traditional plottools interface, offering enhanced functionality for interactive plot modification [12] (Fig. 8.10).

The fundamental syntax for essential plotting utilities follows this structure:

Listing 8.13 Essential plotting utilities syntax

```
% Basic plot customisation syntax
xlabel('X-Axis Label', 'Interpreter', 'latex')
ylabel('Y-Axis Label', 'Interpreter', 'latex')
title('Plot Title', 'FontWeight', 'bold')
legend('Series 1', 'Series 2', 'Location', 'best')
grid on
axis([xmin xmax ymin ymax])
```

The following example demonstrates advanced customisation techniques for scientific visualisation, as shown in Fig. 8.10.

Listing 8.14 Advanced plot customisation for scientific data

```
% Generate sample data
t = linspace(0, 4*pi, 200);
y1 = exp(-0.2.*t).*sin(2.*t);
y2 = exp(-0.2.*t).*cos(2.*t);

% Create figure with specified properties
fig = figure('Units', 'centimeters', 'Position', [10 10
    15 10]);
```

```
% Plot with enhanced styling
plot(t, y1, 'LineWidth', 1.5, 'Color', [0.8500 0.3250
    0.0980])
hold on
plot(t, y2,'-.or', 'LineWidth', 1.5, 'Color', [0 0.4470
    0.7410])

% Customise plot elements
xlabel('Time (s)', 'FontSize', 11)
ylabel('Amplitude', 'FontSize', 11)
title('Damped Harmonic Motion', 'FontWeight', 'bold')
legend('Sin Component', 'Cos Component', 'Location',
    northeast')
grid on
box on

% Adjust axis properties
ax = gca;
ax.LineWidth = 1.2;
ax.TickDir = 'out';
ax.TickLength = [0.02 0.02];
```

For specialised applications in engineering visualisation, the following example demonstrates advanced axis manipulation:

Listing 8.15 Engineering data visualisation example

```
% Generate engineering test data
f = logspace(0, 4, 1000);
magnitude = 20log10(abs(1./(1 + 1jf/100)));

% Create semilogx plot with specific formatting
figure
semilogx(f, magnitude, 'LineWidth', 1.5)
grid on

% Configure axis properties
ax = gca;
ax.XScale = 'log';
ax.XMinorGrid = 'on';
ax.GridLineStyle = ':';
ax.GridAlpha = 0.3;

% Add labels and title
xlabel('Frequency (Hz)')
ylabel('Magnitude (dB)')
title('Frequency Response')

% Set specific axis limits
xlim([1 1e4])
ylim([-40 5])
```

The implementation of these visualisation techniques enables researchers and engineers to create publication-quality figures that effectively communicate complex data relationships. These tools support various scientific and engineering applications, from signal processing to system dynamics analysis [13].

8.3   Data Distribution Plots

Understanding the distribution of data is fundamental in statistical analysis and data science. **Data distribution plots** are powerful tools that enable the visualisation of underlying patterns and characteristics within datasets [14]. MATLAB offers an extensive array of functions for creating various types of data distribution plots, including **histograms**, **box plots**, **violin plots**, and **probability plots**. These visualisations allow users to gain insights into the **central tendency**, **dispersion**, **skewness**, and the presence of **outliers** in their data.

8.3.1   Histograms

Histograms are one of the most common tools for displaying the frequency distribution of numerical data [15]. They partition the data into bins and count the number of observations in each bin, providing a visual impression of the underlying frequency distribution.

The basic syntax for creating a histogram in MATLAB is:

Listing 8.16 Basic syntax for creating a histogram.

```
histogram(data)
```

where `data` is a vector containing the dataset.

For example, to generate a histogram of normally distributed data:

Listing 8.17 Creating a histogram of normally distributed data.

```
data = randn(1000, 1);
figure
histogram(data, 'Normalization', 'probability', '
    FaceColor', 'b', 'EdgeColor', 'w')
xlabel('Value')
ylabel('Probability')
title('Histogram of Normally Distributed Data')
```

In this example, the `randn` function generates a 1000-by-1 vector of normally distributed random numbers. The `histogram` function then creates a histogram of this data, normalised to display probabilities on the y-axis, with customised colours for better visual appeal.

Histograms are invaluable for identifying the shape of the data distribution, detecting skewness, and observing the presence of multiple modes or gaps in the data.

8.3.2 Box Plots

Box plots, also known as **box-and-whisker plots**, are effective for summarising the distribution of data through their quartiles [16]. They provide a graphical depiction of the median, quartiles, and potential outliers.

The basic syntax for creating a box plot in MATLAB is:

Listing 8.18 Basic syntax for creating a box plot.

```
boxplot(data)
```

where data can be a vector or a matrix. For matrices, each column is treated as a separate group.

An example demonstrating the creation of a box plot comparing three different data groups:

Listing 8.19 Creating a box plot for multiple groups.

```
data1 = randn(100, 1);
data2 = randn(100, 1) + 1;
data3 = randn(100, 1) - 1;
figure
boxplot([data1, data2, data3], 'Labels', {'Group 1', '
    Group 2', 'Group 3'})
ylabel('Value')
title('Box Plot of Three Data Groups')
```

This example creates a box plot comparing the distributions of three groups of data. Each group is represented by a box that summarises its distribution, allowing for easy comparison of medians, spreads, and identification of any outliers.

8.3.3 Violin Plots

Violin plots combine the features of box plots and kernel density plots to provide a more detailed view of the data distribution [17]. While MATLAB does not have a built-in violinplot function in its base installation, the **Statistics and Machine Learning Toolbox** includes support for violin plots, or they can be added via custom functions available from the MATLAB File Exchange [18].

An example of creating a violin plot (assuming the Violin Plot function is installed):

Listing 8.20 Creating a violin plot for multiple groups.

```
data = [data1, data2, data3];
violinplot(data, {'Group 1', 'Group 2', 'Group 3'})
ylabel('Value')
title('Violin Plot of Three Data Groups')
```

Violin plots provide richer information about the data distribution, showing the probability density of the data at different values, which can be particularly useful when the data has multiple modes or is not symmetrically distributed.

8.3.4 Probability Plots

Probability plots, such as **normal probability plots**, are used to assess whether a dataset follows a given distribution [19]. MATLAB's `probplot` function creates probability plots for various distributions.

The basic syntax is:

Listing 8.21 Basic syntax for creating a probability plot.

```
probplot('normal', data)
```

For instance, to create a normal probability plot:

Listing 8.22 Creating a normal probability plot.

```
data = randn(1000,1);
figure
probplot(data)
title('Normal Probability Plot')
```

If the data follows a normal distribution, the points on the plot will approximately lie along a straight line. Deviations from this line indicate departures from normality, which is crucial for statistical modelling and hypothesis testing.

8.3.5 Best Practices and Advanced Insights

Employing data distribution plots effectively requires an understanding of both the data and the appropriate visualisation techniques. Recent research emphasises the importance of visual literacy in data science and the need for accurate representations [20].

Advancements in interactive data visualisation tools within MATLAB, such as the `Data Cursor` and live scripts, allow for dynamic exploration of data distributions. Incorporating these tools can enhance the interpretability of complex datasets and facilitate deeper insights.

Furthermore, emerging techniques in data visualisation stress the importance of addressing issues such as overplotting and visual clutter, especially in large datasets [21]. Techniques like adjusting transparency, binning, or using alternative plot types can mitigate these challenges.

An exciting development in the realm of data distribution visualisation is the integration of these plots within **artificial intelligence** and **digital twin** environments [22]. By utilising data distribution plots, engineers and researchers can monitor and analyse the performance of complex systems in real-time, enabling more efficient diagnostics and predictive maintenance strategies.

Understanding the limitations of each plot type is essential. For instance, histograms are sensitive to bin width and position, which can significantly influence the interpretation of the data distribution [15]. Choosing an appropriate binning strategy is critical, and MATLAB provides options to customise the number and edges of bins.

By conscientiously selecting the appropriate data distribution plot and customising it effectively, one can unveil underlying patterns and trends that inform data-driven decision-making across various fields, including artificial intelligence, digital manufacturing, and autonomous systems. Such visual insights are indispensable for academic researchers, engineers, and students engaged in cutting-edge technological advancements.

8.4 Data Distribution Plots

Understanding the distribution of data is fundamental in statistical analysis and data science. **Data distribution plots** are powerful tools that enable the visualisation of underlying patterns and characteristics within datasets [14]. MATLAB offers an extensive array of functions for creating various types of data distribution plots, including **histograms**, **box plots**, **violin plots**, and **probability plots**. These visualisations allow users to gain insights into the **central tendency**, **dispersion**, **skewness**, and the presence of **outliers** in their data.

8.4.1 Histograms

Histograms are one of the most common tools for displaying the frequency distribution of numerical data [15]. They partition the data into bins and count the number of observations in each bin, providing a visual impression of the underlying frequency distribution.

The basic syntax for creating a histogram in MATLAB is:

Listing 8.23 Basic syntax for creating a histogram.

```
histogram(data)
```

where data is a vector containing the dataset.

For example, to generate a histogram of normally distributed data:

Listing 8.24 Creating a histogram of normally distributed data.

```
data = randn(1000, 1);
figure
histogram(data, 'Normalization', 'probability',
   'FaceColor', 'b', 'EdgeColor', 'w')
xlabel('Value')
ylabel('Probability')
title('Histogram of Normally Distributed Data')
```

In this example, the randn function generates a 1000-by-1 vector of normally distributed random numbers. The histogram function then creates a histogram of this data, normalised to display probabilities on the *y*-axis, with customised colours for better visual appeal.

Histograms are invaluable for identifying the shape of the data distribution, detecting skewness, and observing the presence of multiple modes or gaps in the data.

8.4.2   Box Plots

Box plots, also known as **box-and-whisker plots**, are effective for summarising the distribution of data through their quartiles [16]. They provide a graphical depiction of the median, quartiles, and potential outliers.

The basic syntax for creating a box plot in MATLAB is:

Listing 8.25 Basic syntax for creating a box plot.

```
boxplot(data)
```

where data can be a vector or a matrix. For matrices, each column is treated as a separate group.

An example demonstrating the creation of a box plot comparing three different data groups:

Listing 8.26 Creating a box plot for multiple groups.

```
data1 = randn(100, 1);
data2 = randn(100, 1) + 1;
data3 = randn(100, 1) - 1;
figure
boxplot([data1, data2, data3], 'Labels', {'Group 1', '
   Group 2', 'Group 3'})
ylabel('Value')
title('Box Plot of Three Data Groups')
```

This example creates a box plot comparing the distributions of three groups of data. Each group is represented by a box that summarises its distribution, allowing for easy comparison of medians, spreads, and identification of any outliers.

8.4.3   Violin Plots

Violin plots combine the features of box plots and kernel density plots to provide a more detailed view of the data distribution [17]. While MATLAB does not have a built-in `violinplot` function in its base installation, the **Statistics and Machine Learning Toolbox** includes support for violin plots, or they can be added via custom functions available from the MATLAB File Exchange [18].

An example of creating a violin plot (assuming the Violin Plot function is installed):

Listing 8.27 Creating a violin plot for multiple groups.

```
data = [data1, data2, data3];
violinplot(data, {'Group 1', 'Group 2', 'Group 3'})
ylabel('Value')
title('Violin Plot of Three Data Groups')
```

Violin plots provide richer information about the data distribution, showing the probability density of the data at different values, which can be particularly useful when the data has multiple modes or is not symmetrically distributed.

8.4.4   Probability Plots

Probability plots, such as **normal probability plots**, are used to assess whether a dataset follows a given distribution [19]. MATLAB's `probplot` function creates probability plots for various distributions.

The basic syntax is:

Listing 8.28 Basic syntax for creating a probability plot.

```
probplot('normal', data)
```

For instance, to create a normal probability plot:

Listing 8.29 Creating a normal probability plot.

```
data = randn(1000,1);
figure
probplot(data)
title('Normal Probability Plot')
```

If the data follows a normal distribution, the points on the plot will approximately lie along a straight line. Deviations from this line indicate departures from normality, which is crucial for statistical modelling and hypothesis testing.

8.4.5   Best Practices and Advanced Insights

Employing data distribution plots effectively requires an understanding of both the data and the appropriate visualisation techniques. Recent research emphasises the importance of visual literacy in data science and the need for accurate representations [20].

Advancements in interactive data visualisation tools within MATLAB, such as the Data Cursor and live scripts, allow for dynamic exploration of data distributions. Incorporating these tools can enhance the interpretability of complex datasets and facilitate deeper insights.

Furthermore, emerging techniques in data visualisation stress the importance of addressing issues such as overplotting and visual clutter, especially in large datasets [21]. Techniques like adjusting transparency, binning, or using alternative plot types can mitigate these challenges.

An exciting development in the realm of data distribution visualisation is the integration of these plots within **artificial intelligence** and **digital twin** environments [22]. By utilising data distribution plots, engineers and researchers can monitor and analyse the performance of complex systems in real-time, enabling more efficient diagnostics and predictive maintenance strategies.

Understanding the limitations of each plot type is essential. For instance, histograms are sensitive to bin width and position, which can significantly influence the interpretation of the data distribution [15]. Choosing an appropriate binning strategy is critical, and MATLAB provides options to customise the number and edges of bins.

By conscientiously selecting the appropriate data distribution plot and customising it effectively, one can unveil underlying patterns and trends that inform data-driven decision-making across various fields, including artificial intelligence, digital manufacturing, and autonomous systems. Such visual insights are indispensable for academic researchers, engineers, and students engaged in cutting-edge technological advancements.

8.5   Discrete Data Plots

Discrete data plots are indispensable tools for visualising and analysing categorical or discrete variables within datasets [23]. MATLAB offers a comprehensive suite of functions and tools for generating various types of discrete data plots, including **bar charts**, **pie charts**, **stem plots**, and **area plots**. These visualisation techniques enable users to effectively convey the distribution, proportions, and trends of discrete data points, thereby enhancing data interpretation and supporting informed decision-making.

Visualising discrete data appropriately is critical across various disciplines, from engineering to social sciences, where understanding categorical variables is paramount. This section delves into the functionalities and applications of these discrete data plotting tools in MATLAB, offering insights into their usage and highlighting best practices.

8.5.1 Bar Charts

Bar charts are widely used for representing categorical data, allowing for effortless comparison across categories [24]. In MATLAB, the `bar` function creates vertical bar charts, while `barh` generates horizontal bar charts. Advanced features enable the creation of grouped or stacked bar charts, providing a deeper understanding of data relationships.

The basic syntax for a simple bar chart is:

Listing 8.30 Basic syntax for creating a bar chart.

```
bar(x, y)
```

where x represents the categories and y represents the corresponding values.

For example, to compare sales figures across different regions:

Listing 8.31 Creating a bar chart of sales figures.

```
regions = {'North', 'South', 'East', 'West'};
sales = [150, 200, 180, 170];
bar(categorical(regions), sales)
title('Sales Figures by Region')
xlabel('Region')
ylabel('Sales (in units)')
```

Bar charts are highly customisable in MATLAB, with options to modify colours, bar widths, and add error bars. Utilising these features enhances the interpretability and aesthetic appeal of the charts, making them more engaging for the audience.

8.5.2 Pie Charts

Pie charts display data in a circular format, where each slice represents a category's contribution to the whole [25]. MATLAB's `pie` function is employed to generate pie charts, suitable for illustrating proportional data.

The basic syntax is:

Listing 8.32 Basic syntax for creating a pie chart.

```
pie(y)
```

where y is a vector containing the data values.

An example of creating a pie chart with labels is:

Listing 8.33 Creating a pie chart with category labels.

```
market_share = [35, 25, 15, 25];
companies = {'Company A', 'Company B', 'Company C',
 'Company D'};
pie(market_share, companies)
title('Market Share Distribution')
```

Despite their popularity, pie charts can be misleading if overused or used improperly. It is advisable to limit the number of slices and consider alternative plots, such as bar charts, when dealing with numerous categories.

8.5.3 Stem Plots

Stem plots, also known as **lollipop plots**, are effective for displaying discrete data points, especially when highlighting the distribution along an axis [26]. The stem function in MATLAB creates stem plots, which are useful for emphasising individual data points and their positions.

The basic syntax is:

Listing 8.34 Basic syntax for creating a stem plot.

```
stem(x, y)
```

where x and y define the data points.

For instance, to visualise the amplitude of a signal at discrete time intervals:

Listing 8.35 Creating a stem plot of a signal.

```
n = 0:15;
signal = sin(0.2pin);
stem(n, signal)
title('Discrete Signal Representation')
xlabel('n')
ylabel('Amplitude')
```

Stem plots are particularly useful when the data points are discrete and non-uniformly spaced. They provide a clear visual of the data's distribution without the clutter that might arise in a scatter plot for densely packed data.

8.5.4 Area Plots

Area plots represent quantitative data visually by filling the area beneath a line plot [27]. MATLAB's `area` function creates area plots, which can be stacked to show the contribution of individual components to the total.

The basic syntax is:

Listing 8.36 Basic syntax for creating an area plot.

```
area(x, y)
```

where x and y represent the data.

An example demonstrating an area plot:

Listing 8.37 Creating an area plot of cumulative data.

```
months = 1:12;
revenue = cumsum(rand(1,12) * 1000);
area(months, revenue)
title('Cumulative Revenue Over a Year')
xlabel('Month')
ylabel('Revenue (in GBP)')
```

Area plots are effective for showing how quantities accumulate over time or categories, particularly when comparing multiple datasets. They can also be stacked to represent the contribution of each category to the total amount.

8.5.5 Best Practices in Discrete Data Visualisation

Selecting the appropriate type of plot is imperative based on the characteristics of the data and the message intended to be conveyed. Recent advancements in data visualisation emphasise the importance of interactive and dynamic plots [28]. MATLAB's integration with interactive tools, such as the `Live Editor` and the `appdesigner`, enables users to create interactive visualisations that enhance engagement and data exploration.

Adhering to principles of visual perception and cognitive load can significantly improve the effectiveness of data visualisations [29]. Ensuring clarity, simplicity, and proper annotation in plots facilitates better comprehension among academic researchers, professors, and PhD students. Effective use of colour schemes, legends, and labels contribute to the overall quality of the visualisation.

8.6 Vector Fields (`quiver`, `quiver3`)

Visualising vector fields is fundamental for analysing directional data in two and three dimensions. MATLAB provides the `quiver` and `quiver3` functions for creating **2D** and **3D vector field plots**, respectively [30]. These plots display vectors as arrows at specified points, depicting both the magnitude and direction of vector quantities at those locations. Such visualisations are pivotal in fields like fluid dynamics, electromagnetism, and mechanical engineering, where understanding the behaviour of **vector-valued functions** is crucial [25].

The basic syntax for creating a 2D vector field plot using `quiver` is:

Listing 8.38 Basic syntax for creating a 2D vector field plot using `quiver`.

```
quiver(X, Y, U, V)
```

Here, `X` and `Y` define the grid coordinates, while `U` and `V` represent the vector components at each point.

An illustrative example of a 2D vector field plot:

Listing 8.39 Creating a 2D vector field plot with `quiver`.

```
[X, Y] = meshgrid(-2:0.2:2, -1:0.2:1);
U = sin(X) .* cos(Y);
V = -cos(X) .* sin(Y);

figure
quiver(X, Y, U, V)
xlabel('X')
ylabel('Y')
title('2D Vector Field Plot')
```

This code generates a 2D vector field where each arrow represents the magnitude and direction of the vector at each point on the grid defined by `X` and `Y`. The functions `sin` and `cos` are used to define the vector components, creating a rotational field pattern.

For visualising three-dimensional vector fields, MATLAB offers the `quiver3` function:

Listing 8.40 Basic syntax for creating a 3D vector field plot using `quiver3`.

```
quiver3(X, Y, Z, U, V, W)
```

In this syntax, `X`, `Y`, and `Z` define the positions in 3D space, while `U`, `V`, and `W` represent the vector components along the respective axes.

An example of a 3D vector field plot:

Listing 8.41 Creating a 3D vector field plot with `quiver3`.

```
[X, Y, Z] = meshgrid(-2:1:2);
U = Y;
V = -X;
```

```
W = zeros(size(X));

figure
quiver3(X, Y, Z, U, V, W)
xlabel('X')
ylabel('Y')
zlabel('Z')
title('3D Vector Field Plot')
```

In this example, the vector field represents a rotational flow in the XY-plane with no component in the Z-direction.

Understanding and accurately visualising vector fields is critical in modern engineering applications, including **robotics** and **autonomous systems**, where interpreting force fields and motion dynamics is essential for system design and control [31]. The ability to represent these fields in MATLAB facilitates the analysis of complex vector data, supporting the development of advanced technologies in **Industry 4.0** and **digital manufacturing**.

With the advent of **digital twins** and AI-driven simulations, visualisation of vector fields has become even more significant [32]. Integrating vector field visualisations into digital twin models allows engineers to simulate and analyse physical systems in virtual environments, leading to improved performance and predictive maintenance strategies. MATLAB's powerful visualisation capabilities, coupled with its computational tools, make it an indispensable platform for researchers and practitioners working at the forefront of technological innovation.

8.7 Volume Visualisation (`slice`, `isosurface`, `isocaps`)

Visualising volumetric data is crucial for analysing three-dimensional scalar fields, prevalent in domains such as medical imaging, fluid dynamics, and geophysics [33]. MATLAB provides several functions for **volume visualisation**, including slice, isosurface, and isocaps, enabling users to explore and interpret complex 3D datasets [34].

The slice function creates **slice planes** through a volume, allowing examination of cross-sectional views at specified positions [25]. The basic syntax is:

Listing 8.42 Basic syntax for creating slice planes using slice.

```
slice(X, Y, Z, V, xslice, yslice, zslice)
```

Here, X, Y, Z define the coordinates, V is the volumetric data, and xslice, yslice, zslice specify the positions of the slices.

An example of creating slice planes:

Listing 8.43 Creating slice planes through volumetric data.

```
[X, Y, Z] = meshgrid(-2:0.2:2);
V = X .* exp(-X.^2 - Y.^2 - Z.^2);

figure
slice(X, Y, Z, V, 0, [], [])
xlabel('X')
ylabel('Y')
zlabel('Z')
title('Slice Planes')
```

In this example, a slice is taken at X = 0, providing a cross-sectional view of the volumetric data at that plane.

The isosurface function extracts and visualises **isosurfaces**, representing surfaces of constant value within the volume [35].

The basic syntax is:

Listing 8.44 Basic syntax for creating isosurfaces using isosurface.

```
isosurface(X, Y, Z, V, isovalue)
```

An example of generating an isosurface:

Listing 8.45 Creating an isosurface from volumetric data.

```
[X, Y, Z] = meshgrid(-2:0.2:2);
V = X .* exp(-X.^2 - Y.^2 - Z.^2);

figure
p = patch(isosurface(X, Y, Z, V, 0.1));
isonormals(X, Y, Z, V, p)
p.FaceColor = 'red';
p.EdgeColor = 'none';
camlight
lighting gouraud
xlabel('X')
ylabel('Y')
zlabel('Z')
title('Isosurface')
```

This code generates an isosurface corresponding to an **isovalue** of 0.1, applying lighting to enhance visual perception.

The isocaps function creates **isocaps**, which are the end caps of the isosurface, providing additional context and aiding in the perception of enclosed volumes [36].

An example of using isocaps:

Listing 8.46 Adding isocaps to an isosurface plot.

```
[X, Y, Z] = meshgrid(-2:0.2:2);
V = X .* exp(-X.^2 - Y.^2 - Z.^2);
```

```
figure
p1 = patch(isosurface(X, Y, Z, V, 0.1));
isonormals(X, Y, Z, V, p1)
p1.FaceColor = 'red';
p1.EdgeColor = 'none';

p2 = patch(isocaps(X, Y, Z, V, 0.1));
p2.FaceColor = 'interp';
p2.EdgeColor = 'none';

camlight
lighting gouraud
xlabel('X')
ylabel('Y')
zlabel('Z')
title('Isosurface with Isocaps')
```

This example adds isocaps to the isosurface, enhancing the visualisation of the internal structure of the volumetric data.

The use of volume visualisation techniques is increasingly important in the era of big data and complex simulations [33]. In fields such as **artificial intelligence** and **digital twins,** the ability to visualise and interpret volumetric data facilitates better understanding and optimisation of models and systems [37]. For instance, in computational fluid dynamics, visualising flow fields using isosurfaces and slices enables engineers to identify critical regions and improve design [38].

Moreover, integrating real-time volume visualisation in **robotics** and **autonomous systems** enhances perception and decision-making capabilities, contributing to the advancement of intelligent machines [39]. MATLAB's robust visualisation tools empower researchers and professionals to push the boundaries of innovation in these cutting-edge domains.

8.8 Images Displaying

MATLAB provides a **comprehensive suite of tools** for displaying, processing, and analysing images, making it a valuable platform for visualising and interpreting image data. Functions such as imshow, imagesc, image, and subimage allow users to display images in a wide range of formats and layouts. These tools are particularly useful in domains such as computer vision, medical imaging, and remote sensing, where clarity and precision in visualisation are critical (Fig. 8.11).

8.8.1 Basic Image Display Functions

The most commonly used functions for displaying images in MATLAB are [40]:

- **imshow**: Displays an image in a figure window, automatically adjusting the image display range.

Fig. 8.11 Example image display using imshow

The following example demonstrates the basic syntax for displaying an image using imshow:

Listing 8.47 Basic image display using imshow.

```
% Load and display an image using imshow
img = imread('GinkgoLeaves.jpg'); % Load the sample
    image
imshow(img);                      % Display the image
title('Image Display Using imshow'); % Add a title
```

- **imagesc**: Scales image data to the full range of the colormap for better visualisation of data intensities.

  The imagesc function is commonly used to display image data with intensity values scaled to the full range of the current colormap. This is particularly useful for visualising scientific data or images with a wide range of intensity values.

Listing 8.48 Displaying a grayscale image using imagesc.

```
% Load a grayscale image
img = imread('cameraman.tif');

% Display the image using imagesc
figure;
imagesc(img);
colormap(gray); % Apply grayscale colormap
colorbar;       % Add a colour bar
title('Grayscale Image Display Using imagesc');
```

Another example demonstrates how to visualise matrix data using imagesc with a jet colormap:

Listing 8.49 Visualising a matrix using `imagesc` with a jet colormap.

```
% Generate a random matrix
matrixData = rand(100);

% Visualise the matrix using imagesc
figure;
imagesc(matrixData);
colormap(jet); % Apply the 'jet' colormap
colorbar;      % Add a colour bar for reference
title('Matrix Display Using imagesc with Jet Colormap'
    );
```

- **image**: Displays an image as a two-dimensional graphic object, preserving the original data values.

  The `image` function displays an image as a two-dimensional graphic. Unlike `imagesc`, this function does not scale the intensity values, and the displayed image reflects the raw data values.

Listing 8.50 Displaying an image using `image`.

```
% Load a sample image
img = imread('peppers.png');

% Display the image using the image function
figure;
image(img);
title('Image Display Using image (No Scaling)');
```

- **subimage**: Displays multiple images in a single figure. This is particularly useful for comparative analysis.

  The `subimage` function allows users to display multiple images in a single figure for comparative analysis. This is useful for side-by-side visualisation of original and processed images.

Listing 8.51 Displaying multiple images using `subimage`.

```
% Load two sample images
img1 = imread('peppers.png');
img2 = imadjust(img1); % Adjust contrast of the first
    image

% Display the images side by side using subimage
figure;
subplot(1, 2, 1);
subimage(img1);
title('Original Image');

subplot(1, 2, 2);
subimage(img2);
title('Contrast Enhanced Image');
```

8.8.2 *Enhancing Image Contrast*

Enhancing image contrast is essential for improving the visibility of features and details in an image. MATLAB provides several functions for this purpose, including `histeq` and `adapthisteq`. These functions are demonstrated in the following examples.

To improve the visual quality of images, MATLAB offers several functions for contrast enhancement:

- **imadjust**: Adjusts the intensity values of an image to improve contrast.
  Here is an example that uses `imadjust` to enhance contrast:

Listing 8.52 Enhancing contrast using `imadjust`.

```
% Read an image
img = imread('farm.jpg');

% Adjust image contrast
adjusted_img = imadjust(img);

% Display original and adjusted images
subplot(1, 2, 1);
imshow(img);
title('Original Image');

subplot(1, 2, 2);
imshow(adjusted_img);
title('Contrast Enhanced Image');
```

- **histeq**: Performs histogram equalisation to enhance contrast by redistributing intensity values.
  The `histeq` function enhances the contrast of an image by equalising its histogram. This is particularly useful for improving the contrast of images with narrow intensity distributions.

Listing 8.53 Enhancing image contrast using `histeq`.

```
% Load a grayscale image
img = imread('cameraman.tif');

% Perform histogram equalisation
equalised_img = histeq(img);

% Display the original and equalised images
figure;
subplot(1, 2, 1);
imshow(img);
title('Original Image');

subplot(1, 2, 2);
imshow(equalised_img);
title('Histogram Equalised Image');
```

Fig. 8.12 Example image display using imadjust

- **adapthisteq**: Applies contrast-limited adaptive histogram equalisation (CLAHE), which is particularly effective for enhancing local contrast in images with varying lighting conditions (Fig. 8.12).
  The adapthisteq function applies contrast-limited adaptive histogram equalisation (CLAHE) to an image, which enhances local contrast while avoiding over-amplification of noise.

Listing 8.54 Enhancing image contrast using adapthisteq.

```
% Load a grayscale image
img = imread('cameraman.tif');

% Perform adaptive histogram equalisation
adapted_img = adapthisteq(img);

% Display the original and adapted images
figure;
subplot(1, 2, 1);
imshow(img);
```

```
title('Original Image');

subplot(1, 2, 2);
imshow(adapted_img);
title('Adaptive Histogram Equalised Image');
```

- **Combining Contrast Enhancement Techniques**
  It is often useful to combine contrast enhancement techniques for more refined results. The following example applies both `histeq` and `adapthisteq` to the same image for comparison:

Listing 8.55 Combining `histeq` and `adapthisteq` for comparison.

```
% Load a grayscale image
img = imread('cameraman.tif');

% Apply histogram equalisation
histeq_img = histeq(img);

% Apply adaptive histogram equalisation
adapthisteq_img = adapthisteq(img);

% Display the original and enhanced images
figure;
subplot(1, 3, 1);
imshow(img);
title('Original Image');

subplot(1, 3, 2);
imshow(histeq_img);
title('Histogram Equalised');

subplot(1, 3, 3);
imshow(adapthisteq_img);
title('Adaptive Histogram Equalised');
```

8.8.3 Applications in Various Domains

Image visualisation plays a pivotal role in many scientific and industrial fields: - **Computer Vision**: MATLAB simplifies the analysis of object detection, segmentation, and feature extraction tasks through its visualisation functions. - **Medical Imaging**: Functions like `imshow` and `imagesc` are frequently used to display and process X-ray, MRI, and CT scan images. - **Remote Sensing**: Researchers use MATLAB to display satellite imagery and enhance visual features using contrast adjustment techniques.

The following example illustrates how `imagesc` can be applied to visualise remote sensing data:

Listing 8.56 Visualising remote sensing data using `imagesc`.

```
% Generate synthetic remote sensing data
data = peaks(100);

% Display the data with a colour map
imagesc(data);
colormap('jet'); % Apply 'jet' colormap
colorbar; % Add a colour bar to the figure
title('Remote Sensing Data Visualisation');
```

8.8.4   Advanced Image Manipulation

Advanced image manipulation techniques allow users to extract meaningful insights from complex datasets. Functions such as `imrotate`, `imresize`, and `imcrop` enable users to rotate, resize, and crop images, respectively. These operations are often used in pre-processing pipelines for machine learning and image analysis.

The following example demonstrates how to rotate and resize an image:

Listing 8.57 Image rotation and resizing using `imrotate` and `imresize`.

```
% Load an image
img = imread('peppers.png');

% Rotate the image by 45 degrees
rotated_img = imrotate(img, 45);

% Resize the image to half its original size
resized_img = imresize(rotated_img, 0.5);

% Display the original and processed images
figure;
subplot(1, 3, 1);
imshow(img);
title('Original Image');

subplot(1, 3, 2);
imshow(rotated_img);
title('Rotated Image');

subplot(1, 3, 3);
imshow(resized_img);
title('Resized Image');
```

MATLAB's robust image visualisation and manipulation capabilities empower researchers and engineers to process and analyse images effectively. By leveraging tools such as `imshow`, `imadjust`, and `imagesc`, users can enhance image quality, extract relevant information, and apply these techniques across diverse applications.

The inclusion of interactive visualisation tools and advanced manipulation functions ensures that MATLAB remains a leading platform for image processing tasks.

8.8.5 Unique Insights

- Integration with Machine Learning: MATLAB's image processing functions can be seamlessly integrated into machine learning workflows, enabling automated feature extraction and classification.
- Efficient Memory Management: Techniques like rescaling and cropping help optimise memory usage for large datasets, enhancing computational efficiency.
- Domain-Specific Applications: MATLAB's flexibility allows for customisation of visualisation techniques tailored to specific domains, such as using colormaps like hot for thermal imaging or parula for scientific data visualisation.

8.9 Animating Visualisations

Creating animations is a powerful way to visualise dynamic processes and time-varying data. MATLAB provides a comprehensive set of functions and tools for creating engaging and informative animations [41]. The movie function enables users to create a movie from a series of plots or images, while getframe captures the current figure or axis as a movie frame. By combining these functions with loops and other programming constructs, users can create animations that effectively communicate complex ideas and reveal patterns in data.

8.9.1 Creating Animated Plots

MATLAB's animation functions can be used to create **animated plots** that illustrate changes in data over time or in response to varying parameters. The basic syntax for creating an animated plot using the getframe function is:

Listing 8.58 Creating an animated plot.

```
% Initialize an empty array to store the frames
frames = [];

% Create the animation
for i = 1:num_frames
% Plot the data for the current frame
plot(x, y)
% Capture the current frame
```

```
frames(i) = getframe(gcf);

end

% Create a video writer object
video = VideoWriter('animated_plot.avi');
open(video);

% Write the frames to the video
writeVideo(video, frames);

% Close the video writer
close(video);
```

In this example, a loop is used to generate a series of frames, each depicting the data at a specific point in time. The `getframe` function captures each frame, which is stored in the `frames` array. Finally, the frames are written to a video file using the `VideoWriter` and `writeVideo` functions.

8.9.2 Visualising Time-Series Data

Animations are particularly useful for visualising **changes over time** in time-series data. MATLAB provides the `animatedline` function, which allows users to incrementally add data points to a line plot, creating the illusion of motion or growth. Here's an example that showcases the use of `animatedline` for visualising time-series data:

Listing 8.59 Animating time-series data using animatedline.

```
% Create a figure
figure

% Create an animated line object
h = animatedline;

% Set the axis limits
xlim([0, 10])
ylim([-1, 1])

% Define the number of points
num_points = 100;

% Animate the time-series data
for i = 1:num_points
% Generate new data point
x = i/10;
y = sin(x);
% Add the new data point to the animated line
addpoints(h, x, y);

% Update the figure
```

```
drawnow limitrate
pause(0.1)

end
```

In this example, an `animatedline` object is created, and the `addpoints` function is used within a loop to incrementally add new data points to the line plot. The `drawnow` function is used to update the figure, and the `pause` function introduces a small delay between each frame to control the animation speed.

8.9.3 Animating 3D Plots

Animating **3D plots** in MATLAB allows users to explore data from different angles or show changes in parameters over time. The `getframe` function can be used in combination with 3D plotting functions like `surf` and `plot3` to create animated 3D visualisations. Here's an example that demonstrates the animation of a 3D surface plot:

Listing 8.60 Animating a 3D surface plot.

```
% Create a figure
figure

% Define the surface parameters
[X, Y] = meshgrid(-2:0.1:2);

% Initialize an empty array to store the frames
frames = [];

% Create the animation
for i = 1:50
% Calculate the Z values for the current frame
Z = sin(sqrt(X.^2 + Y.^2) - i/10);
% Plot the surface
surf(X, Y, Z)

% Set the view angle
view(-37.5 + i/2, 30)

% Capture the current frame
frames(i) = getframe(gcf);

end

% Create a video writer object
video = VideoWriter('animated_3d_surface.avi');
open(video);

% Write the frames to the video
```

```
writeVideo(video, frames);

% Close the video writer
close(video);
```

In this example, a loop is used to generate a series of frames, each depicting a 3D surface plot with varying Z values and view angles. The `getframe` function captures each frame, which is stored in the `frames` array. Finally, the frames are written to a video file using the `VideoWriter` and `writeVideo` functions.

Animating visualisations in MATLAB allows users to create engaging and informative representations of dynamic processes, time-varying data, and complex 3D structures. By leveraging functions like `movie`, `getframe`, and `animatedline`, users can effectively communicate their findings and insights to a wider audience.

8.10 Formatting and Annotation

Creating plots and charts is only half the battle in effective data visualisation; the real art lies in the **formatting and annotation** that bring clarity and emphasis to the data presented. MATLAB provides a rich set of functions and properties for customising the appearance of graphics and adding informative labels and annotations [42].

8.10.1 Customising Plot Appearance

Meticulous attention to a plot's appearance can greatly enhance its interpretability. MATLAB allows users to customise various aspects such as **line styles**, **colours**, **markers**, and **transparency**, providing flexibility to tailor plots to specific needs [25].

The basic syntax for setting line properties in a plot is:

Listing 8.61 Setting line properties in a plot.

```
x = linspace(0, 2*pi, 100);
y1 = sin(x);
y2 = cos(x);

plot(x, y1, 'LineWidth', 2, 'Color', 'red', 'LineStyle',
    '--')
hold on
plot(x, y2, 'LineWidth', 1.5, 'Color', 'blue', '
    LineStyle', '-.')
hold off
```

In this example, the `plot` function includes additional arguments to specify the line width, colour, and style for each curve. The `hold on` and `hold off` commands are used to overlay multiple plots on the same axes.

Similarly, marker appearance can be customised:

Listing 8.62 Customising marker properties in a plot.

```
x = linspace(0, 2*pi, 10);
y = sin(x);

plot(x, y, 'Marker', 'o', 'MarkerSize', 8, '
    MarkerFaceColor', 'red', 'MarkerEdgeColor', 'black')
```

Here, the `plot` function specifies the marker type, size, face colour, and edge colour, enhancing the visual distinction of data points.

8.10.2 Adding Labels and Titles

Labels and titles are indispensable for providing context and guiding the reader through the data [24]. MATLAB's `xlabel`, `ylabel`, and `title` functions facilitate the addition of descriptive text to plots.

Listing 8.63 Adding labels and a title to a plot.

```
x = linspace(0, 2*pi, 100);
y = sin(x);

plot(x, y)
xlabel('x')
ylabel('sin(x)')
title('Sine Function')
```

In this example, axis labels and a plot title are added, providing essential information about the variables and the nature of the plot.

8.10.3 Adding Legends

When multiple data series are presented in a single plot, a **legend** becomes crucial for differentiation [42]. MATLAB's `legend` function adds a legend to the axes.

Listing 8.64 Adding a legend to a plot with multiple data series.

```
x = linspace(0, 2*pi, 100);
y1 = sin(x);
y2 = cos(x);

plot(x, y1, 'r--', x, y2, 'b-.')
legend('sin(x)', 'cos(x)')
xlabel('x')
ylabel('Value')
title('Sine and Cosine Functions')
```

This code plots both the sine and cosine functions with different line styles and adds a legend to distinguish between them.

8.10.4 Annotating Plots

Annotations can highlight specific data points or regions of interest, adding an extra layer of information to a plot [43]. MATLAB provides the text and annotation functions for this purpose.

An example using the text function:

Listing 8.65 Annotating a plot using the text function.

```
x = linspace(0, 2*pi, 100);
y = sin(x);

plot(x, y)
xlabel('x')
ylabel('sin(x)')
title('Sine Function')

[max_val, max_idx] = max(y);
text(x(max_idx), max_val, ' \leftarrow Maximum',
  'HorizontalAlignment', 'left')
```

In this example, the maximum point of the sine function is annotated with a label. The arrow notation and horizontal alignment enhance the clarity of the annotation.

For more complex annotations, the annotation function can be used to add shapes, arrows, or highlight areas.

8.10.5 Unique Insights and Advanced Formatting

In the realm of **Industry 4.0** and **digital twins,** the precise visualisation of data becomes even more critical. Advanced formatting techniques can improve the interpretability of complex datasets in fields like **artificial intelligence** and **robotics**. For instance, using transparency in plots can help reveal overlapping data points, and utilising customised colour maps can encode additional dimensions of data [28].

Furthermore, interactive features such as data cursors and zooming can be enabled to allow viewers to explore the data more deeply, which is especially beneficial when presenting findings to a technical audience comprising engineers and researchers.

8.11 Advanced Visualisation Techniques

Beyond basic plotting and formatting, MATLAB provides a plethora of **advanced visualisation techniques** that enable the creation of interactive, dynamic, and highly informative graphics. These techniques empower users to delve into complex datasets, effectively communicate findings, and glean deeper insights [44].

8.11.1 Interactive Visualisations

Interactive visualisations allow users to engage with data dynamically, facilitating exploration through **zooming**, **panning**, and data point selection. MATLAB offers a suite of tools dedicated to crafting interactive plots, such as the `zoom`, `pan`, and `datacursormode` functions [45].

The basic syntax to enable interactive features in a plot is:

Listing 8.66 Creating an interactive plot.

```
x = linspace(0, 2*pi, 100);
y = sin(x);

plot(x, y)
xlabel('x')
ylabel('sin(x)')
title('Interactive Sine Plot')

zoom on
pan on
datacursormode on
```

In this example, the `zoom`, `pan`, and `datacursormode` functions activate interactive capabilities, allowing the user to zoom into specific regions, pan across the plot, and display data values by hovering over points.

Moreover, MATLAB's **Live Editor** enhances interactivity by enabling the embedding of controls such as sliders and drop-down menus, facilitating real-time manipulation of variables with immediate visual feedback [25].

8.11.2 Visualising Big Data

Visualising large datasets poses significant challenges due to memory constraints and rendering performance. MATLAB addresses these issues by providing efficient techniques for visualising **big data**, such as data downsampling, binning, and aggregation [46].

An effective method for handling large datasets is to utilise binned plots, which aggregate data points into bins and display statistical summaries:

Listing 8.67 Visualising big data using binning.

```matlab
% Generate a large dataset
x = randn(1e6, 1);
y = randn(1e6, 1);

% Create a binned scatter plot
binscatter(x, y)
xlabel('x')
ylabel('y')
title('Binned Scatter Plot')
```

In this example, the `binscatter` function creates a binned scatter plot, efficiently handling one million data points by aggregating them into bins. This approach reduces rendering time while preserving data patterns.

Additionally, MATLAB supports **tall arrays** for processing data that do not fit into memory, enabling users to work with datasets of virtually unlimited size [47].

8.11.3 Visualising Real-Time Data

The ability to visualise data in real time is crucial in applications such as robotics, autonomous systems, and industrial monitoring. MATLAB provides functions and tools for creating real-time plots, allowing for continuous updates as new data become available [48].

The basic approach to real-time plotting involves updating graphics within a loop:

Listing 8.68 Creating a real-time plot.

```matlab
% Create a figure and an animated line object
figure
h = animatedline;
axis([0, 10, -1, 1])
xlabel('Time (s)')
ylabel('Amplitude')
title('Real-Time Sine Wave')

% Update the plot in a loop
for t = 0:0.1:10
y = sin(t);
addpoints(h, t, y);
drawnow
pause(0.1)
end
```

In this code, the `animatedline` function is used to create a line that can be updated with new data points. The `addpoints` function appends data to the line, and `drawnow` forces MATLAB to update the figure window.

Real-time visualisation is instrumental in **digital manufacturing** and **Industry 4.0,** where monitoring sensor data and system states in real time leads to improved process control and efficiency [49].

8.11.4 Visualising Uncertainty

Communicating uncertainty is essential in data analysis to convey the reliability and variability inherent in measurements. MATLAB offers various techniques for visualising uncertainty, including **error bars**, **confidence intervals**, and **probabilistic distributions** [50].

8.11.4.1 Displaying Error Bars

Error bars represent the variability or uncertainty in data points and are vital for statistical analysis:

Listing 8.69 Displaying error bars.

```
x = 1:5;
y = [2, 4, 5, 3, 6];
err = [0.5, 0.8, 0.3, 0.6, 0.4];

errorbar(x, y, err, 'o-')
xlabel('x')
ylabel('y')
title('Plot with Error Bars')
```

Here, the `errorbar` function plots the data points with vertical error bars representing the uncertainty or standard deviation.

8.11.4.2 Visualising Confidence Intervals

Confidence intervals provide a range within which the true value of a parameter is expected to lie, with a certain level of confidence:

Listing 8.70 Visualising confidence intervals.

```
x = linspace(0, 10, 100);
y = sin(x) + 0.1*randn(size(x));
p = polyfit(x, y, 3);
[y_fit, delta] = polyconf(p, x, [], 'predopt', 'curve');

plot(x, y, 'b.')
hold on
plot(x, y_fit, 'r-')
plot(x, y_fit + delta, 'r--')
```

```
plot(x, y_fit - delta, 'r--')
hold off
xlabel('x')
ylabel('y')
title('Polynomial Fit with Confidence Intervals')
legend('Data', 'Fit', 'Upper Bound', 'Lower Bound')
```

In this example, `polyfit` and `polyconf` are used to perform a polynomial fit and compute confidence intervals, which are then plotted to visualise the uncertainty in the model.

8.11.4.3 Communicating Probabilistic Outcomes

Probabilistic outcomes can be visualised using **density plots** or **violin plots**, which display the distribution of data:

Listing 8.71 Visualising probabilistic outcomes using a density plot.

```
data = randn(1000, 1);

[f, xi] = ksdensity(data);
plot(xi, f, 'LineWidth', 2)
xlabel('Value')
ylabel('Density')
title('Kernel Density Estimate')
```

The `ksdensity` function estimates the probability density function of the data, providing a smooth curve that represents the distribution.

Visualising uncertainty is particularly important in fields such as **artificial intelligence** and **machine learning**, where understanding the confidence and variability of predictions informs better decision-making [51].

8.11.5 Visualising Geographical Data

The visualisation of geographical data is essential in applications ranging from environmental monitoring to logistics planning. MATLAB offers comprehensive tools for creating maps, plotting data on maps, and customising map appearance through the **Mapping Toolbox** [52].

8.11.5.1 Creating Maps

To create a map, MATLAB provides functions like `worldmap` and `axesm`:

Listing 8.72 Creating a simple world map.

```
figure
worldmap('World')
load coastlines
plotm(coastlat, coastlon)
title('World Map with Coastlines')
```

In this example, `worldmap` sets up map axes appropriate for the specified region, and the coastlines data is plotted using `plotm`.

8.11.5.2   Plotting Data on Maps

Data can be plotted on maps using geographic coordinates:

Listing 8.73 Plotting data points on a map.

```
% Define latitude and longitude of data points
lat = [51.5074, 48.8566, 52.5200]; % London, Paris,
    Berlin
lon = [-0.1278, 2.3522, 13.4050];

% Create a map of Europe
figure
worldmap('Europe')
geoshow('landareas.shp', 'FaceColor', [0.8 0.8 0.8])

% Plot data points
plotm(lat, lon, 'ro', 'MarkerSize', 8,
 'MarkerFaceColor', 'r')
title('Major European Cities')
```

Here, the geographic locations of London, Paris, and Berlin are plotted on a map of Europe.

8.11.5.3   Customising Map Appearance

Maps can be customised by changing projections, adding layers, and adjusting visual properties:

Listing 8.74 Customising map appearance.

```
figure
ax = worldmap('World');
setm(ax, 'MapProjection', 'robinson')

% Display land areas with custom face color
geoshow('landareas.shp', 'FaceColor', [0.5 1.0 0.5])

% Add rivers and lakes
```

```
geoshow('worldrivers.shp', 'Color', 'blue')
geoshow('worldlakes.shp', 'FaceColor', 'cyan')

% Add grid and labels
gridm on
mlabel on
plabel on

title('Customised World Map')
```

In this example, the Robinson projection is used for the map. Land areas, rivers, and lakes are displayed with specified colours, and meridian and parallel labels are enabled.

Geospatial visualisation is vital in **digital twins** of geographical systems, where accurate mapping of real-world locations is necessary for simulation and analysis [53].

8.12 Advanced Visualisation Techniques

Beyond basic plotting and formatting, MATLAB provides a plethora of **advanced visualisation techniques** that enable the creation of interactive, dynamic, and highly informative graphics. These techniques empower users to delve into complex datasets, effectively communicate findings, and glean deeper insights [44].

8.12.1 Interactive Visualisations

Interactive visualisations allow users to engage with data dynamically, facilitating exploration through **zooming**, **panning**, and data point selection. MATLAB offers a suite of tools dedicated to crafting interactive plots, such as the `zoom`, `pan`, and `datacursormode` functions [45].

The basic syntax to enable interactive features in a plot is:

Listing 8.75 Creating an interactive plot.

```
x = linspace(0, 2*pi, 100);
y = sin(x);

plot(x, y)
xlabel('x')
ylabel('sin(x)')
title('Interactive Sine Plot')

zoom on
pan on
datacursormode on
```

In this example, the zoom, pan, and datacursormode functions activate interactive capabilities, allowing the user to zoom into specific regions, pan across the plot, and display data values by hovering over points.

Moreover, MATLAB's **Live Editor** enhances interactivity by enabling the embedding of controls such as sliders and drop-down menus, facilitating real-time manipulation of variables with immediate visual feedback [25].

8.12.2   Visualising Big Data

Visualising large datasets poses significant challenges due to memory constraints and rendering performance. MATLAB addresses these issues by providing efficient techniques for visualising **big data**, such as data downsampling, binning, and aggregation [46].

An effective method for handling large datasets is to utilise binned plots, which aggregate data points into bins and display statistical summaries:

Listing 8.76 Visualising big data using binning.

```
% Generate a large dataset
x = randn(1e6, 1);
y = randn(1e6, 1);

% Create a binned scatter plot
binscatter(x, y)
xlabel('x')
ylabel('y')
title('Binned Scatter Plot')
```

In this example, the binscatter function creates a binned scatter plot, efficiently handling one million data points by aggregating them into bins. This approach reduces rendering time while preserving data patterns.

Additionally, MATLAB supports **tall arrays** for processing data that do not fit into memory, enabling users to work with datasets of virtually unlimited size [47].

8.12.3   Visualising Real-Time Data

The ability to visualise data in real time is crucial in applications such as robotics, autonomous systems, and industrial monitoring. MATLAB provides functions and tools for creating real-time plots, allowing for continuous updates as new data become available [48].

The basic approach to real-time plotting involves updating graphics within a loop:

Listing 8.77 Creating a real-time plot.

```
% Create a figure and an animated line object
figure
h = animatedline;
axis([0, 10, -1, 1])
xlabel('Time (s)')
ylabel('Amplitude')
title('Real-Time Sine Wave')

% Update the plot in a loop
for t = 0:0.1:10
y = sin(t);
addpoints(h, t, y);
drawnow
pause(0.1)
end
```

In this code, the `animatedline` function is used to create a line that can be updated with new data points. The `addpoints` function appends data to the line, and `drawnow` forces MATLAB to update the figure window.

Real-time visualisation is instrumental in **digital manufacturing** and **Industry 4.0**, where monitoring sensor data and system states in real time leads to improved process control and efficiency [49].

8.12.4   Visualising Uncertainty

Communicating uncertainty is essential in data analysis to convey the reliability and variability inherent in measurements. MATLAB offers various techniques for visualising uncertainty, including **error bars**, **confidence intervals**, and **probabilistic distributions** [50].

8.12.4.1   Displaying Error Bars

Error bars represent the variability or uncertainty in data points and are vital for statistical analysis:

Listing 8.78 Displaying error bars.

```
x = 1:5;
y = [2, 4, 5, 3, 6];
err = [0.5, 0.8, 0.3, 0.6, 0.4];

errorbar(x, y, err, 'o-')
xlabel('x')
ylabel('y')
title('Plot with Error Bars')
```

Here, the `errorbar` function plots the data points with vertical error bars representing the uncertainty or standard deviation.

8.12.4.2 Visualising Confidence Intervals

Confidence intervals provide a range within which the true value of a parameter is expected to lie, with a certain level of confidence:

Listing 8.79 Visualising confidence intervals.

```
x = linspace(0, 10, 100);
y = sin(x) + 0.1*randn(size(x));
p = polyfit(x, y, 3);
[y_fit, delta] = polyconf(p, x, [], 'predopt', 'curve');

plot(x, y, 'b.')
hold on
plot(x, y_fit, 'r-')
plot(x, y_fit + delta, 'r--')
plot(x, y_fit - delta, 'r--')
hold off
xlabel('x')
ylabel('y')
title('Polynomial Fit with Confidence Intervals')
legend('Data', 'Fit', 'Upper Bound', 'Lower Bound')
```

In this example, `polyfit` and `polyconf` are used to perform a polynomial fit and compute confidence intervals, which are then plotted to visualise the uncertainty in the model.

8.12.4.3 Communicating Probabilistic Outcomes

Probabilistic outcomes can be visualised using **density plots** or **violin plots**, which display the distribution of data:

Listing 8.80 Visualising probabilistic outcomes using a density plot.

```
data = randn(1000, 1);

[f, xi] = ksdensity(data);
plot(xi, f, 'LineWidth', 2)
xlabel('Value')
ylabel('Density')
title('Kernel Density Estimate')
```

The `ksdensity` function estimates the probability density function of the data, providing a smooth curve that represents the distribution.

Visualising uncertainty is particularly important in fields such as **artificial intelligence** and **machine learning**, where understanding the confidence and variability of predictions informs better decision-making [51].

8.12.5 Visualising Geographical Data

The visualisation of geographical data is essential in applications ranging from environmental monitoring to logistics planning. MATLAB offers comprehensive tools for creating maps, plotting data on maps, and customising map appearance through the **Mapping Toolbox** [52].

8.12.5.1 Creating Maps

To create a map, MATLAB provides functions like `worldmap` and `axesm`:

Listing 8.81 Creating a simple world map.

```
figure
worldmap('World')
load coastlines
plotm(coastlat, coastlon)
title('World Map with Coastlines')
```

In this example, `worldmap` sets up map axes appropriate for the specified region, and the coastlines data is plotted using `plotm`.

8.12.5.2 Plotting Data on Maps

Data can be plotted on maps using geographic coordinates:

Listing 8.82 Plotting data points on a map.

```
% Define latitude and longitude of data points
lat = [51.5074, 48.8566, 52.5200]; % London, Paris,
    Berlin
lon = [-0.1278, 2.3522, 13.4050];

% Create a map of Europe
figure
worldmap('Europe')
geoshow('landareas.shp', 'FaceColor', [0.8 0.8 0.8])

% Plot data points
plotm(lat, lon, 'ro', 'MarkerSize', 8,
 'MarkerFaceColor', 'r')
title('Major European Cities')
```

Here, the geographic locations of London, Paris, and Berlin are plotted on a map of Europe.

8.12.5.3   Customising Map Appearance

Maps can be customised by changing projections, adding layers, and adjusting visual properties:

Listing 8.83 Customising map appearance.

```
figure
ax = worldmap('World');
setm(ax, 'MapProjection', 'robinson')

% Display land areas with custom face color
geoshow('landareas.shp', 'FaceColor', [0.5 1.0 0.5])

% Add rivers and lakes
geoshow('worldrivers.shp', 'Color', 'blue')
geoshow('worldlakes.shp', 'FaceColor', 'cyan')

% Add grid and labels
gridm on
mlabel on
plabel on

title('Customized World Map')
```

In this example, the Robinson projection is used for the map. Land areas, rivers, and lakes are displayed with specified colours, and meridian and parallel labels are enabled.

Geospatial visualisation is vital in **digital twins** of geographical systems, where accurate mapping of real-world locations is necessary for simulation and analysis [53].

8.13   Visualisation Best Practices

Effective data visualisation requires careful consideration of various factors to ensure that the intended message is clearly communicated to the target audience. This section explores best practices for creating informative and visually appealing graphics in MATLAB.

8.13.1 Choosing the Right Plot Type

Selecting the most appropriate **plot type** is crucial for effectively conveying the message behind the data. Different plot types, such as line plots, scatter plots, bar graphs, and heatmaps, are suited for different types of data and purposes [54]. For example, line plots are ideal for displaying trends over time, while scatter plots are useful for showing relationships between two variables.

Listing 8.84 Line plot example.

```
x = linspace(0, 2*pi, 100);
y = sin(x);
plot(x, y);
xlabel('x');
ylabel('sin(x)');
title('Sine Function');
```

Listing 8.85 Scatter plot example.

```
x = randn(100, 1);
y = 0.5*x + randn(100, 1);
scatter(x, y);
xlabel('x');
ylabel('y');
title('Scatter Plot');
```

8.13.2 Effective Use of Colour

Colour is a powerful tool in data visualisation that can be used to highlight important information, distinguish between different data series, and improve overall readability [55]. However, it is essential to use colour judiciously and consider factors such as colour blindness and cultural differences in colour perception.

Listing 8.86 Using colour to distinguish data series.

```
x = linspace(0, 2*pi, 100);
y1 = sin(x);
y2 = cos(x);
plot(x, y1, 'r', x, y2, 'b');
xlabel('x');
ylabel('y');
legend('sin(x)', 'cos(x)');
title('Sine and Cosine Functions');
```

8.13.3 Simplifying Complex Visualisations

When dealing with **complex data sets**, it is important to simplify visualisations to ensure that the key insights are clearly communicated [56]. This can be achieved by reducing clutter, using subplots to break down the information, and focusing on the most relevant aspects of the data.

Listing 8.87 Using subplots to simplify complex visualisations.

```
x = linspace(0, 2*pi, 100);
y1 = sin(x);
y2 = cos(x);
y3 = sin(x).*cos(x);

subplot(1,3,1);
plot(x, y1);
title('sin(x)');

subplot(1,3,2);
plot(x, y2);
title('cos(x)');

subplot(1,3,3);
plot(x, y3);
title('sin(x)*cos(x)');
```

8.13.4 Designing for Different Audiences

When creating visualisations, it is crucial to consider the **target audience** and their level of expertise [57]. Visualisations designed for a technical audience may include more complex information and assume a higher level of background knowledge, while those intended for a general audience should prioritize clarity and simplicity.

Listing 8.88 Designing visualisations for different audiences.

```
% Technical audience
x = linspace(0, 2*pi, 100);
y = sin(x);
plot(x, y);
xlabel('x (radians)');
ylabel('sin(x)');
title('Sine Function');
grid on;

% General audience
x = linspace(0, 360, 100);
y = sind(x);
plot(x, y);
xlabel('Angle (degrees)');
```

```
ylabel('Sine of Angle');
title('Sine Function');
```

By following these best practices and considering factors such as plot type, colour usage, simplification, and audience, one can create effective and compelling visualisations in MATLAB that successfully communicate the intended message (Figs. 8.13 and 8.14).

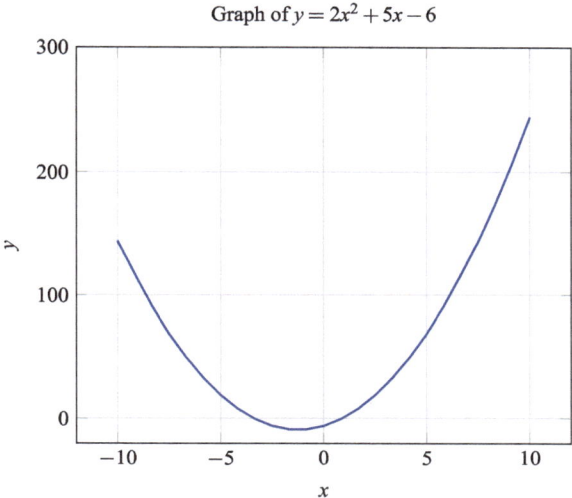

Fig. 8.13 Plot of the quadratic equation $y = 2x^2 + 5x - 6$

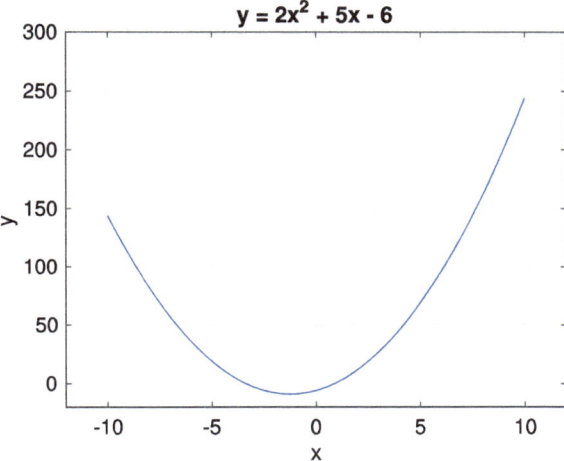

Fig. 8.14 Plotting quadratic curve

8.14 Laboratory

1. Plotting Sine and Cosine Functions

   a. Create a vector **x** with values from 0 to 2pi in increments of 0.1.
   b. Create a vector **y1** by evaluating the sine function on **x**.
   c. Create a vector **y2** by evaluating the cosine function on **x**.
   d. Use the **plot**() function to create a line plot of **y1** vs. **x**.
   e. Use the **hold on** and **plot**() functions to overlay a line plot of **y2** vs. **x** on the same figure.
   f. Add a title, x-label, y-label, and legend to the plot.

Solution:

Listing 8.89 Basic plotting functions.

```
x = 0:0.1:2*pi;
y1 = sin(x);
y2 = cos(x);

figure;
plot(x, y1);
hold on;
plot(x, y2);
title('Sine and Cosine Functions');
xlabel('x');
ylabel('y');
legend('Sine', 'Cosine');
```

2. Plotting Quadratic Curve

$$y = 2x^2 + 5x - 6 \tag{8.1}$$

To draw the curve defined by Eq. 8.1, follow these steps:

a. Let x range from $[-10, 10]$.
b. Display the grid lines on the graph.
c. Add labels to the X-axis and Y-axis.
d. Include a title for the graph.
e. Set the range of the X-axis to $[-12, 12]$.
f. Set the range of the Y-axis to $[-20, 300]$.

Listing 8.90 Plotting Quadratic Curve

```
x = -10:0.1:10;
y = 2*x.^2+5*x-6;
plot(x,y);
grid on
xlabel('x')
ylabel('y')
title(' y = 2x^2 + 5x - 6 ')
axis([-12, 12, -20, 300])
```

3. Customising Plot Properties

  a. Create a vector **x** with values from −10 to 10 in increments of 0.1.
  b. Create a vector **y** by evaluating the Gaussian function on **x**.
  c. Use the **plot()** function to create a line plot of **y** vs. **x**.
  d. Set the line color to red using the **Color** property.
  e. Set the line width to 2 using the **LineWidth** property.
  f. Set the line style to dashed using the **LineStyle** property.
  g. Add a title, x-label, and y-label to the plot.

Listing 8.91 Customising plot properties.

```
    x = -10:0.1:10;
y = exp(-(x.^2)/2) / sqrt(2*pi);

figure;
plot(x, y, 'Color', 'r', 'LineWidth', 2, 'LineStyle', '
    --');
title('Gaussian Function');
xlabel('x');
ylabel('y');
```

4. Subplots and Figure Windows

  a. Create a vector **x** with values from -pi to pi in increments of 0.1.
  b. Create a vector **y1** by evaluating the sine function on **x**.
  c. Create a vector **y2** by evaluating the cosine function on **x**.
  d. Create a new figure window with two subplots (one row, two columns) using **subplot(1, 2, 1)** and **subplot(1, 2, 2)**.
  e. In the first subplot, create a line plot of **y1** vs. **x** with a title "Sine Function".
  f. In the second subplot, create a line plot of **y2** vs. **x** with a title "Cosine Function".
  g. Add x-labels and y-labels to both subplots.

Solution:

Listing 8.92 Subplots and figure windows.

```
    x = -pi:0.1:pi;
y1 = sin(x);
y2 = cos(x);

figure;
subplot(1, 2, 1);
plot(x, y1, '-.*r');
title('Sine Function');
legend('y1 = sin(x)')  ;

xlabel('x');
ylabel('y');
```

```
subplot(1, 2, 2);
plot(x, y2, '--ob');
title('Cosine Function');
legend('y2=cos(x)') ;
xlabel('x');
ylabel('y');
```

5. 3D Plots and Surface Plots

    a. Create vectors **x** and **y** with values from -2 to 2 in increments of 0.1.

    b. Create matrices **X** and **Y** using **meshgrid(x, y)**.

    c. Create a matrix **Z** by evaluating the function $\mathbf{Z} = X^2 + Y^2$ on **X** and **Y**.

    d. Use the $surf()$ function to create a 3D surface plot of **Z**.

    e. Set the colormap to **hot** using **colormap('hot')**.

    f. Add a title, x-label, y-label, and z-label to the plot.

Solution:

Listing 8.93 3D plots and surface plots.

```
    x = -2:0.1:2;
y = -2:0.1:2;
[X, Y] = meshgrid(x, y);
Z = X.^2 + Y.^2;

figure;
surf(X, Y, Z);
colormap('hot');
title('3D Surface Plot');
xlabel('X');
ylabel('Y');
zlabel('Z');
```

6. Image Processing and Image Display

    a. Load an image file (e.g., **image.jpg**) into MATLAB using **img = imread('image.jpg')**.

    b. Display the image using **imshow(img)**.

    c. Convert the image to grayscale using **gray_img = rgb2gray(img)**.

    d. Display the grayscale image using **imshow(gray_img)**.

    e. Apply a Gaussian filter to the grayscale image using **filtered_img = imgaussfilt(gray_img, 2)**.

    f. Display the filtered image using **imshow(filtered_img)**.

    g. Add a title to the filtered image using **title('Filtered Image')**.

Solution:

Listing 8.94 Image processing and image display.

```matlab
    img = imread('image.jpg');
imshow(img);

\begin{lstlisting}[language=matlab, caption={Image
    processing and image display.}]
gray_img = rgb2gray(img);
imshow(gray_img);

filtered_img = imgaussfilt(gray_img, 2);
figure;
imshow(filtered_img);
title('Filtered Image');
```

7. Plot Data Distribution—A Case Study: Salary Analysis

   A company has recorded the salaries paid to its employees this month. The salaries, in British Pounds (£), are as follows:
   8200, 12200, 9000, 6850, 7800, 8500, 7500, 4800, 9200, 7600, 6250, 10800, 11500, 8200, 11000, 12800, 7800, 7250, 7500, 8600, 8200, 6300, 12000, 6500, 7800, 5500, 6600, 7500, 8200, 9000, 6800, 8300, 7800, 8500, 7500, 4800, 9100, 6800, 7600, 15000, 6500, 8200, 20000, 9800, 7800, 18000, 10500, 5700, 7500, 8200, 6000, 6800, 7800, 5500, 7900, 5500, 7500, 7000, 7100, 5800.
   Your task is to:

   a. Calculate the average salary of the employees.
   b. Plot a histogram of the salary distribution.
   c. Count the number of employees with a salary greater than £10,000.

   Please use MATLAB to accomplish these tasks.

Listing 8.95 Plot a histogram of the salary distribution

```matlab
% Salary data in GBP
salaries = [8200, 12200, 9000, 6850, 7800, 8500, 7500,
    4800, 9200, 7600, 6250, 10800, 11500, 8200, 11000,
    12800, 7800, 7250, 7500, 8600, 8200, 6300, 12000,
    6500, 7800, 5500, 6600, 7500, 8200, 9000, 6800,
    8300, 7800, 8500, 7500, 4800, 9100, 6800, 7600,
    15000, 6500, 8200, 20000, 9800, 7800, 18000, 10500,
    5700, 7500, 8200, 6000, 6800, 7800, 5500, 7900,
    5500, 7500, 7000, 7100, 5800];

% Calculate the average salary
averagesalary = mean(salaries);
disp(['The average salary is: ', num2str(averagesalary,
    '%.2f'), 'GBP']);

% Plot the histogram of salary distribution
figure;
```

```
histogram(salaries, 'BinWidth', 1000);
xlabel('Salary (GBP)');
ylabel('Number of Employees');
title('Salary Distribution');

% Count the number of employees with salary greater than
    10,000GBP
highsalarycount = sum(salaries > 10000);
disp(['Number of employees with salary greater than
    10,000GBP: ', num2str(highsalarycount)]);
```

The average salary is: 8590.83GBP
Number of employees with salary greater than 10,000GBP: 13

8. Scatter Plot with Randomly Generated Data
   To create a scatter plot, follow these steps:

   a. Randomly generate a 100×4 array.
   b. Draw a scatter plot with solid diamond markers.
   c. Use the first column for the horizontal coordinates.
   d. Use the second column for the vertical coordinates.
   e. Use the third column (multiplied by 50) to specify the size of the symbols.
   f. Use the fourth column (multiplied by 10) to specify the colour of the symbols
      (Fig. 8.15).

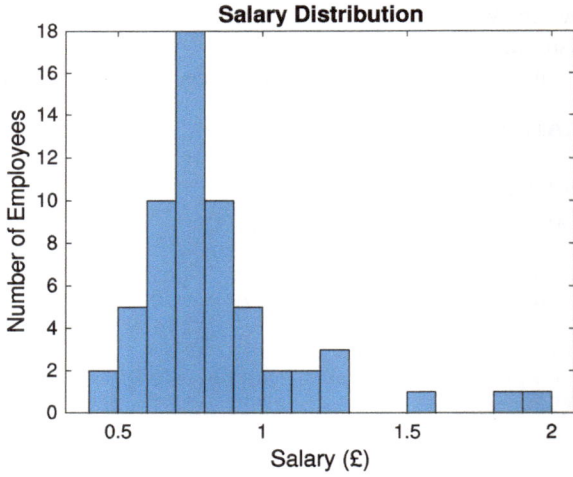

Fig. 8.15 Histogram of salary distribution

Listing 8.96 Scatter Plot with Randomly Generated Data

```
% Generate a 100x4 matrix of random numbers
data = rand(100, 4);

% Extract the first column for the x-coordinates
x = data(:, 1);

% Extract the second column for the y-coordinates
y = data(:, 2);

% Calculate the size of the markers by multiplying the
    third column by 40
sz = 40 * data(:, 3);

% Determine the colour of the markers by multiplying the
    fourth column by 10
color = 10 * data(:, 4);

% Create a scatter plot with solid diamond markers
scatter(x, y, sz, color, 'filled', 'd');
```

9. Vector Diagrams

   Draw two-dimensional and three-dimensional vector diagrams for the equation $z = y^2 - x^2$.

Listing 8.97 Vector Diagrams

```
% Create a grid of points
[x, y] = meshgrid(-3:0.5:3, -3:0.5:3);

% Calculate z using the equation z = y^2 - x^2
z = y.^2 - x.^2;

% Compute gradients for the vector field
[u, v] = gradient(z, 2, 2);

% Create a 2D quiver plot
figure;
quiver(x, y, u, v);
xlabel('x');
ylabel('y');
title('2D Quiver Plot for z = y^2 - x^2');
grid on;
axis equal;

% Compute surface normals for the 3D vector field
[u, v, w] = surfnorm(z);

% Create a 3D quiver plot
figure;
quiver3(x, y, z, u, v, w);
xlabel('x');
```

```
ylabel('y');
zlabel('z');
title('3D Quiver Plot for z = y^2 - x^2');
grid on;
```

10. Volume Visualisation—Three-Dimensional Volume and Slice Planes
    Create the three-dimensional volume $V = x^2 + y^2 + z^2 - 25$ and plot the slice
    planes orthogonal to the x-axis at $x = -3$, $x = 0$, and $x = 3$, as well as the slice
    planes orthogonal to the z-axis at $z = -4$ and $z = 0$.
    Create the three-dimensional volume $V = x^2 + y^2 + z^2 - 25$ and plot the slice
    planes orthogonal to the x-axis at $x = -3$, $x = 0$, and $x = 3$, as well as the slice
    planes orthogonal to the z-axis at $z = -4$ and $z = 0$.

Listing 8.98 Three-Dimensional Volume and Slice Planes

```
% Define the range for x, y, and z
x = -5:0.5:5;
y = -5:0.5:5;
z = -5:0.5:5;

% Create a meshgrid for x, y, and z
[X, Y, Z] = meshgrid(x, y, z);

% Calculate the volume V using the equation V = x^2 + y
    ^2 + z^2 - 25
V = X.^2 + Y.^2 + Z.^2 - 25;

% Create a figure for visualisation
figure;

% Plot the slices orthogonal to the x-axis at x = -3, 0,
    and 3
slice(X, Y, Z, V, [-3, 0, 3], [], []);
hold on;

% Plot the slices orthogonal to the z-axis at z = -4 and
    0
slice(X, Y, Z, V, [], [], [-4, 0]);

% Enhance visualisation
xlabel('x');
ylabel('y');
zlabel('z');
title('Slice Planes of the Volume V = x^2 + y^2 + z^2 -
    25');
colorbar;
shading interp; % Smooth shading
grid on;
axis equal;
view(3); % 3D view
hold off;
```

Explanation

a. **Range Definition**:

- 'x = −5:0.5:5;', 'y = −5:0.5:5;', 'z = −5:0.5:5;': Defines the range for the 'x', 'y', and 'z' axes from −5 to 5 with a step size of 0.5.

b. **Meshgrid Creation**:

- '[X, Y, Z] = meshgrid(x, y, z);': Creates a 3D grid of points.

c. **Volume Calculation**:

- '

$$V = X.^2 + Y.^2 + Z.^2 - 25$$

': Computes the volume 'V'.

d. **Figure Creation**:

- 'figure;': Opens a new figure window.

e. **Slice Plotting**:

- 'slice(X, Y, Z, V, [−3, 0, 3], [], []);': Plots slices orthogonal to the 'x'-axis at 'x = −3', 'x = 0', and 'x = 3'.
- 'slice(X, Y, Z, V, [], [], [−4, 0]);': Plots slices orthogonal to the 'z'-axis at 'z = −4' and 'z = 0'.

f. **Visualisation Enhancement**:

- 'xlabel', 'ylabel', 'zlabel': Labels the axes.
- 'title': Adds a title.
- 'colorbar': Displays a color bar.
- 'shading interp': Applies smooth shading.
- 'grid on': Enables the grid.
- 'axis equal': Sets equal scaling for all axes.
- 'view(3)': Sets the view to a 3D perspective.
- 'hold off;': Releases the plot hold.

11. Draw a Flower in MATLAB

    To draw a flower in MATLAB, you can use polar coordinates to create a simple flower-like pattern. Below is an example MATLAB code to draw a flower with 8 petals.

```
% Define the number of petals
num\_petals = 8;

% Define the theta range
theta = linspace(0, 2*pi, 1000);
```

```matlab
% Define the radius for the flower petals
r = cos(num\_petals * theta);

% Create a new figure
figure;

% Plot the flower using polar coordinates
polarplot(theta, r, 'LineWidth', 2);

% Enhance the visualisation
title(['Flower with ', num2str(num\_petals), ' Petals'])
    ;
ax = gca;
ax.ThetaTickLabel = {}; % Remove theta tick labels
ax.RTickLabel = {}; % Remove radius tick labels
ax.GridAlpha = 0.3; % Make grid lines less prominent
ax.RAxisLocation = 0; % Set the starting angle of the
    radial axis

% Set the aspect ratio to ensure the plot is circular
axis equal;
```

Explanation

a. **Number of Petals**:

- 'num_petals = 8;': Defines the number of petals for the flower.

b. **Theta Range**:

- 'theta = linspace(0, 2*pi, 1000);': Generates 1000 points between 0 and 2π.

c. **Radius Definition**:

- 'r = cos(num_petals * theta);': Defines the radius as a cosine function to create the petal shapes.

d. **Figure Creation**:

- 'figure;': Opens a new figure window.

e. **Plotting the Flower**:

- 'polarplot(theta, r, 'LineWidth', 2);': Plots the flower using polar coordinates with a specified line width.

f. **Enhancing the visualisation**:

- 'title': Adds a title to the plot.
- 'ax = gca;': Gets the current axes.

- 'ax.ThetaTickLabel = ;', 'ax.RTickLabel = ;': Removes the theta and radius tick labels for a cleaner look.
- 'ax.GridAlpha = 0.3;': Makes the grid lines less
- 'ax.GridAlpha = 0.3;': Makes the grid lines less prominent by setting their transparency.
- 'ax.RAxisLocation = 0;': Sets the starting angle of the radial axis to 0 degrees.
- 'axis equal;': Ensures the aspect ratio is equal, making the plot circular.

12. Fourier Series Expansion—a 3D Data Visualisation

3D data visualisation provides learners with intuitive imagery and has a wide range of applications. The previous post mainly introduced the drawing of 3D surfaces, involving 3D surface plotting functions such as meshgrid, mesh, meshc, surf, surfl, and surfc.

3D line plots aim to display the distribution of curves in three-dimensional space, using the MATLAB function plot3. The difference from the two-dimensional line drawing function plot is that coordinate vectors need to be set in the x, y, and z directions; other plotting attributes are essentially the same for both.

Similar functions include the 3D scatter plot scatter3 and the 3D stem plot stem3.

Listing 8.99 Periodic Signal and Fourier Series Expansion in MATLAB (Fig. 8.16)

```
% MATLAB code for Periodic Signal and Fourier Series
    Expansion
clear all;
close all;

%---------------------%
numTerms     = 15; % Number of terms in the series
numPoints    = 512; % Number of points per period
timeConstant = 4; % Total time span
T            = timeConstant / 2;

t1 = linspace(-T/2, T/2, numPoints);
t2 = linspace(T/2, timeConstant - T/2, numPoints);
t3 = [(t1-timeConstant)';
      (t2-timeConstant)';
       t1';
       t2';
      (t1+timeConstant)'];

%--------------------% Construct periodic signals
    %--------------------%
s = zeros(5 * numPoints, 1);
s(1:numPoints) = 1;
s(2*numPoints + 1:3*numPoints) = 1;
s(4*numPoints + 1:5*numPoints) = 1;

y = zeros(numTerms + 1, length(s));
y(numTerms + 1, :) = s - 0.5;
```

```matlab
% Plot setup
figure;
hold on;
grid on;
axis([-2, numTerms + 1, -timeConstant - 1, timeConstant
    + 1, -1, 2]);
set(gca,'XTick', -timeConstant - 1:2:numTerms);
set(gca,'YTick', -timeConstant - 1:1:timeConstant + 1);
set(gca,'ZTick', -1:0.5:2);
set(gcf,'Color','White');

title('Fourier Series Expansion','FontSize',15);
xlabel('Frequency','Rotation', 15);
ylabel('Time','Rotation', -10);
zlabel('Amplitude/Magnitude');

view(-49, 23);

% Period square wave plot
plot3(t3 - (t3 + 2), t3, y(numTerms + 1, :),
 'LineWidth', 2);

fsamp = 1028;
f     = linspace(1, numTerms + 1, fsamp);
A     = 0.5; % Coefficient
freq  = 1:1:numTerms;

% Frequency components plot
plot3(f, timeConstant + f - f + 1, A * sinc(A * f) * 5,
    'LineWidth', 3);

% Magnitude of frequency components plot
mag = A * sinc(A * freq) * 5;
h   = stem3(freq, timeConstant + freq - freq + 1, mag, '
    filled', 'LineWidth', 3);
set(h, 'Marker', 'o', 'MarkerFaceColor', 'g');

Harmonicx   = A * ones(size(t3)); % Harmonic signals
SynthesisX  = A * ones(size(t3)); % Synthesis signals

% Construct the Fourier series
for k = 1:numTerms
    Harmonicx   = 2 * A * sinc(A * k) * cos(2 * pi * t3
        * k / timeConstant);
    SynthesisX  = SynthesisX + Harmonicx;
    y(k, :)     = Harmonicx;
    plot3(k + t3 - t3, t3, y(k, :), 'LineWidth', 1.5);
end

% Final synthesis signal plot
plot3(k + 1 + t3 - t3, t3, SynthesisX - 0.5,
 'LineWidth', 2);
```

Fourier Series Expansion

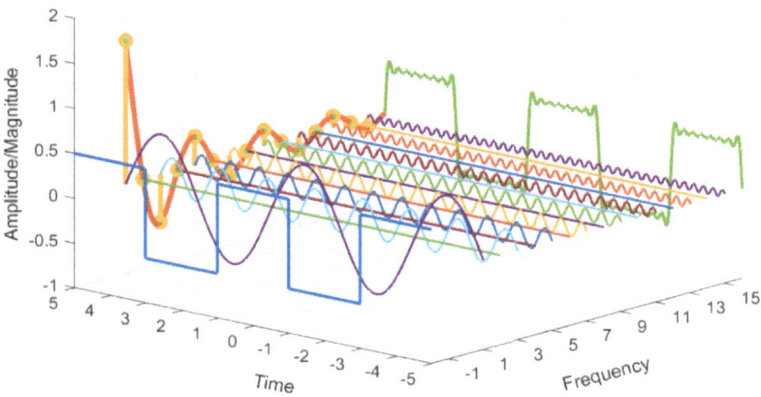

Fig. 8.16 Plot FFT in 3D view

These lab works and exercises cover various aspects of graphics and data visuali-sation in MATLAB, including basic plotting functions, customising plot properties, working with subplots and figure windows, creating 3D plots and surface plots, and image processing and display. The step-by-step instructions, along with the provided code snippets and expected outputs, should help students reinforce their understanding of these concepts through hands-on practice (Fig. 8.16).

8.15 Problems

1. Create a line plot of the function $y = x^2 - 2x + 1$ over the range $-2 \leq x \leq 4$. Annotate the plot with a title, x-label, y-label, and a legend.
2. Plot the vectorized sine function $y = \sin(x)$ over one period $0 \leq x \leq 2\pi$. Customise the line color and style.
3. Generate a scatter plot of random (x,y) points within the range $0 \leq x, y \leq 10$. Add gridlines and use different marker styles and colors to distinguish clusters.
4. Create a bar chart to Visualise student scores in five different courses. Customise the bar colors, add data labels, and rotate the x-tick labels for better readability.
5. Load a image dataset (e.g., from MATLAB's sample data) and display the image. Adjust the colormap and add a descriptive title.
6. Generate a 3D surface plot of the function $z = x^2 + y^2$ over the ranges $-2 \leq x, y \leq 2$. Add appropriate view angles, lighting, and color the surface by z-values.

7. Plot multiple line graphs on the same axes to compare growth trends for different populations over time. Use legends and line styles to differentiate the plots.

8. Load a real-world dataset (e.g., from a CSV file) and create a histogram to Visualise the distribution of a chosen variable. Experiment with different bin widths and overlay a kernel density estimate.

9. Create a filled contour plot of a 2D Gaussian function. Add a colour bar and label the contour levels.

10. Animate a bouncing ball simulation by plotting the ball's position at each time step. Control the animation speed and add a trail effect.

This set of problems covers various plot types (line, scatter, bar, image, 3D, contour), customizations (titles, labels, legends, colors, styles), and advanced topics like animations. The problems involve creating plots from functions, vectors, matrices and real-world data to ensure a comprehensive understanding of MATLAB visualisation capabilities.

8.16 Summary

This chapter covers the key aspects of creating effective data visualisations in MATLAB, from basic plotting to advanced customisation techniques and real-world applications. The sections are organised in a logical progression, allowing readers to build up their skills.

- This chapter covered various **data visualisation** techniques using **MATLAB**, including:

  - Displaying and manipulating **images**
  - Creating **3D surface plots** with customizations
  - Plotting **multiple line graphs** with legends
  - Generating **histograms** and **kernel density estimates**
  - Producing **filled contour plots** with labeled levels
  - Animating **dynamic simulations**, such as a bouncing ball.

- The step-by-step approaches, **sample codes**, and **extensions** provided hands-on examples for mastering these visualisation techniques.

- Concepts like **colormaps**, **view angles**, **lighting**, **legends**, **bin widths**, and **contour labeling** were explored.

- The chapter emphasized the importance of effective **data visualisation** in understanding patterns, trends, and distributions within datasets.

For Undergraduate (UG) Students:

This chapter serves as an excellent introduction to data visualisation using MATLAB. By working through the provided examples and exercises, you will develop essential skills in creating various types of plots, customising their appearance, and interpreting the Visualised data. These techniques are invaluable for exploratory data analysis, communicating research findings, and gaining insights from complex datasets across multiple disciplines.

For Postgraduate (PG) Students and Professional Researchers or Engineers:

The chapter equips you with advanced **data visualisation** tools and techniques that are crucial for research and professional work. The ability to effectively present data through compelling visuals is paramount in scientific communication and decision-making processes. The concepts covered, such as **3D surface plots**, **kernel density estimation**, and **contour plots**, empower you to analyse and interpret intricate datasets more effectively. Additionally, the **animation** capabilities demonstrated provide a powerful means to Visualise and comprehend dynamic systems and simulations.

References

1. MathWorks, Data Distribution Plots. [Online]. https://www.mathworks.com/help/matlab/pie-charts-bar-plots-and-histograms.html. [Accessed: Feb. 17, 2024]
2. MathWorks, "Vector Fields and Volume Visualization in MATLAB," https://www.mathworks.com/help/matlab/vector-fields-and-volume-visualization.html.[Accessed: Feb. 17, 2024]
3. MathWorks, Types of MATLAB Plots. [Online]. https://www.mathworks.com/help/matlab/creating_plots/types-of-matlab-plots.html. [Accessed: Feb. 17, 2024]
4. MathWorks, Create, Store, and Open MATLAB Figures. [Online]. https://www.mathworks.com/help/sltest/ug/create-store-and-access-custom-figures.html. [Accessed: Feb. 17, 2024]
5. MathWorks, "Graphics," [Online]. https://www.mathworks.com/help/matlab/graphics.html. [Accessed: Feb. 17, 2024]
6. MathWorks, "MATLAB Plot Gallery," [Online]. https://www.mathworks.com/products/matlab/plot-gallery.html. [Accessed: Feb. 17, 2024]
7. MathWorks, tiledlayout. [Online]. https://uk.mathworks.com/help/matlab/ref/tiledlayout.html. [Accessed: Feb. 17, 2024]
8. MathWorks, LineSpec - Line style, marker, and color. [Online]. https://uk.mathworks.com/help/matlab/ref/plot.html#btzitot_sep_mw_3a76f056-2882-44d7-8e73-c695c0c54ca8. [Accessed: Feb. 17, 2024]
9. MathWorks, "MATLAB Fundamentals," [Online]. https://www.mathworks.com/help/matlab/. [Accessed: Feb. 17, 2024]
10. MathWorks, Types of MATLAB Plots. [Online]. https://uk.mathworks.com/help/matlab/creating_plots/types-of-matlab-plots.html. [Accessed: Feb. 17, 2024]
11. MathWorks, "MATLAB Graphics Documentation," 2024. [Online]. https://www.mathworks.com/help/matlab/graphics.html
12. MathWorks, "Property Inspector," in MATLAB Documentation, 2024. [Online]. https://www.mathworks.com/help/matlab/ref/inspect.html
13. MathWorks, "Creating Specialized Charts with MATLAB Object-Oriented Programming," in MATLAB Documentation, 2024. [Online]. https://uk.mathworks.com/company/technical-articles/creating-specialized-charts-with-matlab-object-oriented-programming.html
14. MathWorks, "Visualize the distribution of data," [Online]. https://uk.mathworks.com/discovery/data-distribution.html. [Accessed: Oct. 22, 2023]
15. Scott DW (2015) Multivariate density estimation: theory, practice, and visualization, 2nd edn. Wiley, Hoboken, NJ, USA

16. McGill R, Tukey JW, Larsen WA (1978) Variations of box plots. Am Stat. 32(1):12–16
17. Hintze JL, Nelson RD (1998) Violin plots: a box plot-density trace synergism. Am Stat 52(2):181–184
18. MathWorks File Exchange, "Violin Plot," [Online]. https://www.mathworks.com/matlabcentral/fileexchange/45134-violin-plot. [Accessed: Oct. 22, 2023]
19. Chambers JM, Cleveland WS, Kleiner B, Tukey PA (1983) Graphical methods for data analysis. Wadsworth International Group, Belmont, CA, USA
20. Cleveland WS, McGill R (1984) Graphical perception: theory, experimentation, and application to the development of graphical methods. J Am Stat Assoc 79(387):531–554
21. Heer J, Shneiderman B (2012) Interactive dynamics for visual analysis. Commun ACM 55(4):45–54
22. Tuegel EJ, Ingraffea AR, Eason TG, Spottswood SM (2011) Reengineering aircraft structural life prediction using a digital twin. Int J Aerosp Eng 2011:1–14
23. MathWorks, "Data Visualization," [Online]. https://uk.mathworks.com/discovery/data-visualization.html. [Accessed: Oct. 22, 2023]
24. Chapra SC (2018) Applied numerical methods with MATLAB for engineers and scientists, 4th edn. McGraw-Hill Education, New York, NY, USA
25. Attaway S (2017) MATLAB: a practical introduction to programming and problem solving, 4th edn. Butterworth-Heinemann
26. Smith III JO (2013) Introduction to digital filters with audio applications. W3K Publishing, San Francisco, CA, USA
27. Quart R (2016) Data visualization with MATLAB. CRC Press, Boca Raton, FL, USA
28. Tufte ER (2001) The visual display of quantitative information, 2nd edn. Graphics Press, Cheshire, CT, USA
29. Few S (2012) Show me the numbers: designing tables and graphs to enlighten, 2nd edn. Analytics Press, Burlingame, CA, USA
30. MathWorks, "Visualize 2-D Vector Fields," [Online]. https://www.mathworks.com/help/matlab/ref/quiver.html. [Accessed: Oct. 22, 2023]
31. Corke P (2017) Robotics, vision and control: fundamental algorithms in MATLAB, 2nd edn. Springer, Cham, Switzerland
32. Fuller A, Fan Z, Day C, Barrett C (2020) Digital twin: enabling technologies, challenges and open research. IEEE Access 8:108952–108971
33. McCormick BH, DeFanti TA, Brown MD (1987) Visualization in scientific computing. Comput Graph 21(6):1–14
34. MathWorks, "Volume Visualization," [Online]. https://www.mathworks.com/help/matlab/volume-visualization.html. [Accessed: Oct. 22, 2023]
35. Lorensen WE, Cline HE (1987) Marching cubes: a high resolution 3D surface construction algorithm. ACM SIGGRAPH Comput Graph 21(4):163–169
36. MathWorks, "isocaps," [Online]. https://www.mathworks.com/help/matlab/ref/isocaps.html. [Accessed: Oct. 22, 2023]
37. Tao F, Cheng J, Qi Q, Zhang M, Zhang H, Sui F (2018) Digital twin-driven product design, manufacturing and service with big data. Int J Adv Manuf Technol 94(9):3563–3576
38. Versteeg HK, Malalasekera W (2007) An introduction to computational fluid dynamics: the finite volume method, 2nd edn. Pearson Education, Harlow, UK
39. Thrun S, Burgard W, Fox D (2005) Probabilistic robotics. MIT Press, Cambridge, MA, USA
40. MathWorks, "Basic Display," MATLAB Documentation, [Online]. https://uk.mathworks.com/help/images/basic-display-and-exploration.html. [Accessed: Oct. 22, 2023]
41. MathWorks, "Creating Animations in MATLAB," https://www.mathworks.com/help/matlab/creating-animations.html, accessed on Feb. 17, 2024
42. MathWorks, "Graphics," MATLAB Documentation, [Online]. https://www.mathworks.com/help/matlab/graphics.html. [Accessed: Oct. 22, 2023]
43. Few S (2009) Now you see it: simple visualization techniques for quantitative analysis. Analytics Press, Burlingame, CA, USA

44. MathWorks, "Advanced Plotting Functions," MATLAB Documentation, [Online]. https://www.mathworks.com/help/matlab/creating_plots/advanced-plotting-functions.html. [Accessed: Oct. 22, 2023]
45. MathWorks, "Interactively Explore Plotted Data," MATLAB Documentation, [Online]. https://www.mathworks.com/help/matlab/creating_plots/interactively-explore-plotted-data.html. [Accessed: Oct. 22, 2023]
46. MathWorks, "Visualize and Explore Big Data," MATLAB Documentation, [Online]. https://www.mathworks.com/help/matlab/big-data.html. [Accessed: Oct. 22, 2023]
47. MathWorks, "Tall Arrays for Out-of-Memory Data," MATLAB Documentation, [Online]. https://www.mathworks.com/help/matlab/tall-arrays.html. [Accessed: Oct. 22, 2023]
48. MathWorks, "Real-Time Data Streaming and Visualization," MATLAB Documentation, [Online]. https://www.mathworks.com/help/matlab/real-time-data-streaming.html. [Accessed: Oct. 22, 2023]
49. Lee J, Bagheri B, Kao H-A (2015) A cyber-physical systems architecture for industry 4.0-based manufacturing systems. Manuf Lett 3:18–23
50. KentWilliams W (2018) Visualizing uncertainty. Commun ACM 61(4):5–5
51. Gal Y, Ghahramani Z (2016) Dropout as a Bayesian approximation: representing model uncertainty in deep learning. In: Proceedings of the 33rd international conference on machine learning, New York, NY, USA, Jun. 2016, pp 1050–1059
52. MathWorks, "Mapping Toolbox," MATLAB Documentation, [Online]. https://www.mathworks.com/help/map/. [Accessed: Oct. 22, 2023]
53. Batty M (2018) Digital twins. Environ Plann B: Urban Anal City Sci 45(5):817–820
54. Few S (2012) Show me the numbers: designing tables and graphs to enlighten. Analytics Press
55. Stone M (2006) Choosing colors for data visualization. Business Intelligence Network
56. Knafli CN (2015) Storytelling with data: a data visualization guide for business professionals. Wiley
57. Kelleher C, Wagener T (2011) Ten guidelines for effective data visualization in scientific publications. Environ Model Softw 26(6):822–827

Chapter 9
Programming and Algorithm Development

Chapter Learning Outcomes

Upon successful completion of this chapter, you should be able to:

- Understand the fundamental concepts of **programming** and **algorithm development** in MATLAB.
- Write and execute **scripts** and **functions** in MATLAB to solve computational problems.
- Implement **control structures** such as loops and conditional statements for program flow control.
- Utilise **data structures** like vectors, matrices, and cells to organize and manipulate data effectively.
- Develop and apply **algorithms** to solve real-world problems using MATLAB's programming capabilities.
- Employ **debugging** techniques to identify and resolve errors in MATLAB code.

Chapter Key Words

- **Programming**: The process of designing, writing, testing, and maintaining instructions or code that direct a computer to perform specific tasks or operations.
- **Algorithm**: A set of well-defined, step-by-step instructions or rules designed to solve a particular problem or accomplish a specific task within a finite amount of time.
- **Script**: A plain text file in MATLAB containing a sequence of commands and statements that are executed sequentially without requiring user input or returning outputs.
- **Function**: A reusable block of code in MATLAB that performs a specific task and can accept input arguments and return output values, enabling modular and organized programming.

© The Author(s) 2025
Y. Chen and L. Huang, *MATLAB Roadmap to Applications*,
https://doi.org/10.1007/978-981-97-8788-3_9

- **Control Structure**: Programming constructs that allow for the conditional execution of code blocks or repetitive execution of statements based on certain conditions or criteria.
- **Data Structure**: A organized way of storing and managing data in computer memory, such as vectors, matrices, or cells, enabling efficient data manipulation and computation.
- **Debugging**: The process of identifying, locating, and resolving errors or defects in computer programs to ensure correct and intended behavior.
- **Code Reuse**: The practice of using existing code modules (such as functions) in multiple parts of a program or across different programs. It promotes efficiency and reduces code duplication.
- **Maintainability**: The ease with which code can be modified, updated, and debugged without introducing errors. Well-organized and modular code improves maintainability.
- **Computational Problems**: Challenges or tasks that involve mathematical calculations, data manipulation, or algorithmic operations. MATLAB can be used to solve a wide range of computational problems efficiently.

9.1 Introduction to Programming

- Overview of **programming**
  Programming is the process of designing, writing, testing, and maintaining code that instructs computers to perform specific tasks. It involves translating real-world problems into a set of instructions that a computer can understand and execute. **Programming** is a fundamental skill in various fields, including computer science, engineering, data analysis, and scientific research.
- Importance of **algorithms** and **computational thinking**
  Algorithms are a set of well-defined instructions or a sequence of steps that solve a particular problem or perform a specific task. They are the foundation of programming and play a crucial role in computational thinking, which is the process of formulating problems and designing solutions that can be effectively carried out by a computer.
  The importance of algorithms and computational thinking in programming cannot be overstated. Developing efficient and effective algorithms is essential for solving complex problems, Optimising performance, and ensuring the reliability and scalability of software systems.
- MATLAB as a **programming environment**
  MATLAB is a high-level programming language and a powerful numerical computing environment widely used in academia and industry. It provides a comprehensive set of tools for data analysis, algorithm development, and visualisation, making it an ideal environment for scientific and engineering applications.
  MATLAB offers a user-friendly interface, extensive built-in functions, and a flexible scripting language that allows for rapid prototyping and iterative development.

Its matrix-based syntax and powerful visualisation capabilities make it well-suited for tasks such as numerical computation, data analysis, algorithm implementation, and simulation.

The basic syntax for a MATLAB script is:

Listing 9.1 MATLAB Script Syntax

```
% This is a comment

% Variable assignment
x = 5;
y = 10;

% Arithmetic operation
z = x + y;

% Print output
disp(['The result is: ', num2str(z)]);
```

Here's an example of a simple MATLAB script that calculates the area of a circle:

Listing 9.2 Calculate Circle Area

```
% Prompt the user for the radius
radius = input('Enter the radius of the circle: ');

% Calculate the area
area = \pi * radius^2;

% Display the result
disp(['The area of the circle with radius ', num2str(
    radius), ' is ', num2str(area)]);
```

Another example of a MATLAB script that sorts an array of numbers:

Listing 9.3 Sort Array

```
% Create an array of numbers
numbers = [5, 2, 8, 1, 9];

% Sort the array in ascending order
sorted_numbers = \sort(numbers);

% Display the original and sorted arrays
disp(['Original array: ', \mat2str(numbers)]);
disp(['Sorted array: ', \mat2str(sorted_numbers)]);
```

9.2 Algorithms

Algorithms are a fundamental concept in programming and computer science. They are a set of well-defined, step-by-step instructions that are designed to solve a specific problem or perform a particular task. Algorithms are the foundation of any

software program, and their quality and efficiency directly impact the performance
and reliability of the program. When developing algorithms, it is essential to follow
a structured approach to ensure that the algorithm is correct, efficient, and maintainable. Here are some steps to consider when mapping an algorithm to code:

- **Understand the problem**: Before attempting to write an algorithm, it is crucial
  to have a clear understanding of the problem you are trying to solve. This includes
  identifying the inputs, expected outputs, and any constraints or requirements that
  need to be met.
- **Understand the problem**: Before attempting to write an algorithm, it is crucial
  to have a clear understanding of the problem you are trying to solve. This includes
  identifying the inputs, expected outputs, and any constraints or requirements that
  need to be met.
- **Develop a conceptual solution**: Once you have a clear understanding of the problem, develop a conceptual solution by breaking it down into smaller, more manageable steps. This can be done using techniques like flowcharts, pseudocode, or
  natural language descriptions.
- **Choose appropriate data structures**: Select the appropriate data structures, such
  as arrays, lists, or trees, to represent the input, intermediate results, and output of
  the algorithm. The choice of data structures can significantly impact the efficiency
  and performance of the algorithm.
- **Implement the algorithm**: Translate the conceptual solution into code using a
  programming language of your choice. Follow best practices for coding, such as
  using meaningful variable names, commenting your code, and adhering to coding
  standards.
- **Test and debug**: Test your implementation thoroughly to ensure that it produces
  the expected results for various input scenarios. Debug and fix any errors or issues
  that you encounter during testing.
- **Optimise and refine**: Look for opportunities to optimise and refine your algorithm to improve its efficiency, readability, and maintainability. This may involve
  techniques like code refactoring, algorithm analysis, or the use of more efficient
  data structures or algorithms.
- **Document the algorithm**: Properly document your algorithm, including a description of its purpose, inputs, outputs, and any assumptions or limitations. Good documentation will make it easier for others (or your future self) to understand and
  maintain the algorithm.

9.3   From Algorithm to Programming

- Define the algorithm using pseudocode or flowchart notation
- Identify the programming language you will be using
- Translate each step of the algorithm into code using the chosen programming
  language

- Use appropriate control structures (if-else, loops, etc.) to implement the logic of the algorithm
- Test the code thoroughly to ensure that it produces the expected results
- Optimise the code if necessary to improve its efficiency or readability

It is important to note that the process of developing algorithms is often iterative, and you may need to revisit and refine your solution as you gain more insight or encounter new requirements.

Listing 9.4 Bubble Sort Algorithm

```
function sorted_array = bubbleSort(array)
% Get the length of the array
n = length(array);
% Perform bubble sort
for i = 1:n-1
    % Flag to track if any swaps occurred
    swapped = false;

    % Loop through the unsorted portion of the array
    for j = 1:n-i
        % If the current element is greater than the
          next
        if array(j) > array(j+1)
            % Swap the elements
            temp = array(j);
            array(j) = array(j+1);
            array(j+1) = temp;
            swapped = true; % Swaps occurred
        end
    end

    % If no swaps occurred, the array is sorted
    if ~swapped
        break;
    end
end

% Return the sorted array
sorted_array = array;

end
```

Listing 9.5 Linear Search Algorithm

```
function index = linearSearch(array, target)
% Get the length of the array
n = length(array);
% Initialize the index to -1 (not found)
index = -1;

% Loop through the array
for i = 1:n
```

```
      % If the current element matches the target
      if array(i) == target
          % Store the index and exit the loop
          index = i;
          break;
      end
end

end
```

9.4 Programme Organisation

Organising programmes in MATLAB involves structuring code to enhance readability and functionality. Here are several steps and practices:

1. Use Functions: Break down your code into **functions** to perform specific tasks. A function in MATLAB is defined using the function keyword, followed by the output variables, the function name, and input variables in parentheses. Functions should be saved in separate files with the same name as the function.
2. Script Files: Use **script files** for sequences of commands to run together. Scripts do not accept input arguments or return output arguments but can access and modify the workspace.
3. Local Functions: Include **local functions** within script and function files that are only accessible within the file.
4. Subfunctions: Create **subfunctions** within the same file to be used by the primary function. These are functions that appear in the same file as the primary function, after the primary function's end statement. They are only accessible by the primary function and other subfunctions in the same file.
5. Private Functions: Store functions that should be hidden from other functions in a subdirectory named private. They are accessible only to functions in the parent directory. Store functions in a **private** subdirectory to hide them from other functions.
6. Namespaces: Use namespaces to create a scope for functions and classes that prevents name clashes.
7. Code Sections: Use code sections within scripts to organise and run blocks of code independently. This can be done by using %% to start a new section.
8. MATLAB Path: Manage the MATLAB **path** effectively so MATLAB can locate your scripts and functions. Use the MATLAB path effectively so that MATLAB can locate your functions and scripts when they are called.
9. Comments and Documentation: Comment your code using the % symbol for single-line comments and { % ... % } for block comments. Use the help and doc functions to provide documentation for your functions.
10. Version Control: Use version control systems like Git to manage changes and collaborate with others.

Programme Organisation refers to the structured approach of organising and managing MATLAB code to enhance readability, maintainability, and reusability.

Key concepts and terms related to programme organisation in MATLAB include:

- **Functions**: These are segments of code that perform specific tasks and can be called and reused multiple times.
- **Scripts**: MATLAB files that contain a sequence of commands executed in order.
- **Modularity**: The practice of breaking down a programme into smaller, self-contained modules or functions to simplify development and maintenance.
- **Functionality separation**: The division of code into distinct functions, each responsible for a specific task or operation.
- **Function headers**: A section at the beginning of a function that includes the function name, input arguments, and output variables.
- **Function calling**: The process of invoking a function within another function or script to execute a specific task.
- **Variable scope**: The accessibility and visibility of variables within different parts of a MATLAB programme.
- **Local variables**: Variables defined within a specific function or script and accessible only within that scope.
- **Global variables**: Variables defined outside of any function, making them accessible across different functions or scripts.
- **Comments**: Text annotations within code that provide explanations, clarify intent, or document code functionality.
- **Code indentation**: The practice of aligning code blocks to visually represent their hierarchical structure.
- **Code documentation**: The process of adding descriptive comments and annotations to clarify code functionality and usage.
- **Debugging**: The process of identifying and fixing errors or bugs in the code to ensure correct execution.
- **Code versioning**: The practice of managing and tracking different versions of code to facilitate collaboration and code maintenance.
- **MATLAB editor**: The integrated development environment (IDE) provided by MATLAB for writing, editing, and running MATLAB code.

By understanding and applying these concepts, MATLAB programmers can effectively organise and manage their code, resulting in more efficient development, easier maintenance, and improved collaboration.

When discussing MATLAB programming, it is important to be familiar with the following terms:

- **Function Handle**: A MATLAB data type that represents a function. Function handles allow you to call functions indirectly and pass functions as arguments.
- **Path Management**: The process of adding and removing directories from the MATLAB search path, which affects the ability to locate scripts and functions.
- **Workspace**: The set of variables that are currently loaded and can be accessed from the command window or your scripts and functions.
- **M-file**: The file extension (.m) for MATLAB script and function files.

- **Vectorization**: A method of programming that works with vectors and matrices in block operations, which is more efficient than scalar computation in MATLAB.

9.5 Control Flow in MATLAB

Control flow structures are essential in programming, enabling specific code execution based on conditions or repeated conditions. MATLAB, renowned for numerical computations and programming, offers various control flow constructs critical for effective coding. MATLAB provides several mechanisms as Types of Control Flow Constructs in MATLAB:

- **Conditional Statements**
  Conditional statements execute parts of code based on logical conditions. MATLAB uses:

  – **if, elseif, and else**—Execute based on the first true condition (Fig. 9.1).
  – **switch and case**—Execute code for the matching case expression (Fig. 9.2).

  Example:

  ```
  x = 5;
  if x > 0
      disp('x is positive');
  elseif x == 0
      disp('x is zero');
  else
      disp('x is negative');
  end
  ```

- **Loop Statements**
  Loops in MATLAB allow repeating actions:

  – **for loops** - Execute a set number of times.
  – **while loops** - Execute as long as a condition remains true.

  Examples:

  ```
  % Example of a for loop
  for i = 1:5
      disp(['Iteration: ', num2str(i)]);
  end

  % Example of a while loop
  count = 1;
  while count <= 5
      disp(['Count: ', num2str(count)]);
      count = count + 1;
  end
  ```

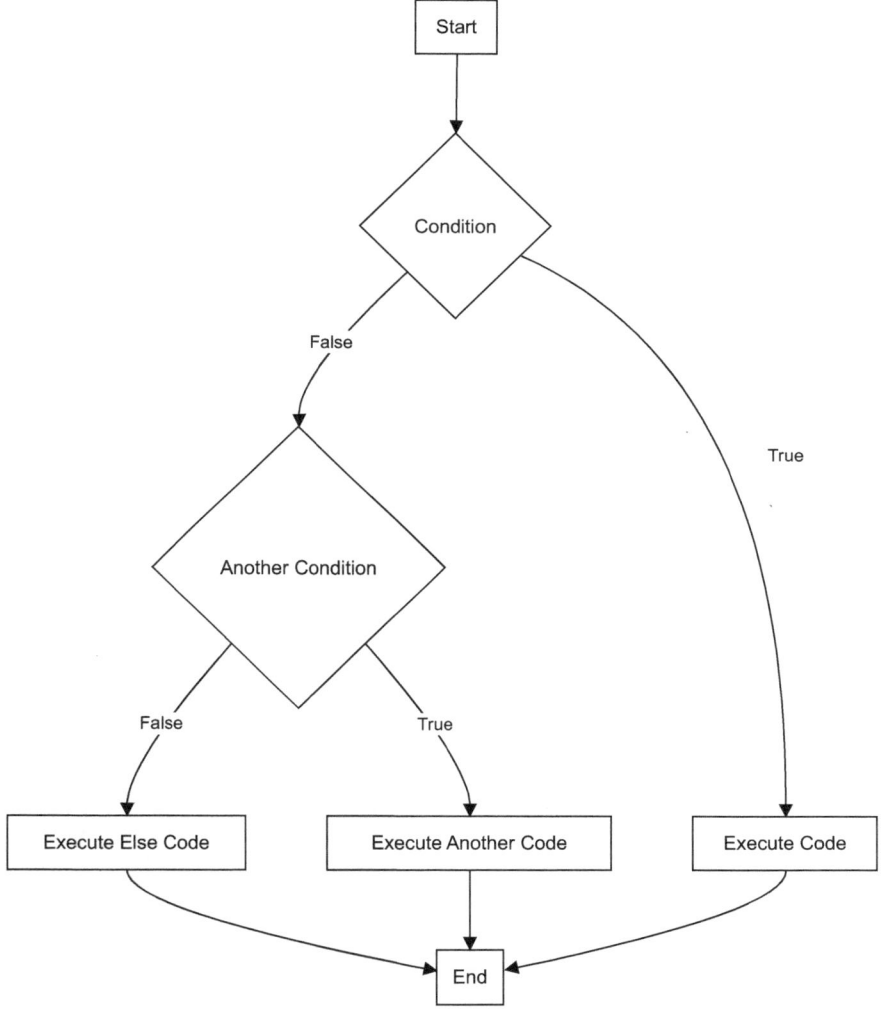

Fig. 9.1 Conditional statements: if statement

- **Jump Statements**
  Control the flow by altering execution:

  - **break** - Exit the loop.
  - **continue** - Skip the loop iteration.
  - **return** - Exit from the current function.

Example:

```
for i = 1:10
    if i == 6
        break; % Exit loop when i is 6
    end
    disp(['i = ', num2str(i)]);
end
```

Effective use of control flow structures enhances problem-solving capabilities in MATLAB, allowing dynamic responses to varying conditions and inputs (Figs. 9.3, 9.4 and 9.5).

9.6   Variable Scope

In programming, **variable scope** refers to the region of a program where a particular variable is accessible and can be used. Understanding variable scope is crucial for writing maintainable and efficient code, as it helps prevent naming conflicts and ensures that variables are used correctly.

In MATLAB, there are two main types of variable scope: **global** and **local**. Global variables are accessible throughout the entire program, while local variables are only accessible within the function or script in which they are defined, also shown in Fig. 9.6.

- **Global Variables**: Global variables are declared outside of any function or script, and they can be accessed and modified from anywhere in the MATLAB workspace. However, overuse of global variables can lead to code that is difficult to maintain and debug, as it becomes harder to track where and how the variables are being modified.
- **Local Variables**: Local variables are declared within a function or script, and they are only accessible within that function or script. When a function is called, it creates its own local workspace, and any variables declared within that function are local to that workspace. Local variables help to encapsulate the functionality of a function and prevent naming conflicts with other parts of the program.

It is generally considered best practice to use local variables whenever possible and to minimise the use of global variables. This approach promotes modular code that is easier to understand, maintain, and debug.

Listing 9.6 Basic variable scope syntax.

```
% Global variable
global x;
x = 10;

% Function using local variable
function y = myFunction(a)
```

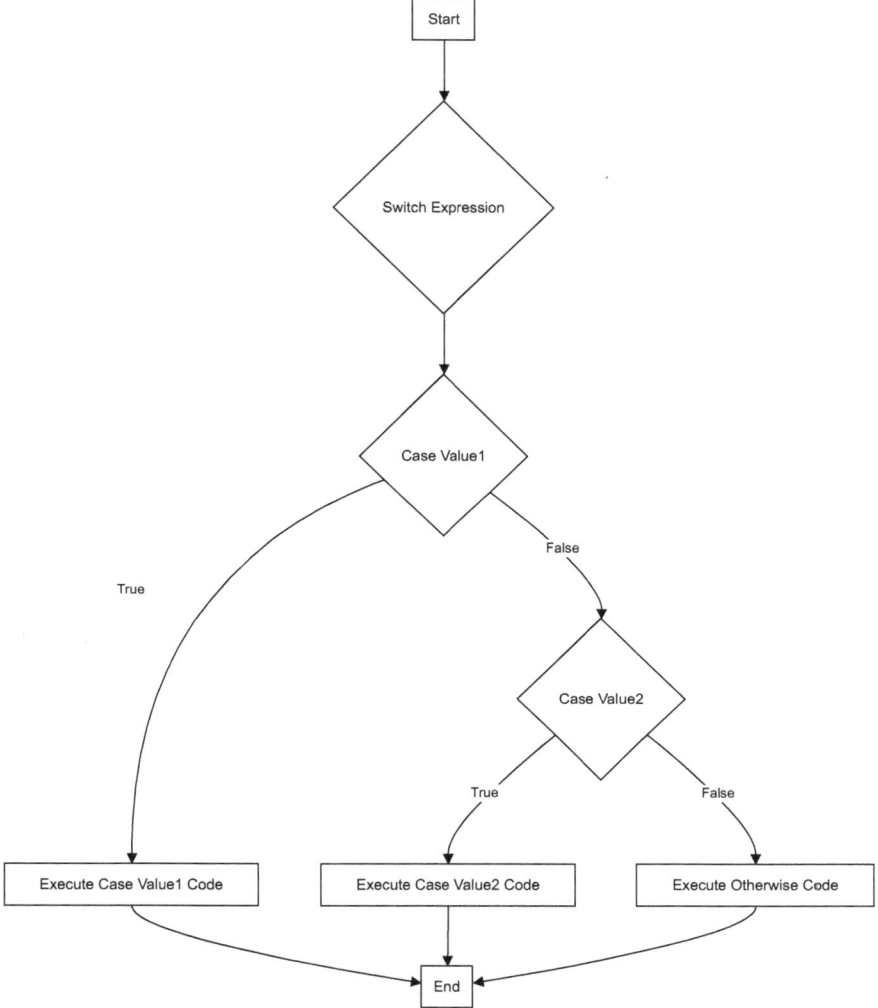

Fig. 9.2 Conditional statements: switch and case

```
y = a + 5; % y is a local variable
end

% Calling the function
result = myFunction(x); % result = 15
```

Listing 9.7 Using nested functions with local variables.

```
function outerFunction()
x = 10; % x is local to outerFunction
```

```
function innerFunction()
    y = x + 5; % y is local to innerFunction, but can
        access x
    disp(y); % Output: 15
end

innerFunction();

end

outerFunction();
```

Fig. 9.3 Loop statements:
for loops

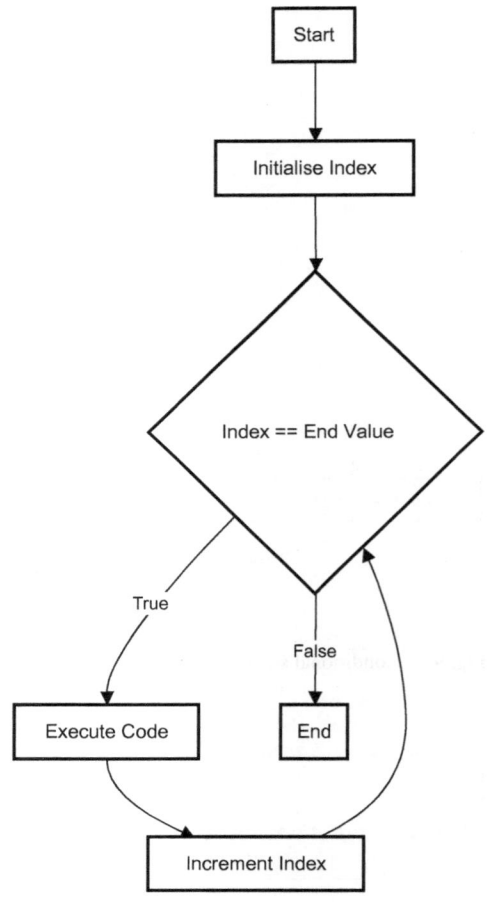

Fig. 9.4 Loop statements: while loops

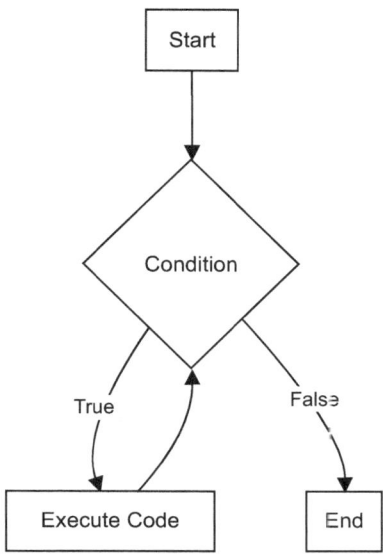

9.6.1 Global and Local Scope

Understanding the scope of variables in MATLAB is crucial for writing effective and error-free code. Variable scope determines where a variable can be accessed or modified within a program. There are two primary types of variable scope in MATLAB: global scope and local scope.

9.6.2 Local Scope

Local scope refers to variables that are accessible only within the function or script where they are declared. These variables are not visible or accessible outside their local context. Local scope ensures that variables do not interfere with each other across different functions or scripts.

- **Function Scope:** Variables declared within a function are local to that function.
- **Script Scope:** Variables declared within a script are local to that script unless explicitly passed to other functions or scripts.

9.6.2.1 Example of Local Scope

Consider the following example to illustrate local scope in MATLAB:

```
function exampleFunction()
    a = 10; % Local variable
    b = 20; % Local variable
```

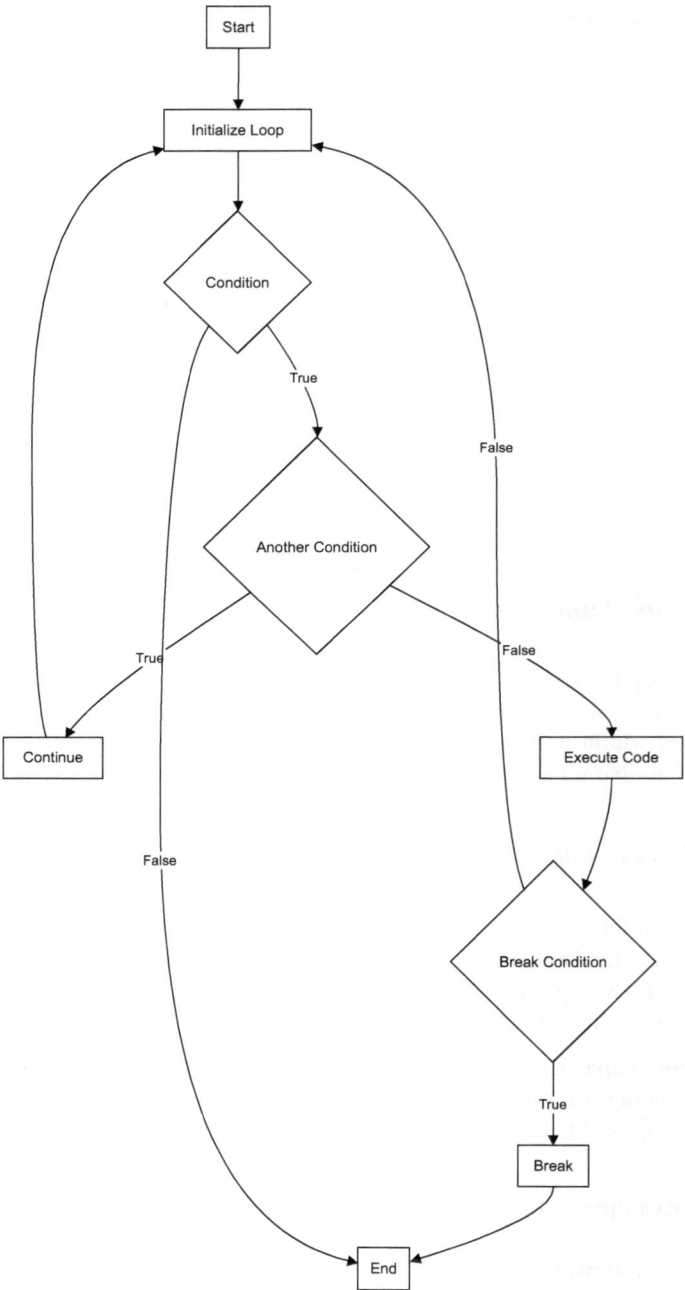

Fig. 9.5 Jump statements

```
    disp(['Inside function, a: ', num2str(a)]);
    disp(['Inside function, b: ', num2str(b)]);
end

exampleFunction();

% Attempting to access a and b outside the function will result in an error
% disp(['Outside function, a: ', num2str(a)]); % Error
% disp(['Outside function, b: ', num2str(b)]); % Error
```

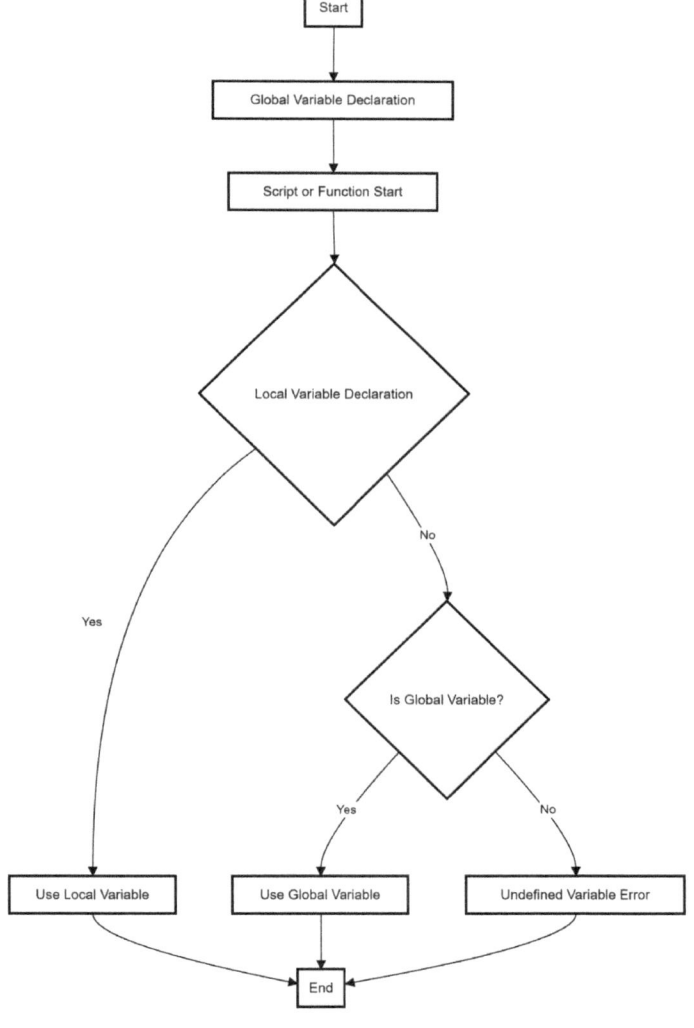

Fig. 9.6 Global and local scope

Output:

```
Inside function, a: 10
Inside function, b: 20
```

9.6.3   Global Scope

Global scope refers to variables that are accessible from any function or script within the MATLAB workspace. These variables are declared using the `global` keyword. Global variables can be useful for sharing data between different functions and scripts, but they should be used sparingly to avoid unintended side effects.

- **Global Variables:** Declared using the `global` keyword and accessible from any function or script that also declares them as global.

9.6.3.1   Example of Global Scope

Consider the following example to illustrate global scope in MATLAB:

```
function setGlobalVariable()
    global gVar;
    gVar = 100; % Set global variable
end

function accessGlobalVariable()
    global gVar;
    disp(['Global variable gVar: ', num2str(gVar)]);
end

setGlobalVariable();
accessGlobalVariable();
```

Output:

```
Global variable gVar: 100
```

9.6.4   Persistent Variables

Persistent variables in MATLAB retain their values between calls to the function in which they are declared. They are declared using the `persistent` keyword. Persistent variables are similar to global variables but are limited to the function scope, providing a way to maintain state information across function calls without exposing the variable globally.

9.6.4.1   Example of Persistent Variables

Consider the following example to illustrate persistent variables in MATLAB:

```
function countCalls()
    persistent count;
    if isempty(count)
        count = 0;
    end
    count = count + 1;
    disp(['Function called ', num2str(count), ' times']);
end

countCalls();
countCalls();
countCalls();
```

Output:

```
Function called 1 times
Function called 2 times
Function called 3 times
```

9.6.5   Nested Functions and Variable Scope

In MATLAB, nested functions are functions that are defined within other functions. These nested functions have access to the variables of their parent functions, which creates a unique variable scope hierarchy. This section provides a detailed explanation of how variable scope operates within nested functions, also shown in Fig. 9.7.

9.6.5.1   Variable Scope in Nested Functions

When a function (the main function) contains another function (the nested function), the nested function can access and modify the variables of the main function. The scope of variables in nested functions follows these rules:

- **Main Function Variables:** Variables declared in the main function are accessible to all nested functions within that main function.

Fig. 9.7 Nested functions
and variable scope

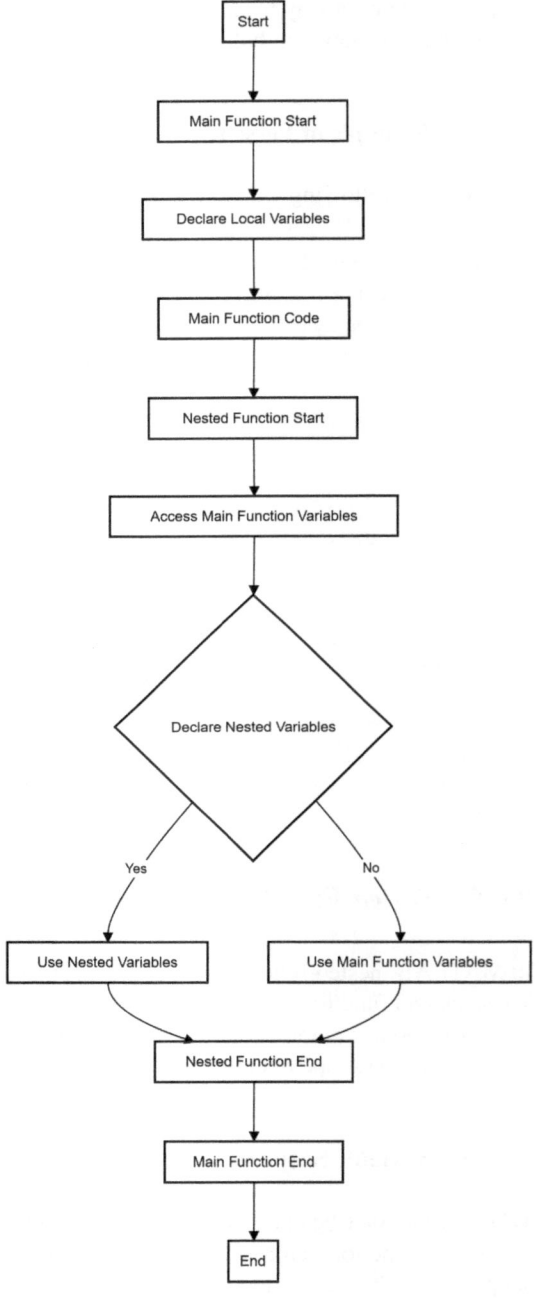

- **Nested Function Variables:** Variables declared in a nested function are local to that nested function and do not affect variables in the main function or other nested functions.
- **Variable Shadowing:** If a nested function declares a variable with the same name as a variable in the main function, the nested function's variable will shadow the main function's variable within its own scope.

9.6.5.2 Example of Nested Functions

Consider the following example to illustrate variable scope in nested functions:

```
function mainFunction()
    x = 10; % Variable in main function
    y = 20; % Variable in main function

    function nestedFunction1()
        disp(['x in nestedFunction1: ', num2str(x)]);
        y = 30; % This shadows the main function's y
        disp(['y in nestedFunction1: ', num2str(y)]);
    end

    function nestedFunction2()
        x = 50; % This shadows the main function's x
        disp(['x in nestedFunction2: ', num2str(x)]);
        disp(['y in nestedFunction2: ', num2str(y)]);
    end

    nestedFunction1();
    nestedFunction2();
    disp(['x in mainFunction: ', num2str(x)]);
    disp(['y in mainFunction: ', num2str(y)]);
end
```

Output:

```
x in nestedFunction1: 10
y in nestedFunction1: 30
x in nestedFunction2: 50
y in nestedFunction2: 20
x in mainFunction: 10
y in mainFunction: 20
```

9.6.5.3 Explanation

- When nestedFunction1 is called, it can access the variable x from the main function. It also declares a local variable y, which shadows the main function's y.
- When nestedFunction2 is called, it declares a local variable x, shadowing the main function's x. It accesses y from the main function because it does not declare a local y.
- The main function's variables x and y remain unchanged after the calls to the nested functions.

9.6.5.4 Persistent and Global Variables

- **Persistent Variables:** Persistent variables retain their values between calls to the function. They are declared using the persistent keyword.
- **Global Variables:** Global variables can be shared among different functions and the base workspace. They are declared using the global keyword.

```
function mainFunction()
    persistent pVar;
    global gVar;

    if isempty(pVar)
        pVar = 0;
    end

    pVar = pVar + 1;
    gVar = pVar;

    function nestedFunction()
        disp(['pVar in nestedFunction: ', num2str(pVar)]);
        disp(['gVar in nestedFunction: ', num2str(gVar)]);
    end

    nestedFunction();
end
```

9.7 Errors and Pitfalls

When programming in MATLAB, it is important to be aware of potential errors and pitfalls that can arise. These errors can be classified into three main categories: **syntax errors**, **logic errors**, and **rounding errors**.

9.7.1 Syntax Errors

Syntax errors occur when the code violates the rules of the MATLAB programming language. These errors prevent the code from running and must be fixed before the program can execute correctly. Common syntax errors in MATLAB include:

- **Incompatible vector sizes**: MATLAB requires vector and matrix operations to be performed on compatible sizes. Attempting to perform operations on vectors or matrices of incompatible sizes will result in a syntax error.
- **Name hiding**: Name hiding occurs when a local variable in a function or script has the same name as a variable in the base workspace or a parent function. This can lead to unintended behavior and should be avoided by using unique variable names.

Listing 9.8 Syntax error example: Incompatible vector sizes.

```
a = [1 2 3];
b = [4 5];
c = a + b; % Syntax error: Incompatible vector sizes
```

Listing 9.9 Syntax error example: Name hiding.

```
x = 10;

function y = myFunction()
x = 5; % This hides the global x
y = x + 2; % y will be 7, not 12
end
```

9.7.2 Logic Errors

Logic errors occur when the code runs without any syntax errors but produces incorrect results due to flaws in the program logic. These errors can be challenging to detect and resolve, as the code executes without any obvious errors. Common logic errors in MATLAB include:

- **Incorrect algorithm implementation**: If the algorithm or mathematical formula is implemented incorrectly in the code, the program will produce incorrect results.
- **Off-by-one errors**: These errors occur when iterative loops or array indexing are off by one iteration or index, leading to incorrect results.

Listing 9.10 Logic error example: Incorrect algorithm implementation.

```
% Incorrect implementation of the mean function
function avg = myMean(arr)
sum = 0;
for i = 1:length(arr)
sum = sum + arr(i); % Correct
```

```
end
avg = sum / (length(arr) + 1); % Logic error: Incorrect
    divisor
end
```

Listing 9.11 Logic error example: Off-by-one error.

```
% Off-by-one error in loop
for i = 1:10 % Should be i = 1:9 to avoid counting 10
    twice
disp(i);
end
```

9.7.3   Rounding Error

Rounding errors are a type of numerical error that can occur when performing arithmetic operations on floating-point numbers. These errors arise due to the finite precision of computer arithmetic and can accumulate, leading to inaccurate results. Rounding errors are particularly common in iterative calculations and computations involving large or small numbers.

To mitigate rounding errors, it is essential to be aware of the limitations of floating-point arithmetic and to use appropriate techniques such as higher precision data types, error analysis, and algorithmic adjustments.

Listing 9.12 Rounding error example.

```
x = 0.1;
sum = 0;
for i = 1:10
sum = sum + x;
end
disp(sum); % Output: 0.999999999999999 (due to rounding
    error)
```

By understanding and addressing these different types of errors, MATLAB programmers can write more robust and reliable code, ensuring accurate and reliable results.

9.8   Debugging and Testing

Debugging and testing are essential processes in software development to ensure the correctness and reliability of programs. MATLAB provides a range of tools and techniques to assist in these processes, including:

- **Debugging** techniques: MATLAB offers a powerful debugger that allows you to step through your code, set breakpoints, and inspect variable values at runtime. This helps identify and fix errors in your code.
- **Error handling** and **exception management**: MATLAB provides mechanisms for handling errors and exceptions, such as try-catch blocks, which allow you to gracefully handle and recover from runtime errors.
- **Unit testing** and **test-driven development**: MATLAB supports unit testing frameworks, such as the built-in matlab.unittest package, which enables you to write and run automated tests for your code. Test-driven development (TDD) is a software development approach that emphasizes writing tests before writing the production code.
- **Code profiling** and **optimisation**: MATLAB offers profiling tools that help identify performance bottlenecks in your code. You can then use this information to optimise your code for better performance.

Listing 9.13 Basic debugging example using the debugger.

```
function result = myFunction(x, y)
% Set a breakpoint here
z = x + y;
% Step through the code
result = z^2;
end
```

Listing 9.14 Exception handling using try-catch block.

```
try
% Code that might throw an exception
x = 1 / 0;
catch ME
% Handle the exception
disp(ME.message);
end
```

Listing 9.15 Unit testing example using matlab.unittest.

```
classdef MyClassTest < matlab.unittest.TestCase
methods (Test)
function testMyFunction(testCase)
% Test the myFunction
result = myFunction(2, 3);
testCase.verifyEqual(result, 25);
end
end
end
```

Listing 9.16 Code profiling example using the Profiler.

```
profile on
% Code to be profiled
```

```
profile off
profview
```

By effectively utilizing these debugging and testing techniques, MATLAB programmers can ensure the correctness, reliability, and performance of their code, leading to more robust and maintainable software solutions.

9.9   Eval and Text Macros

The **eval** function in MATLAB allows you to execute character strings as MATLAB code at runtime. This can be useful for dynamic code generation and evaluation, but it should be used with caution as it can introduce potential security risks and performance issues. MATLAB also provides **text macros** as a safer alternative for dynamic code generation and evaluation.

- **Error trapping with eval and lasterr**: The `lasterr` function can be used in conjunction with `eval` to capture and handle errors that occur during the evaluation of a string as MATLAB code.
- **Eval with try...catch**: The `try...catch` construct can also be used with `eval` to handle errors and exceptions that may occur during the evaluation of a string as MATLAB code.

Listing 9.17 Using eval and lasterr to trap errors.

```
% Create a string with invalid MATLAB code
invalidCode = 'x = 1 / 0;';

% Evaluate the code and trap errors
lastError = [];
try
eval(invalidCode);
catch ME
lastError = ME.message;
end

if ~isempty(lastError)
fprintf('Error: %s\n', lastError);
else
fprintf('Code executed successfully.\n');
end
```

Listing 9.18 Using eval with try...catch.

```
% Create a string with valid MATLAB code
validCode = 'x = 2; y = 3; z = x + y;';

try
eval(validCode);
```

```
fprintf('z = %d\n', z);
catch ME
fprintf('Error: %s\n', ME.message);
end
```

While eval can be a powerful tool in certain situations, it is generally recommended to use it sparingly and with caution, as it can introduce potential security risks and performance issues. Text macros, on the other hand, provide a safer and more efficient alternative for dynamic code generation and evaluation.

9.10 Live Scripts, Code Cells, and Publishing Code

In MATLAB, **Live Scripts** and **Code Cells** provide an interactive environment for writing, executing, and sharing MATLAB code. They support literate programming, allowing you to combine code, output, and explanatory text in a single document.

9.10.1 Live Scripts

Live Scripts are documents that contain executable MATLAB code, formatted text, and output. They enable the creation of self-contained, reproducible documents that can be easily shared and reused. Live Scripts allow for the integration of rich text, equations, images, and hyperlinks alongside the code, which enhances the readability and dissemination of information.

- **Creating a Live Script**: To create a Live Script, navigate to the **Home** tab in MATLAB, and select **New Live Script**.
- **Editing and Running Code**: MATLAB Live Scripts allow users to write code in segments and run these segments individually or collectively, promoting an iterative approach to programming.

Listing 9.19 Creating a Plot in a Live Script

```
% Example: Creating a Plot in a Live Script
x = linspace(0, 2*pi, 100);
y = sin(x);
figure;
plot(x, y);
title('Sine Wave');
xlabel('x');
ylabel('sin(x)');
```

Listing 9.20 Performing Basic Calculations in a Live Script

```
% Example: Performing Basic Calculations in a Live
    Script
```

```
a = 10;
b = 20;
sum = a + b;
fprintf('The sum of %d and %d is %d.\n', a, b, sum);
```

9.10.2 Code Cells

Code Cells are sections of executable MATLAB code within a Live Script or a Live Editor Task. They allow users to run and test code incrementally, making it easier to debug and experiment with different ideas. Code Cells can be created by inserting a cell break, which is typically done by using the **%%** symbol at the beginning of a line.

- **Creating Code Cells**: Insert **%%** at the beginning of a line to start a new cell. Each cell can be run independently, facilitating isolated testing of code segments.
- **Running Code Cells**: Users can run individual cells or the entire script to see the results of specific sections without executing the whole document.

Listing 9.21 Creating and Running Code Cells

```
% Example: Creating and Running Code Cells

%% Cell 1: Define Variables
a = 5;
b = 10;

%% Cell 2: Perform Calculation
sum = a + b;
disp(['Sum: ', num2str(sum)]);
```

Listing 9.22 Using Code Cells for Plotting

```
% Example: Using Code Cells for Plotting

%% Cell 1: Generate Data
x = linspace(0, 10, 100);
y = cos(x);

%% Cell 2: Plot Data
figure;
plot(x, y);
title('Cosine Wave');
xlabel('x');
ylabel('cos(x)');
```

9.10.3 Publishing Code

MATLAB provides the ability to publish Live Scripts and other MATLAB code as HTML, PDF, or other formats, making it easier to share work with others or create reports and documentation. This feature is particularly useful for generating professional reports or presentations that include code, results, and explanations.

- **Publishing Options**: MATLAB supports multiple formats for publishing, including HTML, PDF, and LaTeX.
- **Generating Reports**: Users can publish their scripts directly from the MATLAB environment by using the **Publish** command. This allows for the generation of comprehensive reports that include formatted code, output, figures, and descriptive text.
- **Publishing Options**: MATLAB supports multiple formats for publishing, including HTML, PDF, and LaTeX.
- **Generating Reports**: Users can publish their scripts directly from the MATLAB environment by using the **Publish** command. This generates a document that includes the code, its output, and any accompanying text and images.

Listing 9.23 Publishing a Live Script to HTML

```
% Example: Publishing a Live Script to HTML
% In the MATLAB command window, use:
publish('MyLiveScript.mlx', 'html');
```

Listing 9.24 Publishing a Live Script to PDF

```
% Example: Publishing a Live Script to PDF
% In the MATLAB command window, use:
publish('MyLiveScript.mlx', 'pdf');
```

Live Scripts, Code Cells, and the ability to publish MATLAB code make it easier to create reproducible, shareable documents that combine code, output, and explanations, facilitating collaboration and documentation.

9.11 Files and Folders

MATLAB provides a set of functions and tools for working with files and folders, making it easier to manage and organise your code, data, and other resources.

- **File Operations**: MATLAB offers a variety of functions for performing file operations. These include **fopen**, **fread**, **fwrite**, and **fclose** for reading and writing files, as well as **movefile**, **copyfile**, and **delete** for managing files. These functions allow users to efficiently manipulate files within their MATLAB environment.
  Here are a couple of examples of file operations in MATLAB:

Listing 9.25 Writing data to a file.

```
% Open a file for writing
fileID = fopen('data.txt', 'w');
% Write data to the file
fprintf(fileID, 'Hello, World!\n');
fprintf(fileID, 'This is a sample file.\n');
% Close the file
fclose(fileID);
```

Listing 9.26 Reading data from a file.

```
% Open a file for reading
fileID = fopen('data.txt', 'r');
% Read the contents of the file
data = fscanf(fileID, '%c');
% Close the file
fclose(fileID);
% Display the contents
disp(data);
```

Listing 9.27 Example of moving and deleting a file.

```
% Create a file and then move it
fileID = fopen('temp.txt', 'w');
fprintf(fileID, 'Temporary file');
fclose(fileID);

% Move the file to a new location
movefile('temp.txt', 'newTemp.txt');

% Delete the moved file
delete('newTemp.txt');
```

- **MATLAB Search Path**: The MATLAB search path determines the directories that MATLAB searches for files and functions. Users can add, remove, or reorder directories in the search path using functions like **addpath**, **rmpath**, and **path**. This flexibility allows for better organisation and quicker access to frequently used directories.

  Here are a couple of examples of managing the MATLAB search path:

Listing 9.28 Adding a directory to the search path.

```
% Add a directory to the search path
addpath('C:\MyFunctions');
% Check the current search path
path
```

Listing 9.29 Removing a directory from the search path.

```
% Remove a directory from the search path
rmpath('C:\MyFunctions');
% Check the updated search path
path
```

Listing 9.30 Example of modifying the MATLAB search path.

```
% Add a directory to the search path
addpath('C:\MyMATLABFiles');

% Remove a directory from the search path
rmpath('C:\MyMATLABFiles');

% Display the current search path
disp(path);
```

Listing 9.31 Example of reordering the search path.

```
% Add multiple directories to the search path
addpath('C:\MyMATLABFiles', 'C:\OtherMATLABFiles');

% Reorder the search path
path('C:\OtherMATLABFiles', 'C:\MyMATLABFiles');

% Display the reordered search path
disp(path);
```

In addition to file operations and search path management, MATLAB provides a comprehensive set of functions for working with directories and folders. Functions like cd, dir, mkdir, and rmdir allow you to navigate, list, create, and remove directories programmatically.

By leveraging these file and folder management capabilities in MATLAB, you can effectively organize your code, data, and resources, making your projects more structured and maintainable. These features enhance the overall development experience and facilitate efficient data processing and analysis tasks.

The ability to execute code dynamically and schedule code execution allows for greater flexibility and automation in MATLAB programming, enabling tasks such as dynamic code generation, real-time processing, and scheduled tasks.

9.12   Security in MATLAB Code

Security is an important aspect of MATLAB programming, especially when sharing code or deploying applications. MATLAB provides several features and best practices to ensure the security of MATLAB code. When developing MATLAB code that involves sensitive information or intellectual property, it is essential to consider **security** measures to protect your work and data. MATLAB provides various tools and techniques to help ensure the confidentiality, integrity, and availability of your code and data.

9.12.1   Understanding MATLAB Security

Understanding MATLAB Security is the first step towards writing secure code. It involves being aware of potential vulnerabilities such as buffer overflows, injection attacks, and improper error handling. MATLAB provides built-in functions to help developers write secure code.

- **Using secure functions** is another key aspect of secure MATLAB programming. MATLAB provides a range of secure functions designed to reduce the risk of security breaches. For instance, the `feval` function is a safer alternative to `eval`, as it limits the scope of execution and prevents the execution of arbitrary code.
- **Code Obfuscation Code obfuscation** is the process of making code difficult to understand or reverse-engineer. While not a foolproof method, it can deter casual attempts at code theft or modification. MATLAB does not have built-in obfuscation tools, but developers can implement basic obfuscation techniques such as renaming variables and functions to non-meaningful names. This can help protect your intellectual property and sensitive algorithms.
- **Data Encryption** Sensitive data can be encrypted in MATLAB using built-in encryption functions or external libraries, ensuring that only authorised parties can access and decode the data.
- **Access Control** MATLAB supports various access control mechanisms, such as user authentication and authorization, to restrict access to sensitive code and data.
- **Secure Communication** When transmitting data or code over networks, MATLAB provides tools for secure communication, such as SSL/TLS encryption and digital signatures.
- **Code Signing** MATLAB allows you to digitally sign your code using a code signing certificate. Code signing helps users verify the authenticity and integrity of your code, ensuring that it hasn't been tampered with. Here's an example of signing a MATLAB script:

Listing 9.32 Signing a MATLAB script.

```
% Create a code signing certificate
cert = codeSigningCert('MyCodeSigningCert');
% Sign the MATLAB script
signFile('myscript.m', cert);
```

- **Encrypted Source Code** MATLAB provides the ability to encrypt your source code files, protecting your intellectual property and preventing unauthorized access to your code. You can use the `pcode` function to create an encrypted version of your MATLAB code:

Listing 9.33 Encrypting MATLAB source code.

```
% Encrypt the MATLAB source code
pcode('myscript.m');
```

- **Secure Deployment** When deploying MATLAB applications, it's important to consider security measures to protect your code and data. MATLAB Compiler allows you to package your MATLAB code into standalone applications or shared libraries, which can be deployed securely.
- **Input Validation** Validating user inputs is crucial to prevent potential security vulnerabilities. MATLAB provides functions like `validateattributes` and `inputParser` to validate and sanitize user inputs, ensuring that they meet the expected criteria.

Listing 9.34 Example of input validation.

```matlab
function secureFunction(input)
    % Validate that input is numeric
    validateattributes(input, {'numeric'}, {'nonempty'
        });

    % Proceed with computation
    result = input^2;
    disp(['The square of the input is: ', num2str(
        result)]);
end
```

Listing 9.35 Example using mustBeNumeric.

```matlab
function secureFunction(input)
    % Ensure input is numeric
    arguments
        input {mustBeNumeric}
    end

    % Proceed with computation
    result = input^2;
    disp(['The square of the input is: ', num2str(
        result)]);
end
```

- **Error Handling and Logging** Proper error handling is essential for maintaining the security and stability of your MATLAB code. Use `try-catch` blocks to handle exceptions gracefully and avoid revealing sensitive information in error messages. But be mindful about not over-sharing sensitive details in logs or error messages.
- **File Permissions**
  Proper handling of file permissions is essential to prevent unauthorised access and modifications. Using MATLAB's built-in file functions, users can control access to files. It is advisable to set appropriate permissions when reading from or writing to files.

Listing 9.36 Example of setting file permissions.

```
% Create a file with restricted permissions
fileID = fopen('secureFile.txt', 'w');
fprintf(fileID, 'This is a secure file.');
fclose(fileID);

% Change file permissions to read-only
fileattrib('secureFile.txt', '-w');
```

Listing 9.37 Example of checking file permissions.

```
% Check file permissions
[status, attributes] = fileattrib('secureFile.txt');
disp(attributes);

% Ensure file is read-only
if attributes.UserWrite == 0
    disp('File is read-only');
else
    disp('File is writable');
end
```

- **Safeguarding Against Code Injection**
  Code injection is a significant security threat where an attacker can execute arbitrary code by exploiting vulnerabilities. To mitigate this risk, it is crucial to avoid using **eval** with user inputs and to employ safer alternatives like **str2func**.

Listing 9.38 Unsafe use of eval.

```
% Unsafe example
userInput = 'disp(''This is unsafe!'')';
eval(userInput); % Potential code injection risk
```

Listing 9.39 Safe alternative using str2func.

```
% Safe example
userInput = 'disp';
safeFunction = str2func(userInput);
safeFunction('This is safe!');
```

- **Testing and validation** are essential to ensure the security of MATLAB code. This includes unit testing, where individual components of the code are tested for expected behavior, and integration testing, where components are tested together to ensure they work as intended. MATLAB's built-in testing framework, such as the unittest package, can be Utilised for this purpose.

9.12.2   Example MATLAB Codes for Secure Programming

Listing 9.40 Code obfuscation using the MATLAB Compiler.

```
% Obfuscate a MATLAB function
\textcolor{green}{mbuild} -g -c mysecretfunction.m

% Deploy the obfuscated function
deployedFunction = \textcolor{green}{deploytool}('-build
    ', '-obfuscate', 'mysecretfunction.m');
```

Listing 9.41 Encrypting and decrypting data using MATLAB's built-in encryption functions.

```
% Generate a random encryption key
key = \textcolor{green}{randi}([0, 255], 1, 16);

% Encrypt sensitive data
encryptedData = \textcolor{green}{encrypt}('My sensitive
    data', key);

% Decrypt the data
decryptedData = \textcolor{green}{decrypt}(encryptedData
    , key);
```

Implementing security measures in MATLAB is crucial for protecting sensitive information, intellectual property, and ensuring the integrity and confidentiality of your code and data, especially when working on projects involving critical or proprietary algorithms, data, or systems.

In addition to these security features, it's important to follow general programming best practices, such as keeping your MATLAB installation and toolboxes up to date, using strong passwords, and being cautious when running untrusted code or opening files from unknown sources.

By incorporating security measures into your MATLAB code and following secure coding practices, you can protect your intellectual property, ensure the integrity of your applications, and maintain a secure development environment.

9.13   Graphical User Interfaces

Graphical User Interfaces (GUIs) provide a user-friendly way to interact with MATLAB applications, enabling users to input data, visualise results, and control program execution through a graphical environment. GUIs can greatly enhance the usability and accessibility of MATLAB programs, particularly for non-technical users or those who prefer a visual interface over command-line interactions.

9.13.1 Basic Structure of a GUI

The basic structure of a GUI in MATLAB consists of various components, such as windows, panels, buttons, menus, and other user interface controls. These components are organized in a hierarchical manner, with the main window serving as the parent container for other elements.

Listing 9.42 Creating a simple GUI window.

```
% Create a new figure window
figureHandle = \textcolor{green}{figure};

% Set the window properties
\textcolor{green}{set}(figureHandle, 'Name', 'My GUI', '
   NumberTitle', 'off', 'MenuBar', 'none');
```

Listing 9.43 Adding a button to the GUI.

```
% Create a button
buttonHandle = \textcolor{green}{uicontrol}('Style', '
   pushbutton', 'String', 'Click Me', ...
'Position', [50 50 100 30], 'Callback', @buttonCallback)
   ;

function buttonCallback(~, ~)
disp('Button clicked!');
end
```

9.13.2 A First Example: Getting the Time

Creating a simple GUI to display the current time is a great way to learn the basics of GUI development in MATLAB. This example demonstrates how to create a window with a button and a text box, where clicking the button updates the text box with the current time.

Listing 9.44 Creating the time GUI.

```
function timeGUI()
    % Create the figure window
    fig = uifigure('Name', 'Current Time Display');
    fig.Position = [100 100 300 150];

    % Create a text area to display the time
    timeDisplay = uitextarea(fig);
    timeDisplay.Position = [50 80 200 40];
    timeDisplay.Value = 'Click the button to show time';
    timeDisplay.BackgroundColor = [0.9 0.9 0.9];
```

```
    % Create a button to update the time
    updateButton = uibutton(fig, 'push');
    updateButton.Position = [100 30 100 30];
    updateButton.Text = 'Update Time';

    % Set the button callback function
    updateButton.ButtonPushedFcn = @(btn,event)
        updateTime(timeDisplay);
end

function updateTime(timeDisplay)
    % Get the current time and format it as a string
    currentTime = datestr(now, 'HH:MM:SS');

    % Update the text area with the current time
    timeDisplay.Value = currentTime;
end
```

In this example, the figure function creates the main window, and the uicontrol function is used to create the button and text box components. The Position parameter specifies the location and size of each component within the window.

- The script defines two functions: timeGUI() and updateTime().
- The timeGUI() function is the main function that creates the GUI:

  - It uses uifigure() to create the main window with the title "Current Time Display".
  - The fig.Position property sets the window's size and position on the screen.
  - A text area is created using uitextarea():

    · Its position and size are set with timeDisplay.Position.
    · Initial text and background color are set.

  - A button is created using uibutton():

    · Its position and size are set with updateButton.Position.
    · The button's text is set to "Update Time".

  - The button's callback function is set to call updateTime() when clicked.

- The updateTime() function is called each time the button is pressed:

  - It gets the current time using datestr(now, 'HH:MM:SS').
  - It updates the text area's value with the current time.

- When the script is run:

  - MATLAB executes timeGUI().
  - The GUI window appears with the text area and button.
  - Each time the user clicks the button, updateTime() is called, updating the displayed time.

When you run this code, a window will appear with a button labeled "Update Time" and a text box above it, as shown in Fig. 9.8. Clicking the button will update the text box with the current time in the format "HH:MM:SS", as shown in Fig. 9.9.

This simple example demonstrates how to create a basic GUI with interactive components and how to update the GUI based on user actions. It serves as a foundation for building more complex GUIs with additional functionality and visual elements.

9.13.3 Newton's Method

Newton's method is a powerful numerical technique for finding the roots of a function. It is an iterative method that starts with an initial guess and iteratively refines the approximation until a desired level of accuracy is achieved. In this section, we will explore how to implement Newton's method in MATLAB using a GUI to provide a user-friendly and interactive experience, as shown in Fig. 9.10.

The GUI will consist of several components: a text box for entering the function, a slider for adjusting the initial guess, a button to start the root-finding process, a plot to visualise the function and the convergence of the method, and a text area to display the iterations and the final root.

Fig. 9.8 The initial state of the time GUI

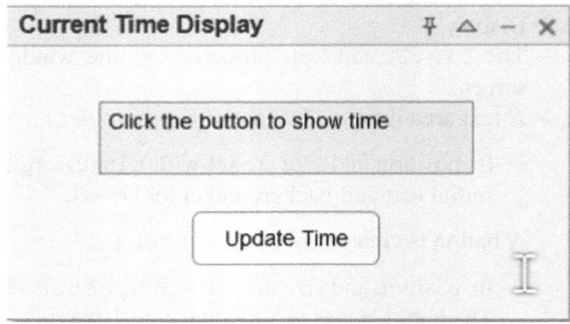

Fig. 9.9 The time GUI after clicking the "Get Time" button

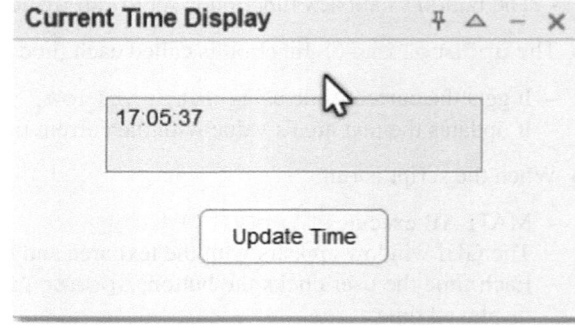

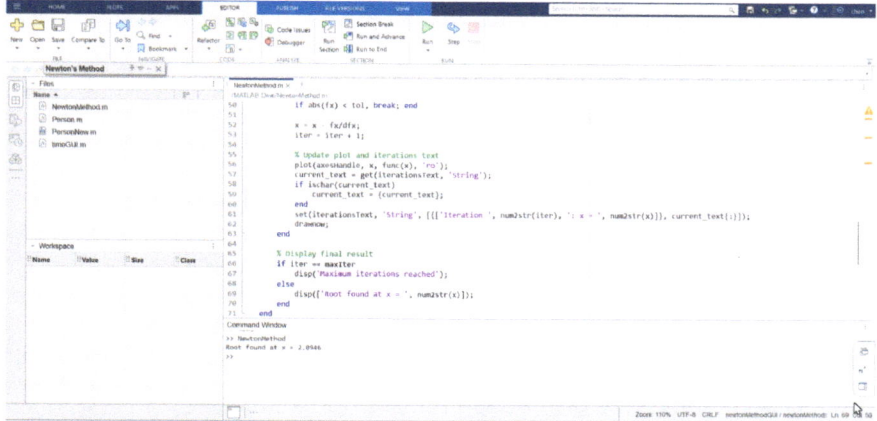

Fig. 9.10 The MATLAB codes for Newton's method GUI

Listing 9.45 Creating the Newton's method GUI.

```
function newtonMethodGUI()
    % Create the main window
    figureHandle = figure('Name', 'Newton''s Method', '
        NumberTitle', 'off', 'MenuBar', 'none', '
        Position', [100 100 800 600]);

    % Create components (text box, slider, button, plot,
        text area)
    functionEdit = uicontrol('Style', 'edit', 'String',
        'x.^3 - 2*x - 5', ...
        'Position', [20 550 200 30]);
    initialGuessSlider = uicontrol('Style', 'slider', '
        Value', 1, 'Min', -10, 'Max', 10, ...
        'Position', [250 550 300 30]);
    startButton = uicontrol('Style', 'pushbutton', '
        String', 'Start', ...
        'Position', [600 550 100 30], 'Callback', {
            @newtonMethod, functionEdit,
            initialGuessSlider});
    axesHandle = axes('Position', [100 100 500 400]);
    iterationsText = uicontrol('Style', 'edit', 'String'
        , '', 'Max', 2, ...
        'Position', [650 100 100 400], '
            HorizontalAlignment', 'left');

    % Store handles in figure's UserData
    setappdata(figureHandle, 'axesHandle', axesHandle);
```

```matlab
        setappdata(figureHandle, 'iterationsText',
            iterationsText);
end

% Newton's method function
function newtonMethod(hObject, ~, functionEdit,
    initialGuessSlider)
    % Get function and initial guess from GUI
    func_str = get(functionEdit, 'String');
    func = str2func(['@(x) ' func_str]);
    x0 = get(initialGuessSlider, 'Value');

    % Get handles from figure's UserData
    figureHandle = gcbf;
    axesHandle = getappdata(figureHandle, 'axesHandle');
    iterationsText = getappdata(figureHandle, '
        iterationsText');

    % Implement Newton's method
    maxIter = 100; tol = 1e-6;
    x = x0; iter = 0;

    % Plot function
    x_range = linspace(-10, 10, 1000);
    y_range = arrayfun(func, x_range);
    plot(axesHandle, x_range, y_range);
    hold(axesHandle, 'on');

    % Numerical differentiation function
    dfunc = @(x) (func(x + 1e-8) - func(x)) / 1e-8;

    while iter < maxIter
        fx = func(x);
        dfx = dfunc(x);

        if abs(fx) < tol, break; end

        x = x - fx/dfx;
        iter = iter + 1;

        % Update plot and iterations text
        plot(axesHandle, x, func(x), 'ro');
        current_text = get(iterationsText, 'String');
        if ischar(current_text)
            current_text = {current_text};
        end
        set(iterationsText, 'String', [{['Iteration ',
            num2str(iter), ': x = ', num2str(x)]},
            current_text{:}]);
        drawnow;
    end

    % Display final result
```

```
        if  iter  ==  maxIter
            disp('Maximum  iterations  reached');
        else
            disp(['Root  found  at  x  =  ',  num2str(x)]);
        end
end
```

In this example, the GUI components are created using the uicontrol function, and the plot axes are created using the axes function, as shown in Fig. 9.11. Specifically,

- The script defines two main functions: newtonMethodGUI() and newtonMethod().
- The newtonMethodGUI() function creates the GUI:

  – It uses figure() to create the main window.
  – GUI components are created using uicontrol():

    · A text edit box for entering the function.
    · A slider for selecting the initial guess.
    · A "Start" button to begin the calculation.
    · A text area to display iteration results.

  – An axes object is created using axes() for plotting.
  – Component handles are stored using setappdata() for later access.

- The newtonMethod() function implements Newton's method:

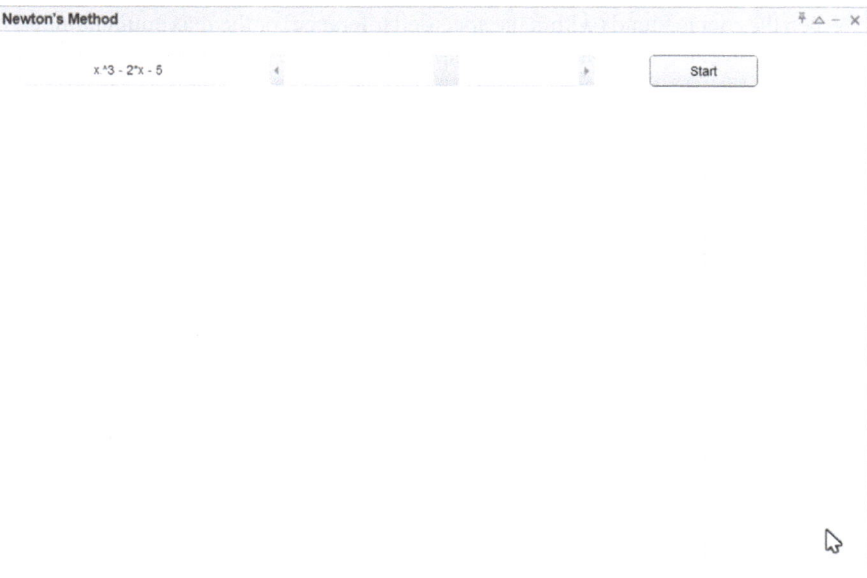

Fig. 9.11 The initial state of the Newton's method GUI

- It's called when the "Start" button is clicked.
- It retrieves the function string and initial guess from GUI components.
- The function string is converted to a function handle using `str2func()`.
- It retrieves necessary handles using `getappdata()`.
- The function is plotted over the range [-10, 10] using `plot()`.
- Newton's method is implemented in a loop:
  - Function value and derivative are calculated at each iteration.
  - New iterate is computed using Newton's formula.
  - The plot is updated with new iterate using `plot()`.
  - The iterations text area is updated with new iterate information.
- The loop continues until convergence or maximum iterations are reached.

- Key aspects of the implementation:
  - `arrayfun()` is used to apply the function to a range of x-values for plotting.
  - Numerical differentiation is used to approximate the derivative.
  - `drawnow()` is used to update the GUI in real-time during iterations.

- When the script is run:
  - MATLAB executes `newtonMethodGUI()`, creating the GUI window.
  - The user can input a function and select an initial guess.
  - Clicking "Start" initiates Newton's method.
  - The GUI updates in real-time, showing the convergence process visually and numerically.

When the root is found (within the specified tolerance) or the maximum number of iterations is reached, the final result is displayed in the MATLAB command window, as shown in Fig. 9.12.

This GUI implementation of Newton's method allows users to interactively explore the root-finding process by adjusting the function and initial guess, and visually observing the convergence behavior. It demonstrates the power of combining numerical methods with graphical user interfaces in MATLAB for educational and visualisation purposes.

9.13.4 Axes on a GUI

Incorporating axes into a GUI is a powerful way to visualise data and results within a MATLAB application. MATLAB's axes function allows you to create and customize plot areas within a GUI window, enabling interactive plotting and data visualisation capabilities.

In this section, we will explore how to add axes to a GUI and demonstrate their use with a simple example of plotting a sine wave.

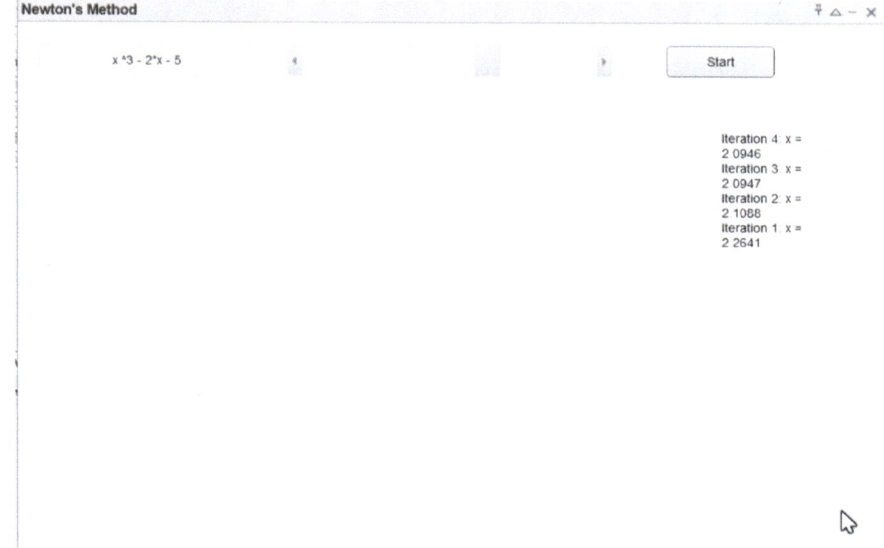

Fig. 9.12 The Newton's method GUI showing the convergence of the method

Listing 9.46 Creating a GUI with axes.

```
% Create the main window
figureHandle = figure('Name', 'Sine Wave Plot', '
    NumberTitle', 'off', 'MenuBar', 'none', ...
    'Position', [200 200 600 400]);

% Create axes
axesHandle = axes('Parent', figureHandle, 'Position'.
    [0.1 0.2 0.8 0.7]);

% Create a button
buttonHandle = uicontrol('Parent', figureHandle, 'Style'
    , 'pushbutton', 'String', 'Plot Sine Wave', ...
    'Position', [450 20 120 30], 'Callback',
        @plotSineWave);

% Callback function to plot the sine wave
function plotSineWave(~, ~)
    % Get the axes handle
    axesHandle = findobj(gcf, 'Type', 'axes');

    % Generate data
    x = linspace(0, 2*pi, 100);
    y = sin(x);

    % Plot the sine wave
    plot(axesHandle, x, y);
```

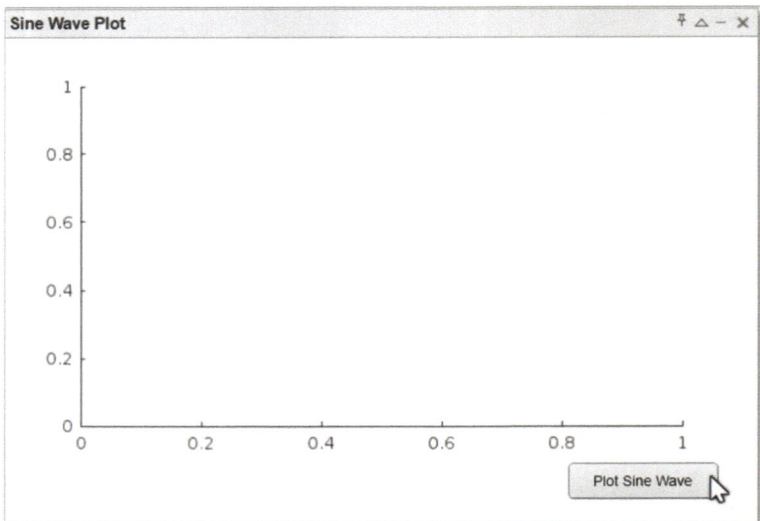

Fig. 9.13 The initial state of the GUI with axes

```
    xlabel(axesHandle, 'x');
    ylabel(axesHandle, 'sin(x)');
    title(axesHandle, 'Sine Wave Plot');
    grid(axesHandle, 'on');
end
```

In this example, the axes function creates a set of axes within the main figure window. The Position property specifies the location and size of the axes relative to the figure's dimensions (using normalized coordinates from 0 to 1).

The uicontrol function creates a button labeled "Plot Sine Wave". When clicked, the plotSineWave callback function is executed, which generates the data for the sine wave using the linspace and sin functions.

The plot function is then used to plot the sine wave on the axes created earlier. The xlabel, ylabel, title, and grid functions are used to add labels, a title, and a grid to the plot, respectively.

When you run this code, a window will appear with a button labeled "Plot Sine Wave", as shown in Fig. 9.13. Clicking the button will generate and display a plot of the sine wave on the axes within the GUI window, as shown in Fig. 9.14.

This example demonstrates how to create and Utilise axes within a GUI for plotting and visualizing data. By combining user interface controls with axes, you can create interactive and visually appealing applications for data analysis, simulation, and visualisation tasks.

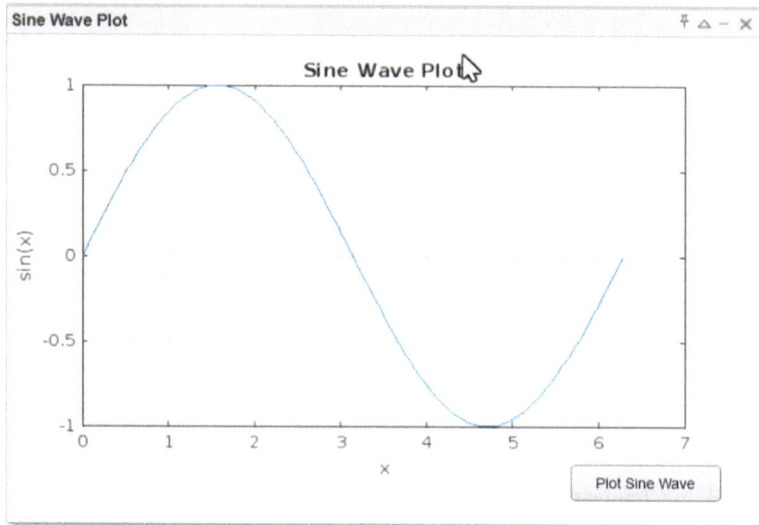

Fig. 9.14 The GUI with the sine wave plotted on the axes

9.13.5   Adding Color to a Button

Customising the appearance of GUI components, such as changing the color of a button, can enhance the visual appeal and usability of an application. MATLAB provides various properties and functions to modify the appearance of GUI elements, allowing developers to create visually appealing and user-friendly interfaces.

In this section, we will explore how to change the colour of a button within a GUI. We will create a simple GUI with a button and demonstrate how to change its colour based on user input.

Listing 9.47 Creating a GUI with a colored button.

```
% Create the main window
figureHandle = figure('Name', 'Colored Button', '
    NumberTitle', 'off', 'MenuBar', 'none', ...
    'Position', [300 300 400 200]);

% Create a text box for color input
colorEdit = uicontrol('Style', 'edit', 'String', 'r',
    ...
    'Position', [20 130 100 30]);

% Create a button
buttonHandle = uicontrol('Style', 'pushbutton', 'String'
    , 'Click to Change Color', ...
    'Position', [150 130 200 30], 'Callback',
        @changeButtonColor);
```

```
% Callback function to change the button color
function changeButtonColor(hObject, ~)
    % Get the figure handle
    figureHandle = ancestor(hObject, 'figure');

    % Find the color edit box
    colorEdit = findobj(figureHandle, 'Style', 'edit');

    % Get the color from the edit box
    color = get(colorEdit, 'String');

    % Change the button color
    set(hObject, 'BackgroundColor', color);
end
```

In this example, we create a figure window and add a text box (colorEdit) for the user to input a color code. We also create a button (buttonHandle) with an initial label "Click to Change Color".

The changeButtonColor callback function is associated with the button's Callback property. When the button is clicked, this function is executed.

Inside the changeButtonColor function, we first retrieve the color code entered by the user in the text box using the get function. We then use the set function to change the BackgroundColor property of the button to the specified color (Fig. 9.15).

To change the button color, simply enter a valid color code (e.g., 'r' for red, 'g' for green, 'b' for blue, or a combination like '[1 0 0]' for red) in the text box and click the button (Fig. 9.16).

This example demonstrates how to change the color of a button within a GUI based on user input. By utilizing the get and set functions, you can modify various properties of GUI components, enabling you to create customized and visually appealing interfaces tailored to your application's needs (Fig. 9.17).

Fig. 9.15 The initial state of the GUI with a default button color

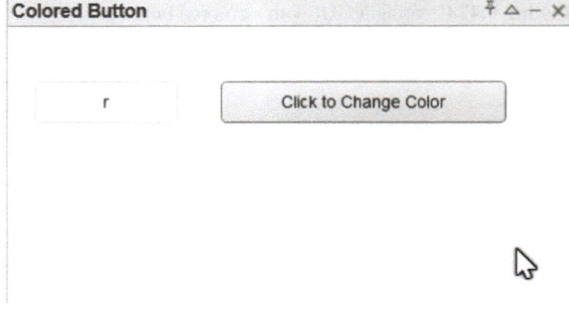

Fig. 9.16 The initial state of the GUI with a default button colour—red

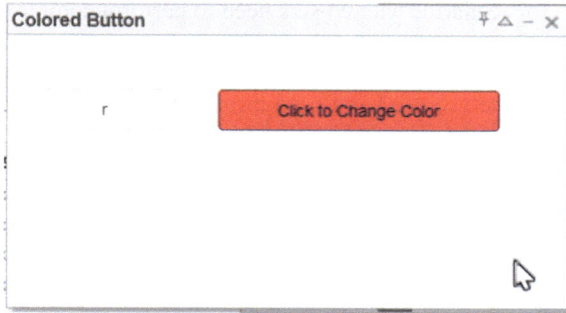

Fig. 9.17 The GUI with the button colour changed to green

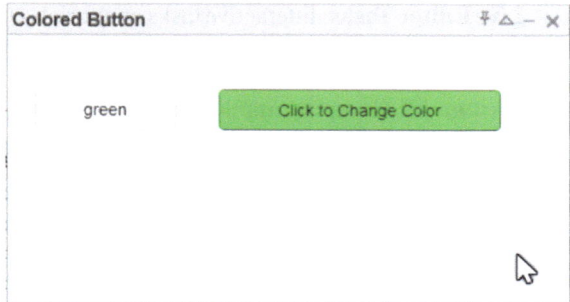

9.14 Apps Building in MATLAB

Creating self-contained applications, embedded Live Editor tasks, and custom user interfaces (UI) is a powerful feature in **MATLAB**. It enables developers to build interactive and user-friendly tools for data analysis, visualisation, simulation, and other computational tasks. **MATLAB** provides a range of tools and techniques for app development, making it easier to transform algorithms and scripts into polished and professional applications.

9.14.1 Types of Apps in MATLAB

MATLAB offers a versatile platform for developing various types of applications, each tailored to specific use cases and deployment scenarios. Here are the main types of apps that can be created in **MATLAB**:

- **Standalone Desktop Apps**: These are self-contained applications designed to run on desktop computers or laptops. They provide a rich user interface and can be deployed as executable files or installed applications, without requiring a full **MATLAB** installation on the target machine. Standalone desktop apps are well-suited for scenarios where users need to perform complex data analysis, simulations, or computations on local machines.

- **Web Apps**: **MATLAB** enables the development of web-based applications that can be accessed and used through a web browser. Web apps offer the advantage of remote access and collaboration, as they can be hosted on servers and accessed by multiple users over the internet or an intranet. Web apps are particularly useful for sharing interactive visualisations, dashboards, or tools with remote teams or stakeholders.
- **Embedded Apps**: **MATLAB** supports the creation of embedded applications that can be integrated into larger systems or workflows. These apps can take various forms, such as:

  - **Live Editor Tasks**: Interactive tasks or tutorials embedded within the **MATLAB** Live Editor environment, allowing users to explore and interact with code, data, and visualisations.
  - **Custom UI Components**: Reusable UI components, such as plots, sliders, or control panels, that can be integrated into existing **MATLAB** applications or third-party software.
  - **System Components**: **MATLAB** code and algorithms can be integrated as components or libraries within larger systems, leveraging **MATLAB**'s performance and computational capabilities.

  Embedded Apps are particularly useful for extending the functionality of existing software or systems, or for creating reusable components that can be shared and integrated across different projects.

  By supporting these different types of apps, **MATLAB** provides a flexible and powerful platform for developing and deploying interactive applications tailored to various use cases and deployment scenarios, meeting the diverse needs of researchers, engineers, and scientists across various domains.

9.14.2 App Development Tools

MATLAB provides a comprehensive set of tools and techniques to facilitate the development of various types of applications. These tools cater to different user preferences and application requirements, offering a range of options for creating and customizing user interfaces (UIs), handling events, and building desktop, web, or embedded apps. The main app development tools in **MATLAB** include:

- **App Designer**: A modern and integrated development environment (IDE) for creating and customizing user interfaces, handling events, and building desktop and web apps. App Designer offers a visual and code-based approach, allowing developers to design UIs using drag-and-drop tools or by writing code directly. It supports the creation of complex layouts, integration of UI components (such as buttons, sliders, and plots), and handling of user interactions through event callbacks. App Designer is the recommended tool for building new applications in **MATLAB**.

- **GUIDE (GUI Development Environment)**: A legacy tool for designing and building graphical user interfaces (GUIs) using a drag-and-drop interface. GUIDE provides a visual canvas for arranging UI components and generating the corresponding **MATLAB** code. While still supported, GUIDE is an older tool, and App Designer is recommended for new GUI development.
- **Coding with UI Components**: **MATLAB** offers a wide range of UI components that can be created and customized programmatically using **MATLAB** code. This approach provides greater flexibility and control over the UI design and behavior compared to visual tools like App Designer or GUIDE. It involves writing code to create and configure UI components, handle events, and manage the application's logic and data flow.
- **MATLAB Compiler and Deployment Tools**: In addition to the UI development tools, **MATLAB** provides tools for packaging and deploying applications to various target environments. The **MATLAB Compiler** allows developers to create standalone applications or components that can be distributed and run on computers without a full **MATLAB** installation. The **MATLAB Web App Server** enables deploying and sharing web apps with remote users over the internet or an intranet. The **MATLAB Production Server** provides a scalable and secure environment for deploying and managing **MATLAB** applications in production environments.

These app development tools in **MATLAB** offer a range of options to cater to different project requirements, user preferences, and deployment scenarios, enabling developers to create powerful and user-friendly applications for data analysis, visualisation, simulation, and other computational tasks.

9.14.3 Creating and Customising UI Components

Building interactive and visually appealing user interfaces is a crucial aspect of app development in **MATLAB**. The platform provides a rich set of UI components and tools for creating and customizing these components, enabling developers to craft intuitive and user-friendly applications. Here are some key aspects of creating and customizing UI components in **MATLAB**:

- **UI Component Library**: **MATLAB** offers a comprehensive library of UI components, including buttons, sliders, text boxes, drop-down menus, tables, plots, and axes. These components can be easily integrated into applications using App Designer, GUIDE, or programmatic coding.
- **Visual Design Tools**: App Designer and GUIDE provide visual design tools that allow developers to drag and drop UI components onto a canvas, arrange them in desired layouts, and customize their properties (such as size, color, and text) through intuitive interfaces.

- **Programmatic Customization**: In addition to visual design tools, **MATLAB** enables programmatic customization of UI components through code. This approach offers greater flexibility and control, allowing developers to dynamically modify component properties, handle events, and integrate UI components with application logic and data processing.
- **Event Handling**: UI components in **MATLAB** support event handling, enabling developers to define callback functions that execute when specific events occur (e.g., button click, slider movement, menu selection). This allows for seamless integration of user interactions with the application's functionality.
- **Layout Management**: **MATLAB** provides layout managers and tools for organizing UI components in various layouts, such as grids, tabbed panels, and nested containers. This ensures that the UI remains responsive and adapts to different screen sizes or window resizing.
- **Theming and Styling**: Developers can customize the appearance of UI components by applying themes, styles, and color schemes, ensuring a consistent and visually appealing look and feel across the application.
- **Data Visualisation**: **MATLAB** excels in data visualisation, offering a wide range of plot types and customization options for creating interactive and dynamic visualisations within the application's UI.

By leveraging these tools and techniques, developers can create highly interactive and user-friendly applications tailored to specific use cases, such as data analysis, simulation, or visualisation tasks. The ability to customize UI components and integrate them with application logic and data processing ensures a seamless and intuitive user experience.

9.14.4 Deploying and Sharing Apps

After developing an application in **MATLAB**, it is often necessary to deploy and share it with others, whether it's for collaboration, distribution, or production use. **MATLAB** provides several options for deploying and sharing applications, each tailored to specific deployment scenarios and target environments. Here are the main deployment and sharing options in **MATLAB**:

- **MATLAB Compiler**: The **MATLAB Compiler** is a powerful tool that allows you to create standalone applications or components that can be distributed and run on computers without a full **MATLAB** installation. It packages your **MATLAB** code, along with the necessary runtime components, into executables or libraries that can be run on various platforms (Windows, Linux, or macOS). This is particularly useful for sharing applications with users who do not have access to **MATLAB** or for deploying applications in production environments.
- **MATLAB Web App Server**: **MATLAB** enables the deployment and sharing of web-based applications through the **MATLAB Web App Server**. This server

allows you to host and serve web apps created with **MATLAB**, making them accessible to remote users over the internet or an intranet. Web apps offer the advantage of cross-platform compatibility, as they can be accessed and used through a web browser on any device, without the need for installation.

- **MATLAB Production Server**: The **MATLAB Production Server** is a scalable and secure environment designed for deploying and managing **MATLAB** applications in production settings. It provides tools for packaging, deploying, and monitoring **MATLAB** applications, as well as integrating them with other enterprise systems and databases. The **MATLAB Production Server** is particularly useful for organizations that need to deploy and manage mission-critical **MATLAB** applications at scale.
- **MATLAB Drive**: **MATLAB Drive** is a cloud-based service that allows you to share and collaborate on **MATLAB** files, including scripts, functions, and apps. It provides version control, commenting, and sharing capabilities, making it easier to collaborate with team members or distribute your work to others.

These deployment and sharing options in **MATLAB** cater to various use cases, from distributing standalone applications to remote users, to hosting web-based apps, to deploying applications in production environments, and facilitating collaboration and sharing among teams or students. By leveraging these tools, developers can ensure that their **MATLAB** applications reach their intended audience and are deployed in the most appropriate and efficient manner.

9.15 Programming for Simulink

9.15.1 Quick Introduction to Simulink

Simulink is a graphical programming environment for modeling, simulating, and analyzing multi-domain dynamic systems. It provides an intuitive **block diagram** environment where users can design and simulate systems using a comprehensive library of pre-built blocks and custom components. This subsection aims to provide a quick introduction to **Simulink** and its key features.

- **Overview of Simulink as a powerful graphical programming environment**: **Simulink** is a powerful tool for modeling and simulating complex systems across various domains, including control systems, signal processing, communications, and more. It enables users to create and modify block diagrams visually, making it easier to understand and analyze system behavior.
- **Importance of understanding algebraic loop concepts in system modeling and simulation**: Algebraic loops are a crucial concept in system modeling and simulation. They occur when there are algebraic constraints between the input and output signals of a block or a system of blocks. Understanding algebraic loops

is essential for accurate simulation and analysis of systems, as they can lead to numerical issues or convergence problems if not handled properly.

- **Introduction to the Simulink interface and its key components**: The **Simulink** interface consists of various components, including the Editor, Library Browser, Model Hierarchy, and Simulation Data Inspector. This subsection introduces these components and their roles in creating, modifying, and analyzing **Simulink** models.
- **Basic steps for creating and running simulations in Simulink**: This part covers the basic steps involved in creating a new **Simulink** model, adding blocks from libraries, connecting blocks, setting parameters, and running simulations. It also discusses common simulation settings and options, such as simulation time, solver settings, and data logging.
- **Exploring the available blocks and libraries in Simulink for system modeling**: **Simulink** provides a vast library of pre-built blocks for various domains, including continuous and discrete systems, signal processing, control design, and more. This subsection explores some of the commonly used block libraries and their applications in system modeling.

By providing an overview of **Simulink**, its interface, and the basic steps for creating and running simulations, this subsection lays the foundation for users to effectively Utilise **Simulink** for system modeling and simulation tasks.

9.15.2 What Is an S-Function

An **S-Function** (System-Function) in **Simulink** is a programming interface that allows users to incorporate custom algorithms, external code, and advanced functionality into their models. S-Functions provide a powerful and flexible way to extend the capabilities of **Simulink** beyond the pre-built blocks and libraries.

- **Exploring the concept of S-Functions in Simulink**: S-Functions are user-defined blocks that can be created using programming languages like C, C++, Fortran, or MATLAB. These blocks can be seamlessly integrated into **Simulink** models, enabling users to implement custom algorithms, complex mathematical operations, and advanced control strategies.
- **Significance and practical applications of S-Functions in custom algorithm implementation and model integration**: S-Functions are particularly useful when the desired functionality is not available in the standard **Simulink** block libraries or when existing blocks need to be customized for specific requirements. They enable users to incorporate proprietary algorithms, hardware-specific code, or third-party software into their **Simulink** models.
- **Understanding the structure and requirements of an S-Function**: An S-Function typically consists of several mandatory and optional callback functions that define its behavior. These functions handle tasks such as initialisation, output

computation, state updates, and termination. This subsection explains the structure of an S-Function and the requirements for each callback.

- **Steps for creating and implementing an S-Function in Simulink**: Creating an S-Function involves several steps, including defining the callback functions, compiling the S-Function code, and integrating it into a **Simulink** model. This part covers the step-by-step process of creating and implementing an S-Function, including best practices and guidelines.
- **Incorporating external code and algorithms into Simulink using S-Functions**: S-Functions allow users to incorporate external code and algorithms from various programming languages into their **Simulink** models. This subsection discusses the techniques and considerations for integrating external code, such as linking to external libraries, handling data exchange, and ensuring compatibility with different platforms and operating systems.

By understanding the concept and capabilities of S-Functions, users can extend the functionality of **Simulink** and create highly customized and specialized models tailored to their specific requirements.

9.15.3 Advanced Simulink Programming Techniques and Applications

This subsection delves into advanced programming techniques and applications of Simulink, enabling users to leverage its full potential for complex system modeling, simulation, and deployment tasks.

- **Modeling Continuous and Discrete Systems**: Explore techniques and best practices for modeling continuous-time and discrete-time systems in Simulink. Understand the use of integration algorithms, fixed-step and variable-step solvers, and handling sampling and quantization effects.
- **Model Verification and Validation**: Discover methods and tools available in Simulink for verifying and validating models. Learn about model checking, formal verification, code generation, and hardware-in-the-loop (HIL) testing.
- **Model Optimisation and Tuning**: Gain insights into strategies and tools for Optimising and tuning Simulink models. Explore techniques for parameter estimation, design exploration, performance analysis, and the use of optimisation algorithms and tools like Simulink Design optimisation.
- **Code Generation and Deployment**: Understand the process of generating code from Simulink models for deployment on various platforms, such as embedded systems, real-time targets, or production environments. Learn about available code generation options, optimisation techniques, and best practices for deploying Simulink models.
- **Advanced Topics and Applications**: Explore advanced topics and applications of Simulink, including model-based design, control system design, signal process-

ing, and real-time simulations. Discover the integration of Simulink with other MATLAB tools and toolboxes for specific application domains.

By mastering these advanced programming techniques and applications, users can unlock the full potential of Simulink for complex system modeling, simulation, and deployment tasks, enabling them to tackle challenging real-world problems more effectively.

9.16 Software Development Tools

This section covers various tools and utilities provided by MATLAB for software development, including debugging, performance optimisation, project organization, source control integration, and toolbox packaging. These tools are essential for developing, testing, and deploying robust and efficient MATLAB applications.

9.16.1 Debugging and Analysis

Debugging is an integral part of the software development process, allowing developers to diagnose and fix issues in their code. MATLAB provides several debugging tools to help identify and resolve problems efficiently.

- **MATLAB Debugger**: The MATLAB Debugger is a powerful tool that enables developers to execute code line by line, set breakpoints, inspect variables, and step through function calls. This tool is particularly useful for understanding code behavior and identifying logical errors.
- **Code Analyzer**: The Code Analyzer is a static analysis tool that checks MATLAB code for potential issues, such as syntax errors, coding style violations, and compatibility problems. It provides suggestions for improving code quality and maintainability.

The basic syntax for setting a breakpoint in the MATLAB Debugger is:

Listing 9.48 Setting a breakpoint.

```
dbstop if condition % Set a conditional breakpoint
dbstop at filename.m:line_number % Set a breakpoint at a
    specific line
```

Here's an example of using the MATLAB Debugger to step through a function:

Listing 9.49 Debugging a function.

```
function result = myFunction(x, y)
dbstop if x < 0 % Set a conditional breakpoint
z = x + y;
result = z^2;
```

```
end

x = -2; y = 3;
result = myFunction(x, y) % Execution will pause at the
    breakpoint
```

9.16.2 Performance and Memory

Optimising code performance and memory usage is crucial for efficient computations, especially when working with large datasets or complex algorithms. MATLAB provides several tools to help developers identify and address performance bottlenecks and memory issues.

- **Profiler**: The Profiler is a tool that analyzes the execution time of MATLAB functions and scripts, helping developers identify performance bottlenecks and optimise their code [1].
- **Memory Profiler**: The Memory Profiler is a tool that tracks memory usage in MATLAB, allowing developers to identify and address memory leaks and optimise memory consumption.

The basic syntax for starting the Profiler is:

Listing 9.50 Starting the Profiler.

```
profile on % Start profiling
% Code to be profiled
profile viewer % Open the Profile Viewer to analyze
    results
```

Here's an example of using the Profiler to analyze the performance of a function:

Listing 9.51 Profiling a function.

```
function result = computeIntensiveTask(n)
    result = zeros(n);
    for i = 1:n
        for j = 1:n
            result(i, j) = heavyComputation(i, j);
        end
    end
end

function y = heavyComputation(x1, x2)
    % Perform some complex calculations
    y = x1^2 + x2^3;
end

profile on
computeIntensiveTask(1000);
profile viewer
```

The Profiler will provide detailed information about the execution time of each function, allowing developers to identify and optimise performance bottlenecks.

9.16.3 Background Processing

MATLAB provides the ability to run code in the background, allowing users to perform other tasks or execute additional code while long-running computations or simulations are in progress.

- **Background Workers**: Background workers enable users to run MATLAB code in a separate process or thread, freeing up the main MATLAB session for other tasks [2].
- **Parallel Computing**: MATLAB supports parallel computing, which allows users to distribute computations across multiple cores or workers, significantly reducing execution time for computationally intensive tasks [3].

The basic syntax for creating and using a background worker is:

Listing 9.52 Creating a background worker.

```
w = parpool; % Create a background worker pool
parfor i = 1:N % Parallel for loop
% Code to be executed in parallel
end
delete(w); % Shut down the worker pool
```

Here's an example of using a background worker to perform a long-running computation:

Listing 9.53 Long-running computation in a background worker.

```
function result = computeLongTask(n)
w = parpool; % Create a background worker pool
parfor i = 1:n
result(i) = heavyComputation(i);
end
delete(w); % Shut down the worker pool
end

function y = heavyComputation(x)
% Perform some complex and time-consuming calculations
pause(0.1); % Simulating a long computation
y = x^3 + sin(x);
end

result = computeLongTask(1000); % Compute in the
    background
% Continue with other tasks or computations in the main
    MATLAB session
```

9.16.4 Projects

As MATLAB projects grow in complexity, it becomes increasingly important to organize and manage code, files, and settings. MATLAB provides project management tools to help developers maintain code structure, share files and settings, and interact with source control systems.

- **MATLAB Projects**: MATLAB Projects is a tool that allows users to organize and manage MATLAB files, settings, and dependencies within a project structure [4].
- **Project Shortcuts**: Project Shortcuts provide a convenient way to access frequently used files, folders, and commands within a MATLAB Project [5].

The basic syntax for creating a new MATLAB Project is:

Listing 9.54 Creating a new MATLAB Project.

```
projectName = 'MyProject'; % Name of the project
rootDir = '/path/to/project/directory'; % Root directory
    for the project
proj = matlab.project.createProject(projectName, rootDir
    );
```

Here's an example of creating and using a MATLAB Project:

Listing 9.55 Using a MATLAB Project.

```
% Create a new project
proj = matlab.project.createProject('MyProject', 'C:\
    Projects\MyProject');

% Add files to the project
addFile(proj, 'main.m');
addFile(proj, 'utils\helper.m');

% Set project properties
proj.RootFolder = 'C:\Projects\MyProject';
proj.SourceFolder = 'src';

% Open the project
openProject(proj);

% Access project files and settings
edit(proj.Files(1).Path); % Open the main.m file
```

9.16.5 Source Control Integration

Source control systems are essential for managing code changes, collaborating with team members, and tracking project history. MATLAB provides integration with popular source control systems, allowing developers to manage their code and collaborate effectively.

- **Git Integration**: MATLAB integrates with Git, a popular distributed version control system, enabling developers to perform common source control operations directly from within the MATLAB environment [6].
- **SVN Integration**: MATLAB also supports integration with Subversion (SVN), a widely used centralised version control system [7].

The basic syntax for initializing a Git repository in MATLAB is:

Listing 9.56 Initializing a Git repository.

```
gitRepo = matlab.project.repository.Git('repo_path');
gitRepo.init(); % Initialize the Git repository
```

Here's an example of using Git integration in MATLAB:

Listing 9.57 Using Git integration in MATLAB.

```
% Initialize a Git repository
gitRepo = matlab.project.repository.Git('C:\Projects\
    MyProject');
gitRepo.init();

% Add files to the repository
gitRepo.add('main.m');
gitRepo.add('utils');

% Commit changes
gitRepo.commit('Initial commit');

% Push changes to a remote repository
gitRepo.addRemote('origin', 'https://github.com/user/
    repo.git');
gitRepo.push('origin', 'master');
```

9.16.6 Testing Frameworks

Testing is a crucial aspect of software development, ensuring that code functions correctly and meets the specified requirements. MATLAB provides testing frameworks that help developers write and run automated tests for their code.

- **MATLAB Unit Testing Framework**: The MATLAB Unit Testing Framework allows developers to write and run unit tests for MATLAB functions and classes [8].
- **Test Automation**: MATLAB supports test automation, enabling developers to run tests automatically as part of the development workflow or continuous integration process [9].

The basic syntax for creating a unit test in MATLAB is:

Listing 9.58 Creating a unit test.

```
classdef MyTest < matlab.unittest.TestCase
% Test methods
methods (Test)
function testMyFunction(testCase)
% Test code for myFunction
% ...
end
end
end
```

Here's an example of a unit test for a simple function:

Listing 9.59 Unit test example.

```
function y = myFunction(x)
y = x^2;
end

classdef MyTest < matlab.unittest.TestCase
methods (Test)
function testMyFunction(testCase)
% Test case 1
x = 2;
expected = 4;
actual = myFunction(x);
testCase.verifyEqual(actual, expected);
% Test case 2
        x = -3;
        expected = 9;
        actual = myFunction(x);
        testCase.verifyEqual(actual, expected);
    end
end

end
```

In this example, we first define a simple function myFunction that takes a scalar input x and returns its square.

We then create a unit test class MyTest that extends the matlab.unittest.TestCase class. Inside the MyTest class, we define a test method testMyFunction that contains two test cases for the myFunction.

In the first test case, we set x to 2, and we expect the output y to be 4. We call the myFunction with x=2 and store the result in actual. We then use the verifyEqual method from the matlab.unittest.TestCase class to assert that the actual value is equal to the expected value of 4.

In the second test case, we set x to -3, and we expect the output y to be 9. We follow a similar process as the first test case, calling the myFunction with x=-3 and using verifyEqual to assert that the actual value matches the expected value of 9.

This unit test class can be executed using MATLAB's Unit Testing Framework, and the test results will indicate whether the myFunction behaves as expected for the provided test cases.

9.16.7 Build Automation

Build automation is the process of automating the compilation, packaging, and deployment of software applications. MATLAB provides tools and interfaces for creating and running build tasks, allowing developers to streamline the build process and ensure consistent and reproducible builds.

- **MATLAB Build Tools**: MATLAB Build Tools provide a set of commands and APIs for creating and running build tasks, such as compiling MATLAB code, packaging applications, and deploying to target environments [10].
- **Continuous Integration**: MATLAB supports integration with popular CI platforms, enabling developers to automatically build, test, and deploy their MATLAB applications as part of the continuous integration workflow.

The basic syntax for defining a build task in MATLAB is:

Listing 9.60 Defining a build task.

```
task = matlab.project.buildtask.CommandTask('TaskName');
task.setCommand('matlab -batch "myScript.m"');
task.addDependencies('myScript.m'); % Add file
    dependencies
```

Here's an example of defining and running a build task to compile a MATLAB application:

Listing 9.61 Running a build task.

```
% Define the build task
compileTask = matlab.project.buildtask.CommandTask('
    CompileApp');
compileTask.setCommand('matlab -batch "deployApp.m"');
compileTask.addDependencies({'deployApp.m', 'main.m', '
    utils*.m'});

% Run the build task
compileTask.run();
```

9.16.8 Continuous Integration (CI)

Continuous Integration (CI) is a software development practice that involves automatically building, testing, and deploying code changes as they are committed to the code repository. MATLAB supports integration with popular CI platforms, enabling developers to continuously develop and integrate their software.

- **CI Platform Integration**: MATLAB can be integrated with various CI platforms, such as Jenkins, Travis CI, and GitHub Actions, to enable automated building, testing, and deployment of MATLAB applications [11].

The basic syntax for defining a CI workflow in MATLAB depends on the specific CI platform being used. However, here's an example of a simple CI workflow using GitHub Actions:

Listing 9.62 GitHub Actions workflow for MATLAB.

```
name: MATLAB CI

on: [push]

jobs:
build:
runs-on: ubuntu-latest

steps:
- uses: actions/checkout@v2
- uses: matlab-actions/setup-matlab@v1

- name: Run tests
  run: matlab -batch "runTests"

- name: Build application
  run: matlab -batch "deployApp"
```

In this example, the CI workflow is triggered on every push to the code repository. It sets up the MATLAB environment, runs unit tests, and builds the application using MATLAB commands.

9.16.9 Toolbox Distribution

MATLAB allows developers to create and distribute toolboxes, which are collections of MATLAB files, data, and documentation organized into a single package. Toolbox distribution enables developers to share their work with others and facilitates code reuse and collaboration.

- **Toolbox Packaging**: MATLAB provides tools and utilities for packaging MATLAB code, data, and documentation into a toolbox file or installer [12].
- **Documentation Generation**: Developers can generate comprehensive documentation for their toolboxes, including function references, examples, and usage instructions, using MATLAB's documentation tools [13].

The basic syntax for creating a toolbox package in MATLAB is:

Listing 9.63 Creating a toolbox package.

```
packager = matlab.project.toolbox.ToolboxPackager('
    ToolboxName.prj');
packager.package(); % Create the toolbox package
```

Here's an example of creating a toolbox package and generating documentation:

Listing 9.64 Toolbox packaging and documentation.

```
% Create a toolbox project
proj = matlab.project.toolbox.ToolboxProject('MyToolbox'
    );
proj.addFiles('main.m', 'utils*.m');

% Generate documentation
doc = help.doc.matlab('MyToolbox', 'C:\Toolboxes\
    MyToolbox');
doc.generateMATLABPage(); % Generate HTML documentation

% Package the toolbox
packager = matlab.project.toolbox.ToolboxPackager(proj);
packager.package(); % Create the toolbox package file
```

9.16.10 Tool Qualification and Certification

In industries where safety and reliability are critical, such as aerospace and automotive, it is essential to qualify and certify the tools used for software development. MATLAB provides certification kits and tools to assist in qualifying MATLAB for use in safety-critical applications.

- **IEC Certification Kit**: The IEC Certification Kit allows users to qualify MATLAB for use in safety-critical applications according to the IEC 61508 and IEC 62304 standards [14].
- **DO Qualification Kit**: The DO Qualification Kit is designed to help users qualify MATLAB for use in airborne software development according to the RTCA DO-178C and RTCA DO-330 standards [15].

The process of qualifying MATLAB using the certification kits typically involves running a set of qualification tests, analyzing the results, and generating qualification reports. The specific steps and requirements may vary depending on the industry and applicable standards.

By following the guidelines and using the certification kits provided by Math-Works, developers can ensure that MATLAB meets the necessary requirements for use in safety-critical and regulated industries.

9.16.11 MATLAB Grader

MATLAB Grader is an interactive platform designed for creating and managing auto-graded MATLAB assignments. It is particularly useful in academic settings for enabling instructors to provide immediate feedback to students on their coding assignments. The platform supports a wide range of problem types and allows seamless integration with Learning Management Systems (LMS). MATLAB Grader facilitates the development of computational thinking and problem-solving skills, which are crucial in fields such as **AI, Industry 4.0**, and **Digital Manufacturing**.

9.16.11.1 Creating Assignments

Creating assignments in MATLAB Grader involves defining the problem statement, providing reference solutions, and setting up test cases for auto-grading. The platform's user-friendly interface assists instructors in designing assignments that can assess both the correctness and efficiency of students' solutions.

Listing 9.65 Creating a simple assignment in MATLAB Grader

```
% Problem Statement: Calculate the square of a number
function y = squareNumber(x)
    % Reference Solution
    y = x^2;
end

% Test cases to validate student submissions
assert(squareNumber(2) == 4);
assert(squareNumber(-3) == 9);
assert(squareNumber(0) == 0);
```

Listing 9.66 Another example of assignment creation

```
% Problem Statement: Check if a number is prime
function isPrime = checkPrime(n)
    % Reference Solution
    isPrime = true;
    if n <= 1
        isPrime = false;
    else
        for i = 2:sqrt(n)
            if mod(n, i) == 0
                isPrime = false;
                break;
            end
        end
    end
end

% Test cases to validate student submissions
```

```
assert(checkPrime(5) == true);
assert(checkPrime(4) == false);
assert(checkPrime(1) == false);
```

9.16.11.2 Managing Assignments

Managing assignments in MATLAB Grader involves monitoring student progress, reviewing submissions, and providing feedback. The platform's dashboard offers insights into common errors and allows instructors to adjust the difficulty level of problems based on student performance.

Listing 9.67 Example of managing assignments

```
% Example of a script to review student submissions
submission = getStudentSubmission('student1', '
    assignment1');
disp(['Student ID: ', submission.studentID]);
disp(['Submission Date: ', submission.date]);
disp(['Code: ', submission.code]);

% Providing feedback based on the submission
if submission.passed
    disp('Great job! Your solution is correct.');
else
    disp('Please review your code and try again.');
end
```

Listing 9.68 Another example of assignment management

```
% Script to generate a report of student performance
students = getAllStudents('assignment1');
for i = 1:length(students)
    submission = getStudentSubmission(students(i).id, '
        assignment1');
    disp(['Student ID: ', students(i).id]);
    disp(['Passed: ', num2str(submission.passed)]);
    disp(['Score: ', num2str(submission.score)]);
end
```

9.16.11.3 Integrating with Learning Management Systems

MATLAB Grader can be integrated with various Learning Management Systems (LMS), such as Moodle and Blackboard, enabling seamless assignment distribution and grading. This integration helps streamline the educational process by providing a unified platform for course management.

Listing 9.69 LMS integration example

```
% LMS integration using LTI (Learning Tools
    Interoperability)
```

```
lmsIntegration = setupLTI('consumer_key', 'shared_secret
    ', 'https://lms.example.com/');

% Example of retrieving assignment grades
grades = getGradesFromLMS(lmsIntegration, 'assignment1')
    ;
for i = 1:length(grades)
    disp(['Student ID: ', grades(i).studentID]);
    disp(['Grade: ', num2str(grades(i).grade)]);
end
```

Listing 9.70 Another LMS integration example

```
% Script to push grades back to the LMS
lmsIntegration = setupLTI('consumer_key', 'shared_secret
    ', 'https://lms.example.com/');
grades = getGradesFromLMS(lmsIntegration, 'assignment1')
    ;

% Update grades in the LMS
for i = 1:length(grades)
    updateGradeInLMS(lmsIntegration, grades(i).studentID
        , grades(i).grade);
end
```

9.16.11.4 Benefits of MATLAB Grader

MATLAB Grader offers several benefits, including:

- **Immediate Feedback**: Students receive instant feedback on their submissions, allowing them to learn from their mistakes and improve their skills.
- **Scalability**: MATLAB Grader can handle a large number of submissions, making it suitable for courses with many students.
- **Integration with LMS**: The platform integrates seamlessly with popula- LMS, facilitating the management of assignments and grades.
- **Customisation**: Instructors can create custom problems and test cases to match the learning objectives of their courses.

These benefits make MATLAB Grader an invaluable tool in modern education, particularly in fields that require strong computational skills.

In conclusion, MATLAB Grader is a powerful tool for both instructors and students. It enhances the learning experience by providing immediate feedback, scalable solutions, and seamless integration with LMS. By leveraging MATLAB Grader, educational institutions can better prepare students for the demands of modern engineering and computational fields.

9.16.12   MATLAB Cody

MATLAB Cody is an **online coding platform** provided by MathWorks that allows users to solve programming challenges and improve their MATLAB skills [16]. It offers a wide range of problems, from basic syntax and data manipulation to advanced algorithms and mathematical modelling. It is a valuable resource for both beginners and experienced users to practice and enhance their coding capabilities. The platform hosts a wide range of problems, from basic syntax to complex algorithmic challenges, allowing users to **learn by doing**.

One of the key features of MATLAB Cody is its **interactive learning environment**. Users can write and test their code directly in the browser, receiving immediate feedback on the correctness and efficiency of their solutions. The platform provides a built-in editor with syntax highlighting and code completion to enhance the coding experience.

- **Key Features of MATLAB Cody**:

  - **Interactive Problems**: Users can solve problems and get immediate feedback on their solutions.
  - **Community Engagement**: Users can see how others solved the same problem, offering diverse perspectives and techniques.
  - **Skill Improvement**: The problems are designed to progressively enhance MATLAB programming skills.
  - **Leaderboard and Achievements**: Users can track their progress and earn badges, making learning engaging and competitive.

  MATLAB Cody offers problems in various **difficulty levels**, ranging from beginner to advanced. Each problem comes with a clear problem statement, sample test cases, and constraints on the solution [17]. Users can submit their solutions and receive points based on the correctness and performance of their code.

  In addition to solving problems, MATLAB Cody encourages **collaboration and learning** among its users. The platform includes a discussion forum where users can ask questions, share insights, and learn from the community [16]. Users can also view other people's solutions after submitting their own, allowing them to explore different approaches and optimizations.

- **Benefits of Using MATLAB Cody**
  Using MATLAB Cody offers several benefits:

  - **Hands-on Practice**: Users can apply theoretical knowledge to practical problems.
  - **Instant Feedback**: Immediate results help users learn from their mistakes and improve their code.
  - **Community Learning**: Viewing solutions from other users can provide new insights and approaches to problem-solving.

These benefits make MATLAB Cody an essential tool for anyone looking to enhance their MATLAB programming skills in an interactive and community-driven environment.

MATLAB Cody is an excellent resource for both beginners and experienced MATLAB programmers looking to enhance their skills and learn new techniques. By solving real-world problems and engaging with the community, users can deepen their understanding of MATLAB and become more proficient in writing efficient and robust code.

9.17 Programming with AI

In this chapter, we will explore the exciting field of programming with **artificial intelligence (AI)** in MATLAB. AI has revolutionized various industries and has become an integral part of modern software development. MATLAB provides powerful tools and frameworks for implementing AI algorithms and applications.

9.17.1 MATLAB AI Chat Playground

The MATLAB AI Chat Playground is a web-based tool that provides a **conversational interface** for interacting with natural language processing (NLP) models directly within MATLAB. It leverages advanced language models trained on a vast corpus of data, including MATLAB code, documentation, and other relevant resources. This innovative feature allows users to ask questions, provide instructions, or describe their programming goals in natural language, and MATLAB will respond with appropriate code suggestions, explanations, or other relevant information [18].

The AI Chat Playground is designed to be **intuitive and user-friendly**, enabling users to communicate with MATLAB in a more natural and conversational manner. It incorporates **contextual understanding**, allowing it to interpret the user's intent and provide relevant responses based on the current context of the user's work.

One of the key advantages of the MATLAB AI Chat Playground is its ability to **generate code snippets** on demand. Users can describe the desired functionality in natural language, and the AI model will generate the corresponding MATLAB code. This feature can significantly accelerate the development process and reduce the time spent on low-level coding tasks [19].

The MATLAB AI Chat Playground also supports **interactive code exploration and debugging**. Users can ask questions about specific code snippets or functions, and the AI model will provide explanations, suggest improvements, or identify potential issues.

By leveraging the power of natural language processing and advanced language models, the MATLAB AI Chat Playground aims to enhance productivity, facilitate code exploration, and provide a more intuitive and natural programming experience for MATLAB users (Fig. 9.18).

9.17.2   ChatGPT

- Introduction to ChatPT [20]

  ChatGPT is a state-of-the-art **conversational AI** developed by Anthropic, a leading artificial intelligence research company. It is based on a large language model that has been trained on a vast amount of data, allowing it to understand and generate human-like text with remarkable fluency and coherence.

  In the context of MATLAB, users can leverage the power of ChatGPT to **generate code snippets, complete functions, or even entire programs** based on natural language prompts. This integration aims to bridge the gap between human language and machine code, enabling a more intuitive and efficient programming experience. One of the key advantages of using ChatGPT with MATLAB is its ability to **understand context and intent**. Users can provide high-level descriptions or requirements in natural language, and ChatGPT will generate the corresponding MATLAB code that fulfills those requirements. This feature can significantly reduce the time and effort required for coding tasks, especially for complex or unfamiliar problems.

- How ChatGPT can help with MATLAB Programming

  Here's an example of using ChatGPT to generate code for a **linear regression** function, using the prompt: **Write a MATLAB function to perform linear regression on a dataset of input features X and output targets y**, as given in Fig. 9.19. And the results as given in Fig. 9.20.

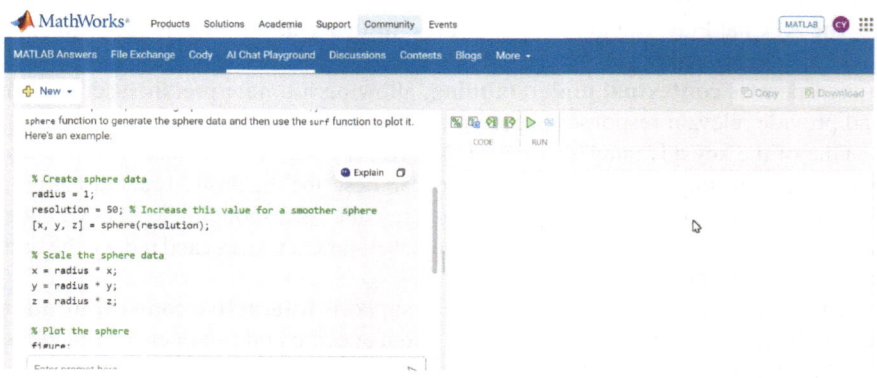

Fig. 9.18 Using the MATLAB AI chat playground

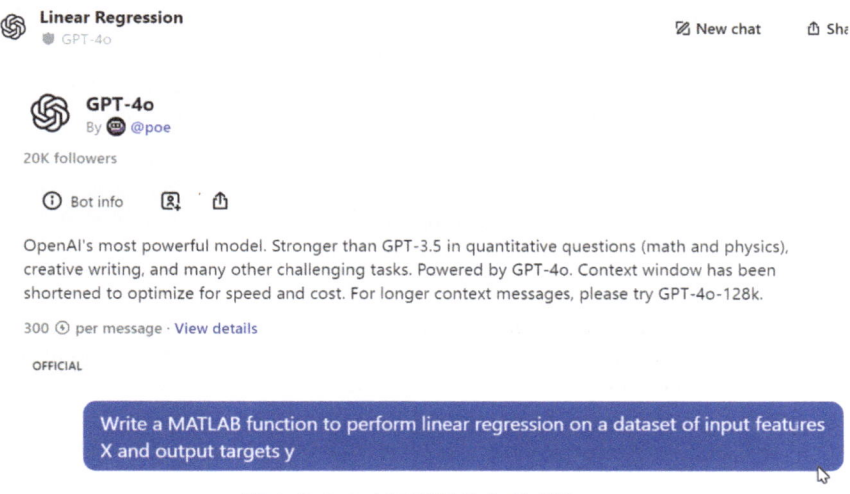

Fig. 9.19 Using ChatGPT with MATLAB—linear regression: prompt

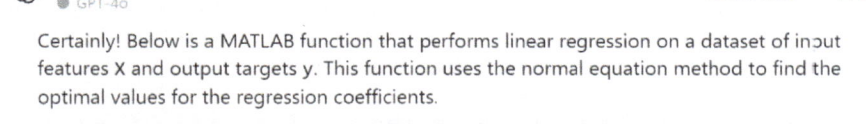

Certainly! Below is a MATLAB function that performs linear regression on a dataset of input features X and output targets y. This function uses the normal equation method to find the optimal values for the regression coefficients.

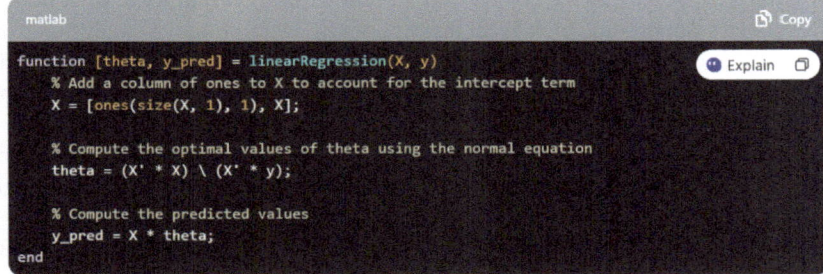

```matlab
function [theta, y_pred] = linearRegression(X, y)
    % Add a column of ones to X to account for the intercept term
    X = [ones(size(X, 1), 1), X];

    % Compute the optimal values of theta using the normal equation
    theta = (X' * X) \ (X' * y);

    % Compute the predicted values
    y_pred = X * theta;
end
```

Fig. 9.20 Using ChatGPT with MATLAB—linear regression: outcome from GPT4o

In this example, the user provides a natural language prompt describing the desired functionality, and ChatGPT generates a complete MATLAB function that performs linear regression on a given dataset using gradient descent optimisation.

ChatGPT can also provide **explanations and insights** into the generated code, helping users understand the underlying logic and algorithms. Additionally, it can suggest improvements, identify potential issues, or provide alternative approaches based on the user's requirements.

By integrating ChatGPT with MATLAB, developers can leverage the power of advanced language models to enhance their productivity, explore new ideas, and focus on higher-level problem-solving tasks while offloading low-level coding tasks to the AI assistant.

9.17.3 Cursor—Pair-Programming with AI

- Introduction to Cursor [21]

  Cursor is an **AI-first code editor** designed for pair-programming with artificial intelligence. It is built on top of Visual Studio Code (VSCode), a popular and widely-used code editor, and aims to enhance the software development experience by seamlessly integrating AI capabilities into the coding workflow.

  One of the core features of Cursor is its ability to **understand the context of a project** and provide intelligent code suggestions tailored to the specific codebase. It achieves this by analysing the entire project repository, enabling the AI to provide relevant and accurate recommendations based on the existing code, files, and documentation.

  Cursor allows developers to **ask questions about their codebase** and receive answers directly within the editor. This feature can save significant time by eliminating the need to search through code or documentation manually. Additionally, developers can refer directly to code definitions, files, and documentation, further streamlining the development process.

  A unique aspect of Cursor is its ability to **follow natural language instructions**. Developers can provide high-level instructions or descriptions of the desired functionality, and Cursor's AI will generate the corresponding code, handling low-level implementation details. This feature promotes a more efficient and intuitive workflow, allowing developers to focus on higher-level problem-solving tasks.

  Cursor also supports **editing code in natural language**. Developers can prompt the AI to modify an entire method or class with a single instruction, streamlining the process of making complex changes to the codebase.

  Furthermore, Cursor introduces a new feature called "Copilot++," which is a **more powerful version of the Copilot AI assistant**. Copilot++ is trained to autocomplete on sequences of edits, making it quick to understand the changes a developer is making. It can suggest mid-line completions and entire diffs, further enhancing the AI-assisted coding experience.

  To ensure a seamless transition for existing VSCode users, Cursor offers **one-click migration** of all VSCode extensions, themes, and keybindings. This feature allows developers to continue using their preferred tools and customizations while benefiting from Cursor's AI-powered capabilities.

  Cursor also addresses privacy and security concerns by providing a **privacy mode** option. In this mode, none of the user's code is stored on Cursor's servers or logs, ensuring that sensitive or proprietary code remains secure and confidential.

  The basic syntax for using Cursor's AI assistant is:

// Ask a question or provide an instruction using natural language cursor.ask ('Instruction or question')

Here's an example of using Cursor to generate code for a simple function:

// Prompt Cursor to create a function that calculates the factorial of a number cursor.ask('Create a function that calculates the factorial of a given number')

// Cursor generates the following code: function factorial(n) if (n === 0) return 1; return n * factorial(n - 1);

In this example, as shown in Fig. 9.21, the user provides a natural language prompt, and Cursor generates a recursive function that calculates the factorial of a given number.

By integrating advanced AI capabilities directly into the code editor, Cursor aims to streamline the software development process, enhance productivity, and provide a more intuitive and efficient coding experience for developers. As shown in Fig. 9.22, Cursor is able to support a few programming languages.

- How Cursor can help with MATLAB Programming

While Cursor is primarily designed for general-purpose programming languages like Python, JavaScript, and TypeScript, it can also be leveraged for MATLAB programming. By integrating with MATLAB's development environment, Cursor enables AI-assisted coding capabilities for MATLAB users.

One of the key advantages of using Cursor with MATLAB is its ability to **understand MATLAB syntax and idioms**. The AI model powering Cursor has been trained on a vast corpus of MATLAB code, allowing it to provide intelligent suggestions and code generation tailored to the MATLAB language.

Cursor can assist MATLAB developers in various tasks, such as **generating code snippets**, **completing partially written functions**, or even **creating entire scripts or programs** based on natural language prompts. This can significantly accelerate the development process and reduce the time spent on low-level coding tasks.

The basic syntax for using Cursor with MATLAB is:

Listing 9.71 Using Cursor with MATLAB.

```
cursor.ask('Ask a question or provide an instruction')
```

Here's an example of using Cursor to generate code for a MATLAB function that calculates the mean of a vector:

Listing 9.72 Generating code for calculating the mean.

```
cursor.ask('Create a function that calculates the mean
    of a vector')

% Cursor generates the following code:
function mean_val = calculate_mean(vec)
sum_val = sum(vec);
length_vec = length(vec);
mean_val = sum_val / length_vec;
end
```

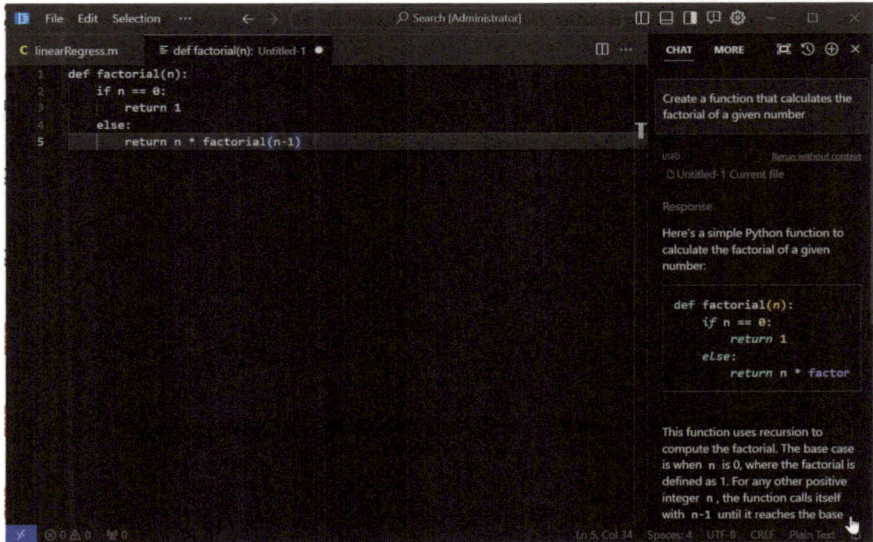

Fig. 9.21 Using cursor for AI-assisted coding—an example of factorial

In this example, the user provides a natural language prompt, and Cursor generates a MATLAB function that calculates the mean of a given vector.

Cursor can also assist with **code refactoring** and **optimisation** tasks in MATLAB. Developers can ask Cursor to suggest improvements or alternative approaches to existing code, leveraging the AI's understanding of best practices and coding patterns.

Additionally, Cursor can provide **explanations and insights** into MATLAB code, functions, and algorithms. Developers can ask questions about specific code snippets or MATLAB features, and Cursor will provide relevant explanations and documentation references, facilitating a deeper understanding of the language and its capabilities. As shown in Fig. 9.22.

By integrating Cursor with MATLAB's development environment, users can benefit from AI-powered code generation, code exploration, and intelligent assistance, streamlining their MATLAB programming workflows and enhancing productivity.

Overall, programming with AI in MATLAB involves leveraging AI models and tools like the MATLAB AI Chat Playground, ChatGPT, and Cursor to enhance the coding experience, automate tasks, and improve productivity (Fig. 9.23).

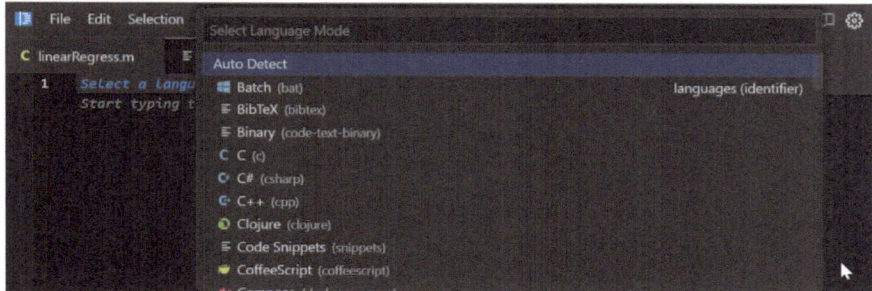

Fig. 9.22 Using cursor for programming languages

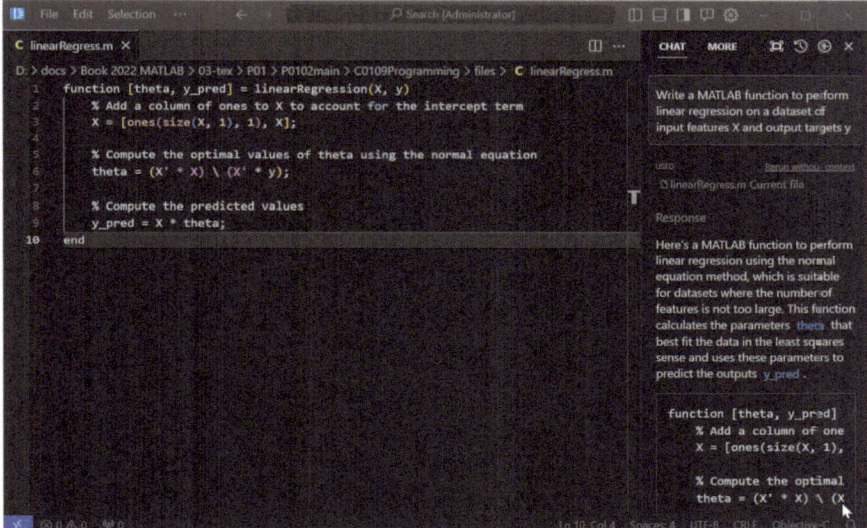

Fig. 9.23 Using cursor for AI-assisted coding—an example of linear regression

9.18   Laboratory

The following laboratory exercises are designed to reinforce the concepts covered in this chapter and provide hands-on experience with programming and algorithm development in MATLAB.

1. Implementing **If-Elseif-Else** Statements

   a. Write a MATLAB script that prompts the user to enter a number and determines whether it is positive, negative, or zero using **if-elseif-else** statements.

Listing 9.73 Determine the sign of a number.

```
num = input('Enter a number: ');

if num > 0
disp('The number is positive.')
elseif num < 0
disp('The number is negative.')
else
disp('The number is zero.')
end
```

b. Implement a MATLAB function that takes three numbers as input and returns the maximum value using **if-elseif-else** statements.

Listing 9.74 Find the maximum of three numbers.

```
function max_val = findMax(num1, num2, num3)
if num1 >= num2 && num1 >= num3
max_val = num1;
elseif num2 >= num1 && num2 >= num3
max_val = num2;
else
max_val = num3;
end
end
```

2. Working with **Switch** Statements

a. Write a MATLAB script that prompts the user to enter a character ('a', 'b', 'c', etc.) and displays the corresponding vowel or consonant using a **switch** statement.

Listing 9.75 Determine if a character is a vowel or consonant.

```
char = input('Enter a character: ', 's');

switch lower(char)
case 'a'
disp('The character is a vowel.')
case 'e'
disp('The character is a vowel.')
case 'i'
disp('The character is a vowel.')
case 'o'
disp('The character is a vowel.')
case 'u'
disp('The character is a vowel.')
otherwise
disp('The character is a consonant.')
end
```

b. Implement a MATLAB function that takes a month number (1-12) as input and returns the corresponding month name using a **switch** statement.

Listing 9.76 Get month name from month number.

```
function month_name = getMonthName(month_num)
switch month_num
case 1
month_name = 'January';
case 2
month_name = 'February';
case 3
month_name = 'March';
% ... (cases for other months)
otherwise
month_name = 'Invalid month number';
end
end
```

3. Code Optimisation and Vectorisation

   a. Write a MATLAB script that calculates the square root of each element in a large vector (e.g., 1 million elements) using both a **for** loop and vectorised operation. Compare the execution times of the two approaches.
   b. Implement a MATLAB function that computes the dot product of two vectors using a **for** loop. Then, optimise the function by using a vectorised operation instead of the loop.
   c. Create a MATLAB script that generates a large random matrix (e.g., 1000 x 1000) and calculates the sum of all elements greater than a specified threshold value using both a nested loop and a vectorised operation. Compare the execution times of the two approaches.

4. Debugging Conditional Statements

   a. Write a MATLAB script that implements a simple calculator with basic arithmetic operations (+, -, *, /). Use **if-elseif-else** statements to perform the appropriate operation based on user input. Debug the script using techniques such as **disp()** statements and the MATLAB debugger.
   b. Implement a MATLAB function that takes a year as input and determines whether it is a leap year or not using **if** statements. Debug the function by creating test cases for different scenarios (e.g., leap years, non-leap years, edge cases).
   c. Create a MATLAB script that prompts the user to enter their age and displays a message indicating their life stage (e.g., child, teenager, adult, senior) using a **switch** statement. Debug the script using techniques such as **fprintf()** statements and the MATLAB debugger.

5. Choosing Appropriate Conditional Statements

   a. Write a MATLAB script that prompts the user to enter a grade (0-100) and displays the corresponding letter grade using both **if-elseif-else** and **switch** statements. Compare the readability and maintainability of the two approaches.

   b. Implement a MATLAB function that takes a string as input and determines whether it is a palindrome (a word, phrase, number, or other sequence of characters that reads the same backward as forward) using **if** statements. Discuss scenarios where **switch** statements might be more appropriate for string comparisons.

   c. Create a MATLAB script that performs different operations based on user input (e.g., calculating the area of a circle, rectangle, or triangle). Use **if-elseif-else** statements for the logical flow and discuss the appropriateness of using **switch** statements in this context.

6. Improving MATLAB Codes with AI

   a. Any suggestions to improve the quality of MATLAB codes

      • Follow best practices for code organization, naming conventions, and commenting
      • optimise performance by vectorizing operations and avoiding unnecessary loops
      • Improve readability and maintainability by breaking down complex logic into smaller functions
      • Ensure proper error handling and input validation
      • Leverage built-in MATLAB functions and data structures when appropriate
      • Consider code modularity and reusability

   b. Based on the above principles, can you help to generate a prompt for the given MATLAB codes improvement?
      Prompt: "Improve the given MATLAB code by addressing the following aspects: code organization, performance optimisation through vectorization, readability and maintainability, error handling and input validation, leveraging built-in MATLAB functions and data structures, and modularization for reusability. Provide suggestions for refactoring the code while preserving its functionality."

   c. Using the prompt above, help to improve the MATLAB codes below:

   Listing 9.77 Original MATLAB code.

   ```
   % Calculate the dot product of two vectors
   a = [1 2 3 4 5];
   b = [6 7 8 9 10];
   dot_product = 0;

   for i = 1:length(a)
   ```

```
dot_product = dot_product + a(i)*b(i);
end

disp(['The dot product of vectors a and b is: '
    num2str(dot_product)]);
```

Improved MATLAB code:

Listing 9.78 Improved MATLAB code.

```
% Function to calculate the dot product of two
    vectors
function dot_product = vectorDotProduct(vec1, vec2
    )
% Input validation
if ~isvector(vec1) || ~isvector(vec2) || length(
    vec1) ~= length(vec2)
error('Input vectors must have the same length.');
end

% vectorised dot product calculation
dot_product = sum(vec1 .* vec2);
end

% Example usage
a = [1 2 3 4 5];
b = [6 7 8 9 10];

result = vectorDotProduct(a, b);
disp(['The dot product of vectors a and b is: '
    num2str(result)]);
```

Explanation:

- The code is modularized into a function **vectorDotProduct** for better reusability.
- Input validation is added to ensure that the input vectors are valid and have the same length.
- The dot product calculation is vectorised using the built-in **sum** function and element-wise multiplication (**.***) operator, improving performance.
- The function returns the dot product value, allowing for flexible usage in other parts of the code.
- An example usage is provided to demonstrate how to call the function.

The improved code follows best practices for code organization, performance optimisation through vectorization, readability and maintainability, error handling and input validation, and leveraging built-in MATLAB functions. It also promotes modularization and reusability by encapsulating the dot product calculation in a separate function.

7. Using AI for Code Generation

   a. Explore the capabilities of AI language models, such as GPT-3 or CodeGPT, in generating MATLAB code snippets or functions based on natural language descriptions or specifications.

      • Provide a natural language description of a desired MATLAB function or code snippet (e.g., "Write a function to calculate the mean and standard deviation of a vector of numbers").
      • Use an AI language model to generate the corresponding MATLAB code.
      • Evaluate the generated code for correctness, efficiency, and adherence to best practices.
      • Refine the natural language description or prompt if the generated code requires improvements.

   b. Investigate the use of AI-assisted code completion tools, such as Tabnine or Codex, that can suggest code snippets or autocomplete MATLAB code based on context and machine learning models.

      • Set up an AI-assisted code completion tool in your MATLAB development environment.
      • Write MATLAB code and observe how the tool suggests code completions or snippets based on your input.
      • Assess the accuracy and usefulness of the suggested code completions in various scenarios.
      • Discuss the potential benefits and limitations of using AI-assisted code completion tools in MATLAB programming.

8. AI for Code Optimisation and Refactoring

   a. Explore the use of AI-powered code optimisation and refactoring tools, such as Microsoft's IntelliCode or Amazon CodeGuru, which can analyze MATLAB code and suggest improvements based on machine learning models.

      • Provide a MATLAB code sample or script that could benefit from optimisation or refactoring.
      • Use an AI-powered code optimisation or refactoring tool to analyze the code and generate suggestions for improvements.
      • Evaluate the suggested improvements in terms of performance, readability, maintainability, and adherence to best practices.
      • Implement the recommended changes and compare the optimised code with the original version.

   b. Investigate the use of AI-powered code summarization tools, which can automatically generate summaries or documentation for MATLAB code based on machine learning models.

- Provide a MATLAB code sample or script that requires documentation or a high-level summary.
- Use an AI-powered code summarization tool to generate a summary or documentation for the provided code.
- Assess the accuracy and usefulness of the generated summary or documentation.
- Discuss the potential benefits and limitations of using AI-powered code summarization tools in MATLAB programming.

9. Can AI solve equations?
   Let's look at lab work 2c, the equations are given in Eq. (3.13).
   Prompt for ChatGPT or other AI chatbot:
   Please solve the following system of linear equations:

$$
\begin{aligned}
4.6x_1 - 2.31x_2 + 8.3x_3 + 29.4x_4 &= 40.34 \\
20.5x_1 + 8.7x_2 + 40.1x_3 - 11.9x_4 &= 1.15 \\
36.4x_1 + 0.92x_2 - 3.7x_3 + 64.3x_4 &= 32.4 \\
7.84x_1 + 40.01x_2 - 2.68x_3 - 7.92x_4 &= 27.55
\end{aligned}
\tag{9.1}
$$

Provide the values of x_1, x_2, x_3, and x_4 that satisfy all the equations simultaneously.

If the AI tool allows for additional instructions or preferences, you can mention the desired method for solving the system of equations, such as Gaussian elimination, matrix inversion, or Cramer's rule.

The choice of the best method to solve a system of linear equations depends on various factors, such as the size of the system, the sparsity of the coefficient matrix, and the computational resources available, specifically,

a. Gaussian Elimination:

- Gaussian elimination is a widely used method for solving systems of linear equations.
- It is efficient for systems with a moderate number of equations and variables.
- The time complexity of Gaussian elimination is approximately $O(n^3)$, where n is the number of equations.
- Gaussian elimination can handle systems with unique solutions, no solutions, or infinitely many solutions.
- It is relatively simple to implement and is numerically stable for well-conditioned systems.

b. Matrix Inversion:

- Matrix inversion involves finding the inverse of the coefficient matrix and multiplying it with the constant terms to obtain the solution.

- It is not the most efficient method for solving large systems of equations due to the computational cost of matrix inversion, which is approximately $O(n^3)$.
- Matrix inversion is sensitive to round-off errors and can be numerically unstable for ill-conditioned matrices.
- It is generally not recommended for solving systems of linear equations, especially when the coefficient matrix is large or ill-conditioned.

c. Cramer's Rule:

- Cramer's rule is a formula that expresses the solution of a system of linear equations in terms of determinants of the coefficient matrix and its submatrices.
- It is mainly used for small systems of equations (e.g., 2x2 or 3x3) due to its computational complexity.
- The time complexity of Cramer's rule is approximately $O(n!)$, which becomes impractical for large systems.
- Cramer's rule is numerically unstable for large systems and can suffer from round-off errors.

Considering the given system of linear equations with 4 equations and 4 variables, Gaussian elimination would be the most suitable choice among the three methods mentioned. It is efficient, numerically stable, and can handle systems of this size effectively.

However, it's worth noting that there are other methods available for solving systems of linear equations, such as LU decomposition, QR decomposition, and iterative methods like Jacobi iteration or Gauss-Seidel iteration. These methods have their own advantages and are particularly useful for large and sparse systems.

In practice, most mathematical software and libraries (like MATLAB, NumPy, or LAPACK) use optimised algorithms based on LU decomposition or other advanced techniques to solve systems of linear equations efficiently and accurately.

10. Function to Compute the Sequence Approximation of $\pi/4$

a. To compute the sequence that approximates $\pi/4$ using the series $\pi/4 \approx 1 - \frac{1}{3} + \frac{1}{5} - \frac{1}{7} + \cdots + \frac{(-1)^n}{2n+1}$, you can write a MATLAB function as shown below.

Listing 9.79 Improved MATLAB code.

```
% Function to calculate the dot product of two
    vectors
function pi_over_4 = compute_pi_over_4 (n)
    % COMPUTE_pi_over_4 Computes an approximation
        of pi/4 using the
    % Gregory-Leibniz series.
    %
    %    pi_over_4 = COMPUTE_pi_over_4 (n) computes
        the approximation
```

```
%     of pi/4 using the first n terms of the
      Gregory-Leibniz series.
%
%     Input:
%           n - Number of terms to use in the
      approximation
%
%     Output:
%           pi_over_4 - Approximation of pi/4

    % Initialise the approximation
    pi_over_4 = 0;

    % Compute the approximation using the series
    for k = 0: 1: n-1
        pi_over_4 = pi_over_4 + ((-1)^k) / (2*k +
           1);
    end
end
```

b. Explanation
   i. **Function Definition**:
      • 'function pi_over_4 = compute_pi_over_4(n)': Defines a function named 'compute_pi_over_4' that takes one input argument, n, which is the number of terms to use in the approximation.
   ii. **Function Documentation**:
      • The comments explain the purpose of the function, its input, and its output. This is good practice for making the code easier to understand and maintain.
   iii. **Initialisation**:
      • 'pi_over_4 = 0;': Initialises the variable 'pi_over_4' to zero. This will hold the cumulative sum of the series.
   iv. **Series Computation**:
      • 'for k = 0:n-1': A 'for' loop that iterates from $k = 0$ to $k = n - 1$.
      • '$pi\_over\_4 = pi\_over\_4 + ((-1)^k)/(2*k + 1);$': Updates the approximation by adding the k-th term of the Gregory-Leibniz series to 'pi_over_4'.

c. MATLAB Live Script for Visualising Moving Average
   The following MATLAB code creates a live script to visualise the moving average of a given stock's price. It includes a sliding bar to adjust the moving average parameter:

```
% Live Script: Visualise Moving Average of Stock
   Price

% Load stock price data
% Assuming 'stockPrices' is a vector containing the
   stock prices
% Example: stockPrices = [100, 101, 102, 100, 99,
   98, 97, 96, 95, 94];
```

```matlab
% Create a UI figure window
fig = uifigure('Name', 'Moving Average Visualisation
    ');

% Create an axis for plotting
ax = uiaxes(fig, 'Position', [50 100 700 400]);

% Create a slider for adjusting the moving average
    window size
slider = uislider(fig, ...
    'Position', [50 50 700 3], ...
    'Limits', [1 50], ...
    'Value', 5, ...
    'MajorTicks', 1:5:50, ...
    'MinorTicks', 1:1:50, ...
    'ValueChangedFcn', @(src, event) updatePlot(ax,
        stockPrices, round(src.Value)));

% Initial plot
updatePlot(ax, stockPrices, round(slider.Value));

% Function to update the plot
function updatePlot(ax, stockPrices, windowSize)
    % Calculate the moving average
    movAvg = movmean(stockPrices, windowSize);

    % Plot the original stock prices
    plot(ax, stockPrices, 'b', 'DisplayName', 'Stock
        Price');
    hold(ax, 'on');

    % Plot the moving average
    plot(ax, movAvg, 'r', 'DisplayName', ['Moving
        Average (Window: ' num2str(windowSize) ')']);
    hold(ax, 'off');

    % Add legend and labels
    legend(ax, 'show');
    title(ax, 'Stock Price and Moving Average');
    xlabel(ax, 'Time');
    ylabel(ax, 'Price');
end
```

- Use a live script to visualise the moving average of the given stock's price.
- Adjust the parameter of the moving average using a sliding bar and observe the changes in the moving average.

11. MATLAB APP Development Requirements

   - Please use the MATLAB APP design function to develop a calculator APP.

Fig. 9.24 MATLAB APP development

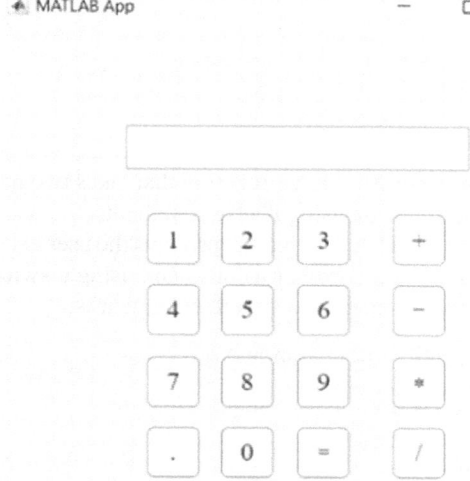

- Specific requirements:
  - The APP should have at least four basic operations: addition, subtraction, multiplication, and division.
  - Compile the APP so that it can run independently outside MATLAB. To compile the app so that it can run independently outside MATLAB:
    1. Open the App Designer and create the app using the code provided. 2. Save the app as a '.mlapp' file. 3. In MATLAB, go to the "HOME" tab and click on "Apps". 4. Select "Application Compiler". 5. Add your '.mlapp' file to the project. 6. Follow the prompts to package the app, which will generate an executable that can run independently of MATLAB.

These lab works aim to introduce students to the emerging field of AI-assisted programming and explore how AI technologies can be leveraged to enhance various aspects of MATLAB programming, such as code generation, optimisation, refactoring, and documentation. Students can gain hands-on experience with AI tools and evaluate their potential benefits and limitations in the context of MATLAB development (Fig. 9.24).

9.19 Problems

1. Write a MATLAB script that prompts the user to enter a number and determines whether it is positive, negative, or zero using **if-else** statements.
   The basic syntax for an **if-else** statement is:

Listing 9.80 If-else statement syntax.

```
if condition
statements
else
statements
end
```

2. Implement a MATLAB function that takes two numbers as input and returns the maximum value using **if-else** statements.

3. Write a MATLAB script that prompts the user to enter a character and determines whether it is a vowel or a consonant using a **switch** statement.
The basic syntax for a **switch** statement is:

Listing 9.81 Switch statement syntax.

```
switch expression
case case1
statements
case case2
statements
...
otherwise
statements
end
```

4. Implement a MATLAB function that takes a year as input and determines whether it is a leap year or not using **if-else** statements. A year is considered a leap year if it is divisible by 4, except for years divisible by 100, which are not leap years unless they are also divisible by 400.

5. Write a MATLAB script that prompts the user to enter a letter grade (A, B, C, D, or F) and calculates the corresponding grade point average (GPA) using **if-elseif-else** statements.
The basic syntax for an **if-elseif-else** statement is:

Listing 9.82 If-elseif-else statement syntax.

```
if condition1
statements
elseif condition2
statements
...
else
statements
end
```

6. Implement a MATLAB function that takes three numbers as input and returns the median value (the middle value when the numbers are arranged in ascending or descending order) using **if-elseif-else** statements.

7. Write a MATLAB script that generates a random integer between 1 and 10, prompts the user to guess the number, and provides feedback using **if-else** statements. The script should continue prompting the user until the correct guess is made.

Fig. 9.25 Line plot with
error band using ChatGPT

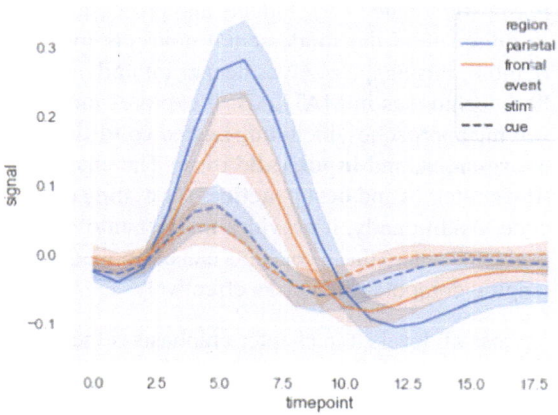

8. Implement a MATLAB function that takes a character as input and determines whether it is an uppercase letter, lowercase letter, or neither using a **switch** statement.
9. Write a MATLAB script that prompts the user to enter their age and displays a message indicating their eligibility to vote based on the following criteria:
   If the age is 18 or above, the user is eligible to vote. If the age is below 18, the user is not eligible to vote.
10. Implement a MATLAB function that takes a string as input and determines whether it is a palindrome (a word, phrase, number, or other sequence of characters that reads the same backward as forward) using **if** statements.
11. Utilise Computational Intelligence Aided Design (CIAD) techniques [22–24] to create a digital twin of a Micro Aerial Vehicle (MAV) robotic system for precision agriculture applications in future farming scenarios.
12. Please collaborate with ChatGPT or a similar AI tool to create a MATLAB function that can generate plots similar to the example figure provided in Fig. 9.25 [25].

9.20 Summary

- For undergraduate (UG) students: This chapter serves as a comprehensive introduction to **conditional statements**, **logical operations**, and **decision-making processes** in MATLAB. Through a diverse set of problems and well-explained examples, students will gain a solid understanding of **if-else** statements, **switch** statements, and their applications in various scenarios. The chapter emphasises the importance of **problem-solving skills** and **algorithmic thinking**, providing a strong foundation for further exploration of programming concepts and techniques in MATLAB.

- For postgraduate (PG) students and professional researchers or engineers:
  Building upon the fundamental concepts introduced in the undergraduate curriculum, this chapter delves deeper into **advanced decision-making** and **control flow** techniques in MATLAB. It explores more complex scenarios and problem-solving approaches, including **nested conditional statements**, **compound logical expressions**, and **input validation**. The chapter also introduces **code optimisation** strategies and **best practices** for writing efficient and maintainable MATLAB code. Additionally, it provides insights into real-world applications and research problems where these concepts can be applied, empowering readers to tackle challenging computational tasks effectively.

Across all levels, the chapter emphasises the following key concepts:

- **Conditional statements**: Understanding and utilising **if-else** and **switch** statements for decision-making in MATLAB programs.
- **Logical operations**: Mastering the use of **logical operators** (AND (&&), OR (||), and NOT (~)) and **relational operators** (>, <, ==, !=, >=, <=) for constructing complex logical expressions.
- **Control flow**: Leveraging **conditional statements** and **loops** to control the flow of program execution based on specific conditions or criteria.
- **Input validation**: Implementing robust techniques to ensure the validity and integrity of user inputs or data sources.
- **Problem-solving skills**: Developing the ability to break down complex problems into smaller, manageable steps and devise efficient algorithmic solutions.
- **Code optimisation**: Exploring strategies for writing efficient, maintainable, and scalable MATLAB code, including **code readability**, **modularity**, and **performance considerations**.

By mastering these concepts and techniques, students, researchers, and professionals will be well-equipped to tackle a wide range of computational challenges, from simple scripting tasks to complex data analysis and simulations, using MATLAB as a powerful and versatile programming environment.

References

1. MathWorks, "Profile Your Code to Improve Performance," https://www.mathworks.com/help/matlab/matlab_prog/profiling-for-improving-performance.html, accessed on Feb. 17, 2024
2. MathWorks, "Run Functions in Background," https://www.mathworks.com/help/matlab/matlab_prog/run-functions-in-the-background.html, accessed on Feb. 17, 2024
3. MathWorks, "Get Started with Parallel Computing Toolbox," https://www.mathworks.com/help/parallel-computing/getting-started-with-parallel-computing-toolbox.html, accessed on Feb. 17, 2024
4. MathWorks, "Projects," https://www.mathworks.com/help/matlab/projects.html, accessed on Feb. 17, 2024

5. MathWorks, "Project Shortcuts," https://www.mathworks.com/help/matlab/ref/matlab.project.project.addshortcut.html, accessed on Feb. 17, 2024
6. MathWorks, "Set Up Git Source Control," https://www.mathworks.com/help/matlab/matlab_prog/set-up-git-source-control.html, accessed on Feb. 17, 2024
7. MathWorks, "Set Up SVN Source Control," https://www.mathworks.com/help/matlab/matlab_prog/set-up-svn-source-control.html, accessed on Feb. 17, 2024
8. MathWorks, "Write Unit Tests," https://www.mathworks.com/help/matlab/write-unit-tests.html, accessed on Feb. 17, 2024
9. MathWorks, "Automated Testing," https://www.mathworks.com/help/slrealtime/automated-testing.html, accessed on Feb. 17, 2024
10. MathWorks, "Overview of MATLAB Build Tool," https://www.mathworks.com/help/matlab/matlab_prog/overview-of-matlab-build-tool.html, accessed on Feb. 17, 2024
11. MathWorks, "Continuous Integration (CI)," https://www.mathworks.com/help/matlab/continuous-integration.html, accessed on Feb. 17, 2024
12. MathWorks, "Create and Share Toolboxes," https://www.mathworks.com/help/matlab/matlab_prog/create-and-share-custom-matlab-toolboxes.html, accessed on Feb. 17, 2024
13. MathWorks, "MATLAB," https://www.mathworks.com/help/matlab/, accessed on Feb. 17, 2024
14. MathWorks, "MathWorks Certification Program," https://www.mathworks.com/learn/training/certification.html, accessed on Feb. 17, 2024
15. MathWorks, "INDUSTRY STANDARDS DO-178C- Achieving certification for airborne systems leveraging Model-Based Design," https://www.mathworks.com/solutions/aerospace-defense/standards/do-178.html, accessed on Feb. 17, 2024
16. MathWorks, "MATLAB Cody," https://uk.mathworks.com/matlabcentral/cody/, accessed May 2024
17. MathWorks, "Problem Types on MATLAB Cody," https://uk.mathworks.com/matlabcentral/cody/problems/, accessed May 2024
18. MathWorks, "The MATLAB AI Chat Playground Has Launched," MATLAB Community. [Online]. https://blogs.mathworks.com/community/2024/11/07/the-matlab-ai-chat-playground-has-launched. [Accessed: Feb. 17, 2024]
19. MathWorks, "MATLAB AI Chat Playground," [Online]. https://www.mathworks.com/matlabcentral/playground. [Accessed: Feb. 17, 2024]
20. "ChatGPT". [Online]. https://chat.openai.com/. [Accessed: Feb. 17, 2024]
21. "Cursor". [Online]. https://cursor.sh/. [Accessed: Feb. 17, 2024]
22. Chen Y, Li Y (2018) Computational intelligence assisted design (In the era of industry 4.0). CRC Press (ISBN 978-1-4987-6066-9)
23. Chen* Y, Li Y (2019) Intelligent autonomous pollination for future farming - a micro air vehicle solution with artificial intelligence and human-in-the-loop. IEEE Access 7(1):119706–119717
24. Chen Y, Zhang G, Jin T, Wu S, Peng B (2014) Quantitative modelling of electricity consumption using computational intelligence aided design. J Cleaner Prod 69:143–152. https://doi.org/10.1016/j.jclepro.2014.01.058
25. Seaborn, "Timeseries plot with error bands," https://seaborn.pydata.org/examples/errorband_lineplots.html, accessed May 2024

Chapter 10
Object-Oriented Programming

Chapter Learning Outcomes

- Understand the fundamental concepts of **object-oriented programming (OOP)** and its advantages.
- Define and create **classes** and **objects** in MATLAB.
- Implement **inheritance, polymorphism**, and **encapsulation** in MATLAB's OOP framework.
- Utilise **properties, methods**, and **constructors** in class definitions.
- Apply **method overloading** and **operator overloading** for custom behavior.
- Leverage **abstract classes** and **interfaces** for code reusability and extensibility.

Chapter Key Words

- **Object-Oriented Programming (OOP)**: A programming paradigm that focuses on creating objects that contain both data (properties) and code (methods) that operate on that data.
- **Class**: A blueprint or template that defines the properties and methods that objects of that class will have, serving as a blueprint for creating objects.
- **Object**: An instance of a class, created from the class blueprint, with its own unique set of property values and access to the methods defined in the class.
- **Inheritance**: A mechanism that allows a new class to be based on an existing class, inheriting its properties and methods, and optionally adding new or modified features.
- **Polymorphism**: The ability of objects of different classes to respond to the same method call in different ways, based on their specific implementation.
- **Encapsulation**: The concept of bundling data and methods within a class, hiding the internal implementation details, and providing controlled access through public interfaces.

© The Author(s) 2025
Y. Chen and L. Huang, *MATLAB Roadmap to Applications*,
https://doi.org/10.1007/978-981-97-8788-3_10

- **Abstract Class**: A class that cannot be instantiated but serves as a base for other classes, providing a common interface and potentially containing abstract methods to be implemented by subclasses.

10.1   Introduction to Object-Oriented Programming

- Overview of **object-oriented programming (OOP)** principles
  Object-oriented programming (OOP) is a programming paradigm that revolves around the concept of **objects**, which are instances of **classes**. Classes are user-defined data types that encapsulate data (properties) and methods (functions) that operate on that data. OOP is based on **four** fundamental principles: **encapsulation, abstraction, inheritance**, and **polymorphism**.

  1. **Encapsulation** is the mechanism of bundling data and methods together within a class, ensuring that the internal implementation details are hidden from the outside world. This promotes code modularity and maintainability by separating the interface from the implementation.
  2. **Abstraction** allows the creation of complex data types by defining objects that capture the essential characteristics of the entity being modelled, while hiding unnecessary details.
  3. **Inheritance** enables the creation of new classes (derived or child classes) based on existing classes (base or parent classes), inheriting their properties and methods. This promotes code reusability and facilitates the creation of hierarchical relationships between classes.
  4. **Polymorphism** allows objects of different classes to be treated as objects of a common superclass, enabling the same method call to behave differently based on the actual object type. This flexibility enhances code extensibility and modularity.

- Benefits of OOP in software development
  Besides **four** fundamental principles, adopting an object-oriented programming approach in MATLAB offers several significant benefits for software development:

  - **Code reusability**: OOP promotes the creation of reusable code components through inheritance and composition. Classes can be designed as building blocks that can be reused across multiple projects, reducing development time and effort.
  - **Modularity**: Objects encapsulate data and behavior, allowing for the separation of concerns and the development of modular, self-contained units. This modular design enhances code organization, maintainability, and scalability.
  - **Code organization and collaboration**: OOP provides a structured approach to organizing code, making it easier for multiple developers to work on different components of a project simultaneously, reducing the risk of conflicts and promoting collaboration.

By leveraging these benefits, MATLAB developers can create more robust, maintainable, and scalable software systems, leading to increased productivity and efficient code management.

- OOP in MATLAB
  MATLAB supports OOP through its class system, which allows the creation and manipulation of objects. Classes in MATLAB are defined using special syntax and can have properties (data members) and methods (function members). Objects are instantiated from classes, and their properties and methods can be accessed and invoked using dot notation.

In the following sections, we will explore various aspects of OOP in MATLAB, including inheritance, method overloading, and advanced OOP concepts, with relevant examples and illustrations.

10.2 Classes and Objects

- Defining **classes**
  Classes are the building blocks of object-oriented programming in MATLAB. They define the blueprint or template for creating **objects**, which are instances of the class. Classes encapsulate data members (**properties**) and function members (**methods**) that operate on that data.

Listing 10.1 Basic syntax for defining a class in MATLAB.

```
classdef ClassName
% Properties
properties
property1
property2
end
% Methods
methods
    function obj = ClassName(arg1, arg2)
        % Constructor
        obj.property1 = arg1;
        obj.property2 = arg2;
    end

    function result = methodName(obj, input)
        % Method implementation
        ...
    end
end

end
```

- Creating **objects**

  Objects are created from classes using the class constructor method. Once instantiated, an object's properties can be accessed and modified, and its methods can be invoked using dot notation.

Listing 10.2 Example of creating and using an object in MATLAB.

```
% Create an object
myObj = ClassName(value1, value2);

% Access properties
myObj.property1 = newValue;
value = myObj.property2;

% Call a method
result = myObj.methodName(input);
```

- **Properties** and **methods**

  Properties and **methods** are fundamental components of classes in object-oriented programming (OOP) with MATLAB. They define the data members and behavior of objects created from a class, respectively.

  Properties: Properties represent the attributes or characteristics of an object. They are the data members of a class and can be of various data types, including primitive types (e.g., double, char, logical) and user-defined types (e.g., other classes, structs).

  Properties are declared within the properties block of a class definition, and each property is given a name and an optional initial value.

Listing 10.3 Example of property declaration in a class.

```
classdef ClassName
properties
PropertyName1 = initialValue1
PropertyName2
end

% ... methods ...

end
```

In the example above, PropertyName1 is initialised with initialValue1, while PropertyName2 will be initialised to its respective default value (e.g., 0 for numeric types, false for logical types, and empty matrices or strings for other types).

Properties can be accessed and modified using dot notation on an object instance of the class.

Listing 10.4 Example of accessing and modifying properties.

```
% Create an object
obj = ClassName();

% Access and modify properties
obj.PropertyName1 = newValue;
value = obj.PropertyName2;
```

Methods: Methods are the function members of a class that define the behavior and operations that can be performed on objects of that class. They can access and manipulate the object's properties, as well as perform other computations or actions.

Methods are declared within the methods block of a class definition, and they can take input arguments and return output values, similar to regular MATLAB functions.

Listing 10.5 Example of method declaration in a class.

```
classdef ClassName
properties
Property1
Property2
end

methods
    function result = MethodName(obj, input1, input2)
        % Method implementation
        result = computeResult(obj.Property1, obj.
            Property2, input1, input2);
    end
end

end
```

In the example above, the MethodName takes two input arguments (input1 and input2) and operates on the object's properties (Property1 and Property2) to compute a result. Methods can also modify the object's properties as needed.

Methods are invoked using dot notation on an object instance, similar to accessing properties.

Listing 10.6 Example of invoking a method.

```
% Create an object
obj = ClassName();

% Invoke a method
result = obj.MethodName(arg1, arg2);
```

Properties and methods work together to encapsulate the data and behavior of objects, promoting code modularity, reusability, and maintainability in object-oriented programming.

- **Constructors** and **destructors**

  Constructors are special methods in a class that are automatically called when an object of that class is created. They are used to initialise the properties (data members) of the object with desired values or perform any necessary setup operations. If no constructor is defined, MATLAB provides a default constructor.

  In MATLAB, constructors are defined as regular methods within the methods block of a class definition, but they have the same name as the class itself. Constructors can take input arguments, which are used to initialise the object's properties.

Listing 10.7 Example of a constructor in MATLAB.

```
classdef ClassName
properties
Property1
Property2
end
methods
    function obj = ClassName(value1, value2)
        % Constructor
        obj.Property1 = value1;
        obj.Property2 = value2;
    end
end
end
```

In the example above, the ClassName has two properties, Property1 and Property2. The constructor ClassName takes two input arguments, value1 and value2, which are used to initialise these properties when creating an object of this class.

If a class does not define a constructor explicitly, MATLAB provides a default constructor that creates an object with all properties initialised to their respective default values (e.g., 0 for numeric types, false for logical types, and empty matrices or strings for other types).

Constructors can also perform additional operations beyond initializing properties, such as opening files, establishing connections to external systems, or performing complex computations required during object creation.

Listing 10.8 Example of a constructor with additional operations.

```
classdef ClassName
properties
Data
FileHandle
end

methods
    function obj = ClassName(filename)
        % Constructor
        obj.FileHandle = fopen(filename, 'r');
        obj.Data = fscanf(obj.FileHandle, '%g');
    end
end
end
```

In this example, the constructor ClassName takes a filename as input and opens the file using the fopen function. It then reads the data from the file using fscanf and stores it in the Data property of the object.

Constructors play a crucial role in ensuring that objects are properly initialised and ready for use immediately after creation, promoting code robustness and maintainability.

Destructors are special methods that are automatically called when an object is no longer needed and is about to be destroyed. They are used to perform any necessary cleanup or release resources held by the object, such as closing files, releasing memory, or disconnecting from external systems. Destructors are defined using the same syntax as regular methods but with a specific name (delete).

In MATLAB, destructors are defined using the special method name delete. The syntax for defining a destructor is similar to other methods, but without input arguments and with the method name delete.

Listing 10.9 Example of a destructor in MATLAB.

```
classdef ClassName
properties
ResourceHandle
end

methods
    function obj = ClassName()
        % Constructor
        obj.ResourceHandle = openResource();
    end

    function delete(obj)
        % Destructor
        closeResource(obj.ResourceHandle);
    end
end

end
```

In the example above, the ClassName has a property ResourceHandle that holds a handle to an external resource. The constructor ClassName initialises this handle by calling the openResource function. The destructor delete is responsible for closing the resource by calling the closeResource function with the ResourceHandle.

It is important to note that destructors are automatically called when an object goes out of scope or is explicitly removed from memory using the delete function in MATLAB. However, it is generally a good practice to explicitly call the delete function to ensure proper cleanup and resource release, especially for objects that hold significant resources or external connections.

10.3   Inheritance and Hierarchies

- **Superclasses and Subclasses**

    Inheritance is a fundamental concept in object-oriented programming that allows a new class, called a **subclass,** to be derived from an existing class, known as a **superclass.** The subclass inherits properties and methods from the superclass, facilitating code reuse and promoting a hierarchical organization of classes.
    The basic syntax for creating a subclass in MATLAB is:

Listing 10.10 Syntax for creating a subclass.

```
classdef SubClassName < SuperClassName
% Properties and methods specific to SubClassName
end
```

Here's an example of a subclass DerivedClass inheriting from a superclass BaseClass:

Listing 10.11 Example of inheritance.

```
classdef BaseClass
properties
BaseProperty
end

methods
    function obj = BaseClass(value)
        obj.BaseProperty = value;
    end

    function result = BaseMethod(obj)
        result = obj.BaseProperty * 2;
    end
end
end

classdef DerivedClass < BaseClass
properties
DerivedProperty
end

methods
    function obj = DerivedClass(baseValue,
        derivedValue)
        obj@BaseClass(baseValue);   % Call superclass
            constructor
        obj.DerivedProperty = derivedValue;
    end
end

end
```

In this example, the DerivedClass inherits the BaseProperty and BaseMethod from the BaseClass. Additionally, it has its own DerivedProperty and constructor, which calls the superclass constructor using obj@BaseClass(baseValue).

- Overriding Methods and Property Inheritance
When a subclass inherits properties and methods from its superclass, it can choose to override the inherited methods to provide its own implementation. This is known as **method overriding**. Similarly, a subclass can redefine the properties it inherits from the superclass, a concept called **property overriding**.
The syntax for overriding a method in a subclass is as follows:

Listing 10.12 Syntax for method overriding.

```
classdef  SubClassName  <  SuperClassName
methods
function  result  =  OverriddenMethod(obj,  ...)
%  New  implementation  for  the  overridden  method
end
end
end
```

Here's an example that demonstrates method overriding:

Listing 10.13 Example of method overriding.

```
classdef  BaseClass
methods
function  result  =  CalculateResult(obj,  value)
result  =  value  *  2;
end
end
end

classdef  DerivedClass  <  BaseClass
methods
function  result  =  CalculateResult(obj,  value)
result  =  value^2;  %  Overridden  implementation
end
end
end
```

In this example, the DerivedClass overrides the CalculateResult method inherited from BaseClass with its own implementation.

Property overriding follows a similar concept but applies to properties instead of methods. To override a property in a subclass, simply redeclare the property with the desired characteristics.

- **Abstract Classes and Interfaces**
Abstract classes and **interfaces** are closely related concepts in object-oriented programming, but they have distinct characteristics and use cases.

Abstract Classes

An abstract class is a class that cannot be instantiated directly but serves as a blueprint for other classes to inherit from. Abstract classes can contain both concrete (implemented) and abstract (unimplemented) methods and properties. Subclasses of an abstract class must provide implementations for all inherited abstract methods and properties.

The key features of abstract classes are:

– They cannot be instantiated directly.
– They can have both abstract and concrete methods and properties.
– Subclasses must implement all abstract members (methods and properties).
– They provide a common base for related classes, defining a common interface and shared implementation.
– They support code reuse and inheritance.

Abstract classes are useful when you want to define a common interface and shared implementation for a group of related classes, while allowing subclasses to provide specific implementations for certain methods or properties.

Interfaces

An interface, on the other hand, is a construct that defines a contract or set of methods that a class must implement. Interfaces consist solely of abstract methods without any implemented methods or properties. Classes that implement an interface must provide concrete implementations for all the methods defined by the interface.

The key features of interfaces are:

– They cannot be instantiated.
– They consist of only abstract methods (no method implementations).
– They cannot have properties or data members.
– Classes that implement an interface must provide implementations for all the interface methods.
– A class can implement multiple interfaces.
– They promote loose coupling and code modularity.

Interfaces are useful when you want to define a contract or a set of methods that multiple unrelated classes need to implement, without specifying any implementation details. They promote code modularity and allow for the creation of interchangeable components.

Comparison and Use Cases

While abstract classes and interfaces share some similarities, they have distinct use cases:

– Abstract classes are more suitable when you have a group of related classes that share some common implementation and need to define a common interface.

- Interfaces are more suitable when you want to define a contract or set of methods that unrelated classes need to implement, without specifying any implementation details.
- A class can inherit from only one abstract class, but it can implement multiple interfaces.
- Abstract classes can have both abstract and concrete members, while interfaces can only have abstract methods.

In practice, abstract classes and interfaces can be used together to achieve a higher level of abstraction and code modularity. For example, an abstract class can implement one or more interfaces, and concrete classes can inherit from the abstract class while implementing the interfaces.

10.4 Encapsulation and Access Control

Encapsulation is a fundamental principle of object-oriented programming that bundles data and methods into a single unit, called a class. It provides a way to hide the implementation details of an object and expose only a well-defined interface for interacting with the object. **Access control** mechanisms are used to control the visibility and accessibility of class members (properties and methods) from outside the class.

- **Private, Protected, and Public Access Modifiers**
  In MATLAB, classes support three access modifiers: **private**, **protected**, and **public**.

  - **Private**: Private members are accessible only within the class in which they are defined. They cannot be accessed from outside the class or from subclasses.
  - **Protected**: Protected members are accessible within the class in which they are defined and from any subclasses that inherit from that class.
  - **Public**: Public members are accessible from anywhere, both within the class and from outside the class.

The basic syntax for defining properties and methods with access modifiers in MATLAB is:

Listing 10.14 Syntax for access modifiers.

```
classdef ClassName
properties (AccessModifier)
PropertyName % Property with specified access modifier
end

methods (AccessModifier)
    function result = MethodName(obj, ...)  % Method
        with specified access modifier
        % Method implementation
                .
```

```
          end
    end
end
```

Here's an example that demonstrates the use of access modifiers:

Listing 10.15 Example of access modifiers.

```
classdef BankAccount
properties (Private)
balance = 0; % Private property
end

methods
    function obj = BankAccount(initialBalance)
        obj.balance = initialBalance;  % Access
            private property
    end

    function deposit(obj, amount)
        obj.balance = obj.balance + amount;  % Access
            private property
    end

    function withdrawn = withdraw(obj, amount)
        if obj.balance >= amount
            obj.balance = obj.balance - amount;  %
                Access private property
            withdrawn = amount;
        else
            withdrawn = 0;
        end
    end

    function showBalance(obj)
        disp(['Current balance: $' num2str(obj.balance
            )]);  % Access private property
    end
end

end
```

In this example, the balance property is declared as **private**, meaning it can only be accessed and modified within the BankAccount class. The class methods (deposit, withdraw, and showBalance) can access and modify the private balance property, but code outside the class cannot directly access or modify the balance.

- **Getters and Setters for Property Access**
  Getters and **setters** are methods that provide controlled access to the properties of a class. Getters are used to retrieve the value of a property, while setters are used to modify the value of a property. They are commonly used to enforce data validation, maintain class invariants, or perform additional operations when accessing or modifying properties.

Here's an example that demonstrates the use of getters and setters:

Listing 10.16 Example of getters and setters.

```
classdef Person
properties (Private)
name
age
end

methods
    function obj = Person(name, age)
        obj.name = name;
        obj.setAge(age);
    end

    function setAge(obj, age)
        if age >= 0
            obj.age = age;
        else
            error('Age cannot be negative.');
        end
    end

    function age = getAge(obj)
        age = obj.age;
    end
end
end
```

In this example, the Person class has two private properties: name and age. The setAge method is a **setter** that validates the age value before setting it to the age property. The getAge method is a **getter** that retrieves the value of the age property. By using getters and setters, the class maintains control over accessing and modifying its properties, allowing for data validation and additional logic if needed.

- **Friend Classes and Packages**

  In MATLAB, the concept of **friend classes** and **packages** is not directly supported. However, you can achieve similar functionality by carefully structuring your code and using appropriate access modifiers.

  For example, you can create a separate class that acts as a "friend" and provide it with access to the private or protected members of another class by passing the necessary objects or data as arguments to its methods.

Listing 10.17 Example of a "friend" class.

```
classdef BankAccount
properties (Private)
balance = 0;
end
methods
    function obj = BankAccount(initialBalance)
        obj.balance = initialBalance;
    end

    function withdrawn = withdraw(obj, amount)
        if obj.balance >= amount
            obj.balance = obj.balance - amount;
            withdrawn = amount;
        else
            withdrawn = 0;
        end
    end
end

end

classdef BankManager
methods (Static)
function showBalance(account)
disp(['Current balance: $' num2str(account.balance)]);
end
end
end
```

In this example, the BankManager class has a static method showBalance that takes a BankAccount object as an argument and displays its balance. By passing the BankAccount object to the BankManager class, it can access the private balance property, effectively acting as a "friend" class.

While MATLAB does not have a built-in mechanism for friend classes or packages, you can achieve similar functionality by carefully managing access modifiers and passing objects or data between classes as needed.

10.5 Polymorphism and Overloading

Polymorphism is a core concept in object-oriented programming that allows objects of different classes to respond differently to the same method call. It enables code reusability and flexibility by providing a consistent interface for interacting with objects of different types. **Overloading** is a specific form of polymorphism that allows methods or operators to have multiple implementations with different signatures (i.e., different numbers or types of parameters).

- **Method Overloading and Dynamic Method Dispatch**

In MATLAB, **method overloading** is supported, which means that a class can have multiple methods with the same name but different parameter lists. When a method is called, MATLAB determines the appropriate implementation based on the number and types of arguments passed to the method. This process is known as **dynamic method dispatch**.

The basic syntax for method overloading in MATLAB is:

Listing 10.18 Syntax for method overloading.

```
classdef ClassName
methods
function result = MethodName(obj, param1, param2, ...)
    % Method signature 1
% Method implementation 1
end

function result = MethodName(obj, param1, ...)   %
    Method signature 2
        % Method implementation 2
    end

    % Additional overloaded method signatures
end
end
```

Here's an example that demonstrates method overloading:

Listing 10.19 Example of method overloading.

```
classdef Calculator
methods
function result = add(obj, a, b)
result = a + b;
end

function result = add(obj, a, b, c)
        result = a + b + c;
    end
end

end

% Using the overloaded methods
calc = Calculator();
result1 = calc.add(2, 3); % Calls add(obj, a, b)
result2 = calc.add(2, 3, 4); % Calls add(obj, a, b, c)
```

In this example, the Calculator class has two add methods with different parameter lists. When the add method is called with two arguments, the first implementation add(obj, a, b) is executed. When the add method is called with three arguments, the second implementation add(obj, a, b, c) is executed.

- **Operator Overloading**

  Operator overloading is a form of polymorphism that allows you to define how operators (e.g., $+$, $-$, $*$, $/$) should behave when applied to objects of a specific class. This allows you to create intuitive and natural syntax for working with objects of your custom classes.

  In MATLAB, operator overloading is achieved by defining specific methods with predefined names that correspond to the desired operators. For example, to overload the $+$ operator for a class, you need to define a method named plus. Here's an example that demonstrates operator overloading:

Listing 10.20 Example of operator overloading.

```
classdef Vector
properties
x
y
end

methods
    function obj = Vector(x, y)
        obj.x = x;
        obj.y = y;
    end

    function result = plus(obj1, obj2)
        result = Vector(obj1.x + obj2.x, obj1.y + obj2
            .y);
    end
end

end

% Using the overloaded + operator
v1 = Vector(1, 2);
v2 = Vector(3, 4);
v3 = v1 + v2; % Calls v3 = plus(v1, v2)
disp(v3) % Output: Vector with properties: x: 4, y: 6
```

In this example, the Vector class overloads the $+$ operator by defining the plus method. When two Vector objects are added using the $+$ operator, the plus method is called, and a new Vector object is returned with the sum of the corresponding x and y components.

- **Benefits and Considerations of Polymorphic Design**

  Polymorphism and overloading provide several benefits in object-oriented programming:

  - **Code reusability**: Polymorphic code can work with objects of different classes as long as they implement the required interface, promoting code reuse and reducing duplication. This allows you to write more generic and flexible code that can handle different types of objects seamlessly, without the need for extensive conditional statements or type-checking.

- **Flexibility and extensibility**: New classes can be added to a system without modifying existing code, as long as they implement the required interface. This makes it easier to extend and evolve the codebase over time, as new classes can be seamlessly integrated with existing polymorphic code.
- **Intuitive and natural syntax**: Operator overloading allows for more intuitive and natural syntax when working with custom objects. It enables you to define how standard operators should behave with your custom classes, making the code more readable and expressive.
- **Abstraction and encapsulation**: Polymorphism supports the principles of abstraction and encapsulation by allowing objects of different classes to be treated as instances of a common superclass or interface. This hides the implementation details of each class and promotes a more modular and maintainable codebase.
- **Separation of concerns**: Polymorphism helps separate the concerns of different components in a system. Each class can focus on its specific responsibilities, while polymorphic code can interact with objects of different classes through a common interface, promoting loose coupling and easier maintenance.

However, it's important to use polymorphism and overloading judiciously and follow best practices to maintain code clarity and avoid confusion. Some considerations and potential drawbacks include:

- **Complexity**: Overuse or misuse of polymorphism and overloading can lead to code that is difficult to understand and maintain, especially if the inheritance hierarchies or method signatures become overly complex.
- **Performance overhead**: Polymorphic method dispatch and virtual function calls can introduce a performance overhead compared to static method calls, although modern compilers often optimize these operations.
- **Naming conventions**: Overloaded methods and operators should follow consistent naming conventions and have clear and meaningful names to avoid confusion and ambiguity.
- **Documentation and readability**: Polymorphic code may require more documentation and comments to explain the intended behavior and interactions between different classes, especially for complex systems or libraries intended for wide use.

Overall, polymorphism and overloading are powerful tools in object-oriented programming, but they should be used judiciously and with careful consideration of code clarity, maintainability, and performance implications.

10.6   Events, Listeners, and Callbacks

In object-oriented programming, **events** are occurrences or actions that happen within an object or application, such as user interactions, data changes, or system notifications. **Listeners** are objects or methods that listen for and respond to specific events. **Callbacks** are functions or methods that are executed when an event occurs, allowing for custom behaviour to be implemented in response to the event.

- Defining and triggering events
  In MATLAB, events are defined as properties of classes or app components. These properties can be assigned event handlers (callbacks) that specify the actions to be taken when the event occurs. Events can be triggered by various actions, such as user input, data changes, or system occurrences.
  The basic syntax for defining an event in MATLAB is:

Listing 10.21 Defining an event.

```
classdef MyClass < handle
events
EventName
end
% Other class properties and methods
end
```

Here's an example that demonstrates triggering an event:

Listing 10.22 Triggering an event.

```
classdef MyClass < handle
events
ValueChanged
end

properties
    Value = 0;
end

methods
    function obj = MyClass()
        % Constructor
    end

    function setValue(obj, newValue)
        obj.Value = newValue;
        \textcolor{green}{notify(obj, 'ValueChanged')}
            % Trigger the ValueChanged event
    end
end
end
```

In this example, the MyClass defines an event called ValueChanged. When the setValue method is called and the Value property is updated, the notify function is used to trigger the ValueChanged event.

- Adding and removing listeners
  To respond to events, you need to add **listeners** that execute specific callbacks when the event occurs. Listeners are added to an object using the addlistener function, and they can be removed using the removelistener function.
  The basic syntax for adding a listener is:

Listing 10.23 Adding a listener.

```
listener = addlistener(obj, 'EventName', @callback);
```

Here's an example that demonstrates adding and removing a listener:

Listing 10.24 Adding and removing a listener.

```
myObj = MyClass();

% Add a listener
listener = \textcolor{green}{addlistener(myObj, '
    ValueChanged', @myCallback)};

% Trigger the event (will execute myCallback)
myObj.setValue(10);

% Remove the listener
\textcolor{green}{removelistener(myObj, 'ValueChanged'
    , listener)};
```

In this example, a listener is added to the myObj object, which listens for the ValueChanged event and executes the myCallback function when the event is triggered. After triggering the event, the listener is removed using the removelistener function.

- Implementing callbacks for custom behaviour
  Callbacks are functions or methods that are executed when an event occurs, allowing you to implement custom behaviour in response to the event. Callbacks can be defined as anonymous functions or separate named functions.
  Here's an example that demonstrates implementing a callback function:

Listing 10.25 Implementing a callback function.

```
function myCallback(src, event)
% Access properties of the source object
value = src.Value;
% Implement custom behaviour
disp(['Value changed to: ', num2str(value)]);

end
```

In this example, the myCallback function is defined to handle the ValueChanged event. It receives two input arguments: src (the source object that triggered the event) and event (an event data structure). Inside the callback function, you can access properties of the source object and implement custom behaviour, such as displaying a message or performing additional computations.

10.7   Advanced OOP Concepts

• **Encapsulation and Information Hiding**

Encapsulation is a fundamental principle of object-oriented programming that involves bundling data and methods within a class, and controlling access to the class members through access modifiers (public, private, and protected). Information hiding is a related concept that ensures the internal implementation details of a class are hidden from the outside world, promoting code modularity and maintainability.

The basic syntax for setting access modifiers in MATLAB is:

Listing 10.26 Setting access modifiers.

```
classdef MyClass
properties (Access = private)
% Private properties
end

properties (Access = public)
    % Public properties
end

methods (Access = private)
    % Private methods
end

methods (Access = public)
    % Public methods
end

end
```

Here's an example that demonstrates encapsulation and information hiding:

Listing 10.27 Encapsulation and information hiding.

```
classdef BankAccount
properties (Access = private)
balance = 0;
end

methods
    function obj = BankAccount(initialBalance)
        obj.balance = initialBalance;
    end

    function deposit(obj, amount)
        obj.balance = obj.balance + amount;
    end

    function withdrawn = withdraw(obj, amount)
        if amount <= obj.balance
            obj.balance = obj.balance - amount;
            withdrawn = amount;
```

```
            else
                withdrawn = 0;
            end
        end

        function getBalance(obj)
            disp(['Current balance: ', num2str(obj.balance
                )]);
        end
    end

end
```

In this example, the BankAccount class encapsulates the balance property and provides public methods (deposit, withdraw, and getBalance) to interact with the account. The balance property is marked as private, hiding its implementation details from external code.

- **Composition** and **Aggregation**

  Composition and **aggregation** are two ways to establish relationships between objects in object-oriented programming. Composition represents a strong "part-of" relationship, where the lifespan of the composed objects is tied to the lifespan of the containing object. Aggregation, on the other hand, represents a weaker "has-a" relationship, where the aggregated objects can exist independently of the containing object.

  The basic syntax for composition and aggregation in MATLAB involves creating class properties that hold instances of other classes:

Listing 10.28 Composition and aggregation.

```
classdef Car
properties
engine % Composition (strong ownership)
wheels % Aggregation (weak ownership)
end

methods
    function obj = Car()
        obj.engine = Engine();
        obj.wheels = Wheel.empty(0, 4); % Array of 4
            Wheel objects
    end
end

end
```

Here's an example that demonstrates composition and aggregation:

Listing 10.29 Composition and aggregation example.

```
classdef Car
properties
engine
```

```
wheels
end

methods
    function obj = Car()
        obj.engine = Engine();
        obj.wheels = Wheel.empty(0, 4);
        for i = 1:4
            obj.wheels(i) = Wheel();
        end
    end

    function start(obj)
        obj.engine.start();
    end

    function drive(obj)
        for i = 1:4
            obj.wheels(i).rotate();
        end
    end
end

end
```

In this example, the Car class has a composition relationship with the Engine class (the car owns the engine, and the engine's lifespan is tied to the car) and an aggregation relationship with the Wheel class (the car has wheels, but the wheels can exist independently).

- **Static** and **dynamic** binding

  In object-oriented programming, **static binding** (also known as early binding) and **dynamic binding** (also known as late binding) refer to the way methods are resolved and executed at runtime.

  Static binding occurs when the method to be executed is determined at compile-time based on the declared type of the object reference. In contrast, dynamic binding occurs when the method to be executed is determined at runtime based on the actual type of the object being referenced.

  The basic syntax for static and dynamic binding in MATLAB involves creating classes with methods that have the same name but different implementations:

Listing 10.30 Static and dynamic binding.

```
classdef BaseClass
methods
function output = doSomething(obj)
output = 'BaseClass';
end
end
end

classdef DerivedClass < BaseClass
```

```
methods
function  output  =  doSomething(obj)
output  =  'DerivedClass';
end
end
end
```

Here's an example that demonstrates static and dynamic binding:

Listing 10.31 Static and dynamic binding example.

```
% Static binding
baseObj = BaseClass();
disp(baseObj.doSomething()); % Output: 'BaseClass'

% Dynamic binding
derivedObj = DerivedClass();
disp(derivedObj.doSomething()); % Output: '
    DerivedClass'

% Dynamic binding with reference to base class
baseRef = derivedObj;
disp(baseRef.doSomething()); % Output: 'DerivedClass'
```

In this example, when using static binding (baseObj.doSomething()), the method executed is determined by the declared type of the object reference (BaseClass). When using dynamic binding (derivedObj.doSomething() and baseRef.doSomething()), the method executed is determined by the actual type of the object being referenced (DerivedClass).

- **Object Serialisation** and **Deserialisation**

  Object serialisation is the process of converting an object's state (properties and data) into a sequence of bytes that can be stored or transmitted. **Deserialisation** is the reverse process of reconstructing the object from the serialised data.

  In MATLAB, serialisation and deserialisation can be achieved using the saveobj and loadobj functions, respectively. These functions support various formats, including MAT-files and portable archives (.pxat files).

Listing 10.32 Object serialisation and deserialisation.

```
% Serialisation
obj = MyClass();
% ... (set object properties)
saveobj(obj, 'myObject.mat'); % Save object to MAT-
    file

% Deserialisation
loadedObj = loadobj('myObject.mat'); % Load object
    from MAT-file
```

Here's an example that demonstrates object serialisation and deserialisation:

Listing 10.33 Object serialisation and deserialisation example.

```matlab
classdef Person
    properties
        Name
        Age
    end

    methods
        function obj = Person(name, age)
            if nargin == 2
                obj.Name = name;
                obj.Age = age;
            end
        end
    end
end

% Create an object
person = Person('John Doe', 30);

% Serialise the object
save('person.mat', 'person');

% Clear the workspace
clear person;

% Deserialise the object
load('person.mat');

% Display the loaded object's properties
disp(['Name: ', person.Name, ', Age: ', num2str(person
    .Age)]);
```

In this example, a Person object is created and serialised to a MAT-file using the saveobj function. After clearing the workspace, the object is deserialised from the MAT-file using the loadobj function, and its properties are displayed.

10.8 OOP Design Patterns

- **Introduction to Design Patterns Design patterns** are reusable solutions to common software design problems. They provide proven approaches for structuring code and organizing classes and objects to address recurring design issues in an elegant and maintainable way. Design patterns promote code reusability, extensibility, and flexibility, making it easier to develop robust and scalable object-oriented systems

- **Common Design Patterns in MATLAB**
Singleton Pattern
MATLAB supports several well-known design patterns, including:
The **Singleton** pattern ensures that a class has only one instance and provides a global point of access to it. This pattern is useful when you need to control the instantiation of a class and provide a single point of coordination for a specific set of operations or data.

Listing 10.34 Singleton pattern syntax.

```
classdef Singleton
properties (Access = private)
persistent uniqueInstance
end

methods (Access = private)
    function obj = Singleton()
        % Constructor code
    end
end

methods (Static)
    function instance = getInstance()
        if isempty(Singleton.uniqueInstance)
            Singleton.uniqueInstance = Singleton();
        end
        instance = Singleton.uniqueInstance;
    end
end

end
```

Here's an example that demonstrates the Singleton pattern:

Listing 10.35 Singleton pattern example.

```
% Get the Singleton instance
instance1 = Singleton.getInstance();
instance2 = Singleton.getInstance();

% Both instances are the same object
isequal(instance1, instance2) % Output: true
```

Factory Pattern The **Factory** pattern provides an interface for creating objects in a super-class, while allowing subclasses to alter the type of objects that will be created. This pattern promotes loose coupling by eliminating the need to specify the exact class of the object being created.

Listing 10.36 Factory pattern syntax.

```
classdef ProductFactory
methods (Static)
function product = createProduct(type)
```

```
switch type
case 'A'
product = ProductA();
case 'B'
product = ProductB();
otherwise
error('Invalid product type');
end
end
end
end
```

Here's an example that demonstrates the Factory pattern:

Listing 10.37 Factory pattern example.

```
% Create a product of type A
productA = ProductFactory.createProduct('A');
disp(class(productA)); % Output: 'ProductA'

% Create a product of type B
productB = ProductFactory.createProduct('B');
disp(class(productB)); % Output: 'ProductB'
```

Observer Pattern

The **Observer** pattern defines a one-to-many dependency between objects, so that when one object changes state, all its dependents are notified and updated automatically. This pattern is commonly used to implement event-driven systems or for decoupling the subject (observable) from its observers.

Listing 10.38 Observer pattern syntax.

```
classdef Subject
properties
observers = ObserverList()
end
methods
    function attach(obj, observer)
        obj.observers.addObserver(observer);
    end

    function detach(obj, observer)
        obj.observers.removeObserver(observer);
    end

    function notify(obj, data)
        obj.observers.notifyObservers(data);
    end
end

end
```

Here's an example that demonstrates the Observer pattern:

Listing 10.39 Observer pattern example.

```matlab
% Create a subject
subject = Subject();

% Create observers
observer1 = Observer1();
observer2 = Observer2();

% Attach observers to the subject
subject.attach(observer1);
subject.attach(observer2);

% Notify observers with some data
subject.notify('Hello, observers!');
```

10.9   OOP Applications and Best Practices

- **OOP in Scientific Computing and Data Analysis**
  OOP plays a pivotal role in **scientific computing** and **data analysis**. By encapsulating data and associated functionalities into modular, reusable **objects**. OOP enhances code organisation and readability, which is essential when dealing with complex algorithms and large datasets [1]. It enables the creation of custom data structures and processing pipelines tailored to specific domains or applications, promoting code reuse and facilitating collaborative development.

  OOP's ability to model real-world entities and relationships makes it particularly suitable for simulating complex systems, performing numerical computations, and implementing sophisticated data analysis techniques. For instance, in control systems engineering, OOP can be utilised to design controllers that are modular and easily extendable.

  – PID Controller Design Example
    To illustrate the application of OOP in scientific computing, consider the design of a **Proportional-Integral-Derivative (PID) controller** using MATLAB's object-oriented features. The PID controller is a fundamental component in control systems, widely used in industry to regulate processes and maintain desired output levels [2].

    Class Definition

    A class named `PIDController` can be defined to encapsulate the properties and methods associated with the PID controller. This class includes parameters for the proportional (Kp), integral (Ki), and derivative ($Kd?$) gains, as well as methods to compute the control signal based on the error input.

Listing 10.40 PIDController Class Definition.

```
classdef PIDController
properties
Kp % Proportional gain
Ki % Integral gain
Kd % Derivative gain
PreviousError = 0
Integral = 0
end
methods

function obj = PIDController(Kp, Ki, Kd)
% Constructor: initialise PID gains
obj.Kp = Kp;
obj.Ki = Ki;
obj.Kd = Kd;
end
function [u, obj] = computeControl(obj, setpoint,
    measurement, dt)
% Compute the PID control signal
error = setpoint - measurement;
obj.Integral = obj.Integral + error * dt;
derivative = (error - obj.PreviousError) / dt;
u = obj.Kp * error + obj.Ki * obj.Integral + obj.Kd
    * derivative;
obj.PreviousError = error;
end
end
end
```

Explanation

In the PIDController class:

·   **Properties** include the PID gains (Kp, Ki, Kd), the previous error, and the
    integral term, all essential for computing the control signal.
·   The **constructor** method initialises the PID gains when creating an instance
    of the class.
·   The **method** computeControl calculates the control output u based on
    the current error, integrating over time step dt.
·   By returning the updated object obj, the method ensures that the state of the
    controller (integral and previous error) is maintained between computations.

Usage

To use the PIDController, an instance is created with specified gains, and
the control signal is computed within a simulation loop.

Listing 10.41 Using the PIDController in a Simulation.

```matlab
% Define PID gains
Kp = 2.0;
Ki = 1.0;
Kd = 0.5;

% Create PID controller instance
pid = PIDController(Kp, Ki, Kd);

% Simulation parameters
dt = 0.01; % Time step
t = 0:dt:10; % Simulation time
setpoint = 1.0; % Desired value
measurement = 0; % Initial measurement
u = zeros(size(t)); % Control signal
y = zeros(size(t)); % System output

% Simple system model: First-order process
tau = 1.0; % Time constant

for i = 1:length(t)
% Compute control signal
[u(i), pid] = pid.computeControl(setpoint,
    measurement, dt);

% Update system (simple discretised first-order lag)
measurement = measurement + dt * ( -measurement + u(
    i) ) / tau;
y(i) = measurement;

end

% Plot results
figure;
plot(t, y, 'b-', t, setpoint * ones(size(t)), 'r--')
    ;
xlabel('Time (s)');
ylabel('Output');
legend('System Output', 'Setpoint');
title('PID Controller Response');
```

Discussion

As shown in Fig. 10.1, this example demonstrates how OOP facilitates the encapsulation of the PID controller logic, making it reusable and easily maintainable. The `PIDController` class can be extended to include additional features, such as anti-windup mechanisms or filter implementations for the derivative term [3]. Furthermore, the object-oriented approach allows for multiple instances of controllers to be created, each with different parameters, which is beneficial in multi-variable control systems.

– Unique Insights

Fig. 10.1 An example of
OOP for PID controller
design

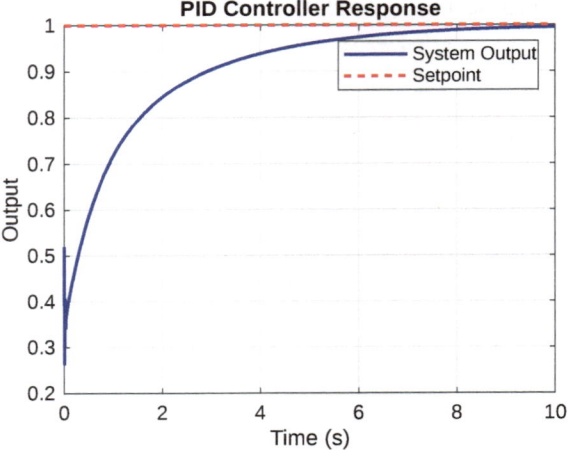

Integrating OOP into scientific computing promotes better software engineering
practices within the research community. Recent trends emphasise the importance of **reproducible research** and code maintainability [4]. By adopting OOP,
researchers can develop more organised codebases, facilitating collaboration
and reducing errors.

Moreover, the combination of OOP and MATLAB's extensive numerical
libraries enhances the development of sophisticated data analysis pipelines. For
instance, in **machine learning** and **artificial intelligence**, OOP enables the creation of modular components such as data loaders, preprocessors, models, and
evaluators, improving the scalability and flexibility of analytical workflows [5].
Additionally, the emergence of **digital twins** and **cyber-physical systems**
necessitates robust simulation tools that can model complex interactions
between physical and virtual entities [6]. OOP provides the structural foundation to build such simulations, allowing for modularity and extensibility, which
are essential in rapidly evolving technological landscapes.

- **OOP for GUI Development and Event Handling**
  OOP significantly enhances the development of **Graphical User Interfaces
  (GUIs)** and event handling in MATLAB. By employing OOP principles, developers can create modular, reusable, and extensible GUI components, leading to more
  maintainable and robust applications [7]. OOP facilitates the encapsulation of GUI
  elements and their associated behaviours into **classes** and **objects**, streamlining
  the management of complex GUIs.

  – Digital Clock Example Using OOP
    To illustrate the application of OOP in GUI development, consider the creation
    of a **digital clock** using MATLAB's object-oriented features. The digital clock
    will display the current time and update every second, demonstrating event

handling and GUI component management within a class structure.

Class Definition

The digital clock can be encapsulated in a class named `DigitalClock`, which inherits from the `handle` class to allow for reference behaviour. The class manages the GUI components, timer events, and updates to the display.

Listing 10.42 DigitalClock Class Definition.

```
%
classdef DigitalClock < handle
properties (Access = private)
Figure % Figure window
TimeDisplay % Text UI component for time display
Timer % Timer object
end
methods
function obj = DigitalClock()
% Constructor: initialise GUI and start timer
obj.createGUI();
obj.startTimer();
end
function delete(obj)
% Destructor: clean up resources
obj.stopTimer();
if isvalid(obj.Figure)
close(obj.Figure);
end
end
end
methods (Access = private)
function createGUI(obj)
% Create the GUI components
obj.Figure = figure('Name', 'Digital Clock', '
    NumberTitle', 'off', ...
'MenuBar', 'none', 'ToolBar', 'none', ...
'CloseRequestFcn', @(src, event)delete(obj));
obj.TimeDisplay = uicontrol('Style', 'text', ...
'FontSize', 36, 'FontWeight', 'bold', ...
'Units', 'normalized', 'Position', [0.2 0.4 0.6
    0.2]);
end
function startTimer(obj)
% Start the timer to update time every second
obj.Timer = timer('ExecutionMode', 'fixedRate',
    Period', 1, ...
'TimerFcn', @(src, event)obj.updateTime());
start(obj.Timer);
end
function stopTimer(obj)
% Stop and delete the timer
if ~isempty(obj.Timer) && isvalid(obj.Timer)
```

```
stop(obj.Timer);
delete(obj.Timer);
end
end
function updateTime(obj)
% Update the time display
currentTime = datestr(now, 'HH:MM:SS');
obj.TimeDisplay.String = currentTime;
end
end
end
```

Explanation

In the `DigitalClock` class:

- **Properties** store handles to the figure window, time display text, and timer object, encapsulated with private access to prevent external modification.
- The **constructor** method `DigitalClock()` sets up the GUI and initiates the timer upon object creation.
- The **destructor** method `delete(obj)` ensures that resources such as the timer and figure window are properly released when the object is deleted.
- Private methods `createGUI()`, `startTimer()`, `stopTimer()`, and `updateTime()` handle specific tasks related to GUI creation and event handling.
- The **timer object** triggers the `updateTime()` method every second, updating the displayed time.

Usage

To utilise the digital clock, an instance of the `DigitalClock` class is created:

Listing 10.43 Instantiate the Digital Clock.
```
clock = DigitalClock();
```

This command launches the GUI window displaying the current time, which updates every second. The clock will continue to run until the figure window is closed, at which point the `delete` method is invoked to clean up resources.

- Discussion

As given in Fig. 10.2, the digital clock example showcases how OOP in MAT-LAB simplifies GUI development and event handling. By encapsulating the GUI elements and logic within a class, the code achieves greater modularity and reusability. The use of **event-driven programming** with timer events demonstrates effective handling of asynchronous operations, enhancing the responsiveness of the application [1].

The adoption of OOP in GUI development aligns with modern programming practices, promoting scalability and maintainability. Recent research emphasises the importance of integrating OOP with advanced GUI development tools

Fig. 10.2 An example of OOP for GUI—digital clock

such as MATLAB's **App Designer**, which offers a rich set of components and a user-friendly interface for building sophisticated applications [8].

Furthermore, the combination of OOP and GUI development facilitates the implementation of **Model-View-Controller (MVC)** design patterns, enabling a clear separation of data models, user interfaces, and control logic [9]. This approach enhances code organisation and supports collaborative development efforts, which is particularly beneficial in complex engineering and research projects.

The integration of OOP with event handling also opens avenues for implementing customised behaviours and extending existing GUI components. By leveraging inheritance and polymorphism, developers can create specialised interfaces tailored to specific application requirements, fostering innovation in fields such as robotics, autonomous systems, and real-time data analysis.

- **Best Practices for Code Organisation, Documentation, and Maintenance**
  To ensure the long-term maintainability and extensibility of object-oriented code, it is essential to follow best practices for code organization, documentation, and maintenance. These practices include:

  - **Modular Design**: Divide functionality into well-defined, self-contained classes and methods, promoting code reuse and easier debugging.
  - **Encapsulation**: Encapsulate data and implementation details within classes, providing a clear interface for external access and modification.

– **Inheritance and Polymorphism**: Leverage inheritance and polymorphism to create hierarchies of classes and promote code reuse and flexibility.
– **Consistent Naming Conventions**: Follow a consistent naming convention for classes, properties, and methods to improve code readability and maintainability.
– **Code Documentation**: Document classes, properties, and methods using MATLAB's built-in documentation tools (e.g., `help` and `doc`) to facilitate understanding and collaboration.
– **Version Control**: Use a version control system (e.g., Git) to track changes, collaborate with others, and manage code revisions effectively.
– **Testing and Debugging**: Implement unit tests and leverage MATLAB's debugging tools to identify and fix issues early in the development process.

By following these best practices, you can develop robust, maintainable, and extensible object-oriented code in MATLAB.

10.10 Laboratory

This section provides several hands-on lab exercises to reinforce the concepts and applications of Object-Oriented Programming (OOP) in MATLAB. Each exercise is designed to challenge your understanding and problem-solving skills.

1. **Creating a Simple Class**

   a. Create a new class called `Rectangle` with properties `length` and `width`.
   b. Implement a constructor method that initialises the `length` and `width` properties.
   c. Create a method called `area` that calculates and returns the area of the rectangle.
   d. Create a method called `perimeter` that calculates and returns the perimeter of the rectangle.
   e. Create an instance of the `Rectangle` class and test the `area` and `perimeter` methods.

Listing 10.44 Rectangle class definition.

```
classdef Rectangle
properties
length
width
end

methods
    function obj = Rectangle(l, w)
        obj.length = l;
        obj.width = w;
    end
```

```
    function area = area(obj)
        area = obj.length * obj.width;
    end

    function perimeter = perimeter(obj)
        perimeter = 2 * (obj.length + obj.width);
    end
end
end
```

2. **Inheritance and Polymorphism**

   a. Create a new class called Square that inherits from the Rectangle class.
   b. Modify the Square class constructor to accept only one parameter (side length) and initialise both length and width with this value.
   c. Override the area and perimeter methods in the Square class to provide more efficient implementations.
   d. Create instances of both Rectangle and Square classes and test their area and perimeter methods.

3. **GUI Development with OOP**

   a. Create a new GUI application with a figure window and a panel.
   b. Add a uicontrol (e.g., a button or a slider) to the panel.
   c. Create a custom class that inherits from matlab.ui.component container.ComponentContainer and encapsulates the panel and the uicontrol.
   d. Implement event handling in the custom class to respond to user interactions with the uicontrol.
   e. Test the GUI application by creating an instance of the custom class and adding it to the figure window.

4. **Data Analysis with OOP**

   a. Create a new class called DataSet with properties to hold numerical data and metadata.
   b. Implement methods to load data from a file, perform basic statistical analysis (e.g., mean, variance, correlation), and visualise the data.
   c. Create a subclass called TimeSeries that inherits from DataSet and adds functionality for handling time-series data, such as resampling and smoothing.
   d. Load a dataset, create instances of DataSet and TimeSeries, and test their methods.

5. **File I/O with OOP**

   a. Create a new class called FileManager with methods to read and write different file formats (e.g., .txt, .csv, .mat).
   b. Implement error handling and validation mechanisms to ensure data integrity during file operations.

    c. Create a subclass called `LoggedFileManager` that inherits from `FileManager` and adds logging functionality to track file operations.

    d. Test the `FileManager` and `LoggedFileManager` classes by reading and writing different types of files.

6. Class Definitions and Usage

- Create a class `Rectangle` with two properties: `length` and `width`. These two properties can only be accessed by the class itself and its subclasses.
- The class should include two methods: one to calculate the circumference and another to calculate the area.
- Create a subclass of `Rectangle` named `Parallelogram`, which represents a parallelogram. The `Parallelogram` class should have an additional property for the angle.
- Instantiate objects of `Rectangle` and `Parallelogram`, and obtain their circumference and area.

The following MATLAB code demonstrates the creation of the `Rectangle` and `Parallelogram` classes, and how to use these classes to calculate circumference and area.

```
classdef Rectangle
    properties (Access = protected)
        length
        width
    end

    methods
        function obj = Rectangle(length, width)
            if nargin > 0
                obj.length = length;
                obj.width = width;
            end
        end

        function circumference =
            calculateCircumference(obj)
            circumference = 2 * (obj.length + obj.
                width);
        end

        function area = calculateArea(obj)
            area = obj.length * obj.width;
        end
    end
end

classdef Parallelogram < Rectangle
    properties
        angle
    end
```

```
    methods
        function obj = Parallelogram(length, width,
            angle)
            obj@Rectangle(length, width);
            if nargin > 0
                obj.angle = angle;
            end
        end

        function area = calculateArea(obj)
            area = obj.length * obj.width * sin(
                deg2rad(obj.angle));
        end
    end
end

% Create Rectangle object
rect = Rectangle(5, 10);
rectCircumference = rect.calculateCircumference();
rectArea = rect.calculateArea();

fprintf('Rectangle Circumference: %.2f\n',
    rectCircumference);
fprintf('Rectangle Area: %.2f\n', rectArea);

% Create Parallelogram object
para = Parallelogram(5, 10, 30);
paraCircumference = para.calculateCircumference();
paraArea = para.calculateArea();

fprintf('Parallelogram Circumference: %.2f\n',
    paraCircumference);
fprintf('Parallelogram Area: %.2f\n', paraArea);
```

For each lab exercise, provide step-by-step guidance, example code snippets, and expected outputs to assist students in completing the tasks successfully.

10.11 Problems

This section presents a collection of problems designed to challenge and reinforce your understanding of Object-Oriented Programming (OOP) concepts in MATLAB. Each problem is accompanied by a step-by-step approach, sample code, sample output, and suggestions for extensions and variations.

1. **Bank Account Management**
   - Create a class called BankAccount with properties for account number, account holder's name, and balance.
   - Implement methods for depositing, withdrawing, and checking the balance.

- Incorporate error handling to prevent negative balances and invalid transactions.
- Create subclasses for different account types (e.g., `SavingsAccount`, `CheckingAccount`) with additional functionalities like interest calculation or overdraft fees.

2. **Geometric Shapes**

- Create an abstract class called `Shape` with a method for calculating the area.
- Derive concrete subclasses for different shapes (e.g., `Circle`, `Rectangle`, `Triangle`) and implement their respective area calculation methods.
- Create a function that takes an array of `Shape` objects and calculates the total area of all shapes.
- Extend the problem by adding methods for calculating perimeter, volume (for 3D shapes), or other relevant properties.

3. **Student Record Management**

- Create a class called `Student` with properties for name, ID, and grades.
- Implement methods for adding/updating grades, calculating the grade point average (GPA), and generating a report card.
- Create a subclass called `GraduateStudent` that inherits from `Student` and includes additional properties and methods specific to graduate students (e.g., thesis, research projects).
- Design a system to manage a collection of students, perform operations like sorting or filtering based on specific criteria, and generate statistical reports.

4. **Employee Payroll System**

- Create an abstract class called `Employee` with properties for name, ID, and a method for calculating the monthly salary.
- Derive concrete subclasses for different employee types (e.g., `Hourly Employee`, `SalariedEmployee`, `ContractEmployee`) and implement their respective salary calculation methods.
- Incorporate features like overtime pay, bonuses, or deductions based on employee type or performance.
- Design a system to manage a company's payroll, generate paychecks, and produce reports for HR or financial analysis.

5. **Library Management System**

- Create a class called `Book` with properties for title, author, publication date, and availability status.
- Implement methods for checking out, returning, and reserving books.
- Create a class called `Library` that manages a collection of `Book` objects and provides functionality for searching, sorting, and filtering books based on various criteria.
- Incorporate features like user accounts, late fees, and notifications for overdue books or reserved items.

6. **Vehicle Rental System**

- Create an abstract class called `Vehicle` with properties for make, model, year, and a method for calculating the daily rental rate.
- Derive concrete subclasses for different vehicle types (e.g., `Car`, `Truck`, `Motorcycle`) and implement their respective rental rate calculation methods.
- Create a class called `RentalCompany` that manages a fleet of `Vehicle` objects and provides functionality for renting, returning, and tracking vehicle availability.
- Incorporate features like customer accounts, discounts for long-term rentals, and additional services like insurance or GPS navigation.

7. **Social Media Platform**

- Create a class called `User` with properties for username, profile information, and a list of followers/following.
- Implement methods for posting updates, liking or commenting on posts, and managing followers/following lists.
- Create a class called `SocialMedia` that manages a collection of `User` objects and provides functionality for searching, filtering, and generating news feeds based on user interactions.
- Incorporate features like hashtags, privacy settings, and notifications for new posts or interactions.

8. **Online Shopping Cart**

- Create a class called `Product` with properties for name, description, price, and quantity.
- Implement a class called `ShoppingCart` that manages a collection of `Product` objects and provides functionality for adding, removing, and updating quantities.
- Create a class called `Order` that represents a customer's order, including the shopping cart contents, shipping information, and payment details.
- Incorporate features like discounts, promotions, and order tracking.

9. **Flight Reservation System**

- Create a class called `Flight` with properties for airline, origin, destination, departure time, and available seats.
- Implement methods for booking seats, canceling reservations, and checking seat availability.
- Create a class called `FlightReservationSystem` that manages a collection of `Flight` objects and provides functionality for searching and booking flights based on user preferences (e.g., dates, destinations).
- Incorporate features like seat selection, frequent flyer programs, and notifications for flight changes or delays.

10. **Weather Monitoring System**

- Create a class called `WeatherStation` with properties for location, temperature, humidity, and pressure.
- Implement methods for recording and retrieving weather data, as well as generating reports or visualisations.
- Create a class called `WeatherMonitoringSystem` that manages a network of `WeatherStation` objects and provides functionality for analyzing and predicting weather patterns based on collected data.
- Incorporate features like real-time data updates, weather alerts, and integration with external weather services or APIs.

10.12 Summary

- This chapter covered the principles of **object-oriented programming (OOP)** and its implementation in MATLAB through several practical examples.
- Key **OOP concepts** such as **classes, objects, properties, methods, inheritance**, and **polymorphism** were introduced and demonstrated.
- Various problem domains were explored, including **banking systems, event management, library management, social media platforms**, and **weather monitoring systems**.
- Each example provided a **step-by-step approach, sample code, sample output**, and suggestions for **extensions and variations**.
- The examples showcased the benefits of **code organization, reusability**, and **maintainability** offered by OOP in solving complex problems.
- For undergraduate (UG) students, this chapter serves as an introduction to the fundamental concepts of object-oriented programming and its implementation in MATLAB. The practical examples and step-by-step approaches provide a solid foundation for understanding how to model real-world problems using classes, objects, and their relationships. The chapter emphasizes the importance of code organization, reusability, and maintainability, which are essential skills for developing robust and scalable software applications.
- For postgraduate (PG) students, this chapter offers an in-depth exploration of object-oriented programming principles and their application in solving complex problems across various domains. The diverse range of examples, such as banking systems, event management, library management, social media platforms, and weather monitoring systems, demonstrates the versatility and power of OOP in tackling challenges in different fields. PG students can leverage the provided code samples as a starting point for further research and development, incorporating advanced features, optimizations, and integrations as needed.
- For professional researchers or engineers, this chapter serves as a valuable resource for implementing object-oriented programming techniques in MATLAB for research or industrial applications. The examples showcase best practices in

software design, code organization, and scalability, which are crucial for developing robust and maintainable systems. The chapter's focus on extensions and variations encourages professionals to explore additional features, integrate with databases, implement user authentication and authorization, and tailor the solutions to meet specific project requirements.

References

1. Attaway S (2016) MATLAB: a practical introduction to programming and problem solving, 4th edn. Butterworth-Heinemann, Oxford, UK
2. Franklin GF, Powell JD, Emami-Naeini A (2015) Feedback control of dynamic systems, 7th edn. Pearson Education Limited, Harlow, UK
3. Åström KJ, Murray RM (2008) Feedback systems: an introduction for scientists and engineers. Princeton University Press, Princeton, NJ, USA
4. Wilson G et al (2014) Best practices for scientific computing. PLoS Biol 12(1):1–7
5. Murphy KP (2012) Machine learning: a probabilistic perspective. MIT Press, Cambridge, MA, USA
6. Grieves M, Vickers J (2017) Digital twin: mitigating unpredictable, undesirable emergent behaviour in complex systems. In: Kahlen F-J, Flumerfelt S, Alves A (eds) Transdisciplinary perspectives on complex systems. Springer, Cham, Switzerland, pp 85–113
7. MathWorks, "Develop Graphics Objects (Handle Classes)," [Online]. https://www.mathworks.com/help/matlab/handle-classes.html. [Accessed: Oct. 2, 2024]
8. MathWorks, "App Designer," [Online]. https://www.mathworks.com/products/matlab/app-designer.html. [Accessed: Oct. 2, 2024]
9. Bruegge B, Dutoit AH (2009) Object-oriented software engineering using UML, Patterns, and Java, 3rd edn. Prentice Hall, Upper Saddle River, NJ, USA

Appendix A
Solutions to Chapter Problems

These solutions provide a starting point for addressing the given problems and can be further extended and customized based on specific requirements or additional features desired in the book. The solutions aim to illustrate the problem-solving process, showcase MATLAB's capabilities, and reinforce the concepts and techniques covered in the chapters.

A.1   Solutions to Chapter 1

1. **Creating a Simple Calculator**
   - **Problem Statement and Background**: This problem involves creating a simple calculator program in MATLAB that can perform basic arithmetic operations (addition, subtraction, multiplication, and division) on two numbers provided by the user. It also requires implementing error handling to gracefully handle invalid input or division by zero.
   - **Step-by-Step Approach**:
     a. Prompt the user to enter the first number using the 'input' function.
     b. Prompt the user to enter the second number using the 'input' function.
     c. Prompt the user to enter the operation ('+', '-', '*', or '/') using the 'input' function.
     d. Use a switch-case statement or an if-elseif-else structure to perform the appropriate operation based on the user's input.
     e. Inside each case or condition, perform the corresponding arithmetic operation on the two numbers.

Y. Chen and L. Huang, *MATLAB Roadmap to Applications*,
https://doi.org/10.1007/978-981-97-8788-3

f. Include error handling to check for division by zero and display an appropriate error message.

g. Display the result of the operation to the user using the 'fprintf' or 'disp' function.

- **Sample Code**:

Listing A.1 Simple Calculator

```
% Prompt user for input
num1 = input('Enter the first number: ');
num2 = input('Enter the second number: ');
operation = input('Enter the operation (+, -, *, /)
    : ', 's');

% Perform operation based on user input
switch operation
case '+'
result = num1 + num2;
case '-'
result = num1 - num2;
case '*'
result = num1 * num2;
case '/'
if num2 == 0
fprintf('Error: Division by zero is not allowed.\n'
    );
return
end
result = num1 / num2;
otherwise
fprintf('Invalid operation.\n');
return
end

% Display result
fprintf('Result: %.2f %s %.2f = %.2f\n', num1,
    operation, num2, result);
```

- **Sample Output**:

```
Enter the first number: 10
Enter the second number: 5
Enter the operation (+, -, *, /): *
Result: 10.00 * 5.00 = 50.00
```

- **Extensions and Variations**:

  – Enhance the calculator to support additional operations like exponentiation, modulus, or trigonometric functions.
  – Implement a graphical user interface (GUI) for the calculator using MATLAB's built-in tools.

– Allow the user to enter multiple operations in a single expression and evaluate the expression according to the order of operations.

2. **Data Analysis and Visualisation**

- **Problem Statement and Background**: This problem involves loading a dataset from a provided file (e.g., a CSV file containing weather data or stock prices), performing data cleaning and preprocessing steps, and then analyzing the data by computing summary statistics and visualizing the results using appropriate plots.

- **Step-by-Step Approach**:

  a. Load the dataset from the provided file using MATLAB's file reading functions (e.g., 'readtable' for CSV files).
  b. Inspect the loaded data and identify any potential issues, such as missing values or outliers.
  c. Perform data cleaning and preprocessing steps as necessary:

    – Handle missing values (e.g., remove rows or columns with missing values, or impute missing values using appropriate techniques).
    – Remove outliers or data points that fall outside of expected ranges.
    – Normalize or scale the data, if required, to ensure consistent ranges across different features.

  d. Compute summary statistics for the cleaned dataset, such as mean, median, standard deviation, and quartiles, using MATLAB's built-in functions (e.g., 'mean', 'median', 'std').
  e. Visualize the data using appropriate plotting techniques:

    – For time-series data, use line plots or scatter plots to visualize the data over time.
    – For continuous data, use histograms or kernel density estimates to visualize the distribution of values.
    – For categorical data, use bar charts or pie charts to visualize the frequencies or proportions of different categories.

  f. Customize the plots by adding titles, labels, legends, and other visual elements to enhance clarity and readability.

- **Sample Code**:

Listing A.2 Data Analysis and Visualisation

```
% Load data from a CSV file
data = readtable('weather_data.csv');

% Handle missing values
data = rmmissing(data);

% Compute summary statistics
meanTemp = mean(data.Temperature);
```

```
medianHumidity = median(data.Humidity);
stdPressure = std(data.Pressure);

% Visualize data
figure;
subplot(2, 2, 1);
histogram(data.Temperature);
title('Temperature Distribution');

subplot(2, 2, 2);
plot(data.Date, data.Humidity);
title('Humidity over Time');
xlabel('Date');
ylabel('Humidity');

subplot(2, 2, 3);
scatter(data.Pressure, data.Temperature);
title('Temperature vs. Pressure');
xlabel('Pressure');
ylabel('Temperature');
```

- **Sample Output**: [Relevant Visualisations and plots will be included here, showcasing the data analysis and Visualisation results.]
- **Extensions and Variations**:
  - Explore and apply more advanced data cleaning and preprocessing techniques, such as feature scaling, dimensionality reduction, or handling imbalanced datasets.
  - Implement additional Visualisations or interactive plots using MATLAB's built-in tools or third-party libraries.
  - Perform statistical hypothesis testing or predictive modeling on the cleaned dataset using MATLAB's statistical and machine learning toolboxes.

3. **Implementing a Simple Algorithm**

   - **Problem Statement and Background**: This problem requires implementing a sorting algorithm (e.g., bubble sort, insertion sort) in MATLAB as a function, and writing a script that generates a random array of numbers and calls the sorting function to sort the array. The correctness of the implementation should be verified by comparing the sorted array with the expected output.
   - **Step-by-Step Approach**:

     a. Choose a sorting algorithm to implement, such as bubble sort or insertion sort.
     b. Define a MATLAB function that takes an array as input and returns the sorted array using the chosen sorting algorithm.
     c. Within the function, implement the logic of the sorting algorithm using loops, conditional statements, and array manipulations.
     d. Write a script that generates a random array of numbers using MATLAB's 'rand' or 'randi' functions.

    e. Call the sorting function with the random array as input and store the sorted output in a new array.

    f. Verify the correctness of the sorted array by:

        – Checking if the sorted array is in non-decreasing order (for ascending sort).

        – Comparing the sorted array with the expected output obtained by sorting the original array using MATLAB's built-in 'sort' function.

- **Sample Code (Bubble Sort Implementation)**:

Listing A.3 Bubble Sort Implementation

```
function sorted_array = bubble_sort(arr)
n = length(arr);
for i = 1:n-1
for j = 1:n-i
if arr(j) > arr(j+1)
temp = arr(j);
arr(j) = arr(j+1);
arr(j+1) = temp;
end
end
end
sorted_array = arr;
end

% Generate random array
random_array = randi(100, 1, 20);

% Sort the array
sorted_array = bubble_sort(random_array);

% Verify correctness
sorted_expected = sort(random_array);
if isequal(sorted_array, sorted_expected)
disp('Sorting successful!');
else
disp('Sorting failed.');
end
```

- **Sample Output**:

```
Sorting successful!
```

- **Extensions and Variations**:

    – Implement and compare the performance of different sorting algorithms, such as quicksort, merge sort, or shell sort.

    – Extend the sorting function to handle sorting arrays of structs or cell arrays based on specific fields or elements.

    – Implement parallel sorting algorithms using MATLAB's Parallel Computing Toolbox for improved performance on large datasets.

4. **Solving Systems of Equations**

- **Problem Statement and Background**: This problem involves writing a MAT-LAB function that takes a system of linear equations as input (in the form of coefficient matrices and constant vectors), and uses MATLAB's built-in functions to solve the system of equations and return the solution vector. The function should be tested with multiple sets of linear equations, including cases with unique solutions, no solutions, and infinitely many solutions.

- **Step-by-Step Approach**:

   a. Define a MATLAB function that takes the coefficient matrix (A) and the constant vector (b) as input arguments.
   b. Within the function, use MATLAB's backslash operator ('\') to solve the system of linear equations: 'x = A \ b;'.
   c. Check for specific cases:

      – If the system has a unique solution, the solution vector 'x' will be returned.
      – If the system has no solution (inconsistent equations), MATLAB will return a warning or error message.
      – If the system has infinitely many solutions (underdetermined system), MATLAB will return a particular solution, and additional steps may be required to find the general solution.

   d. Handle potential warnings or error messages and provide appropriate feedback to the user.
   e. Return the solution vector 'x' or an appropriate message indicating the nature of the solution.

- **Sample Code**:

Listing A.4 Solving Systems of Equations

```
function [x, sol_type] = solve_linear_system(A, b)
try
x = A \ b;
sol_type = 'Unique Solution';
catch ME
if strcmp(ME.identifier, 'MATLAB:singular')
% System has no solution (inconsistent equations)
x = [];
sol_type = 'No Solution';
else
% System has infinitely many solutions (
    underdetermined)
x = A \ b;
sol_type = 'Infinitely Many Solutions';
end
end
end
```

```
% Test with a system of linear equations
A = [1 2 3; 4 5 6; 7 8 9];
b = [6; 15; 24];

[x, sol_type] = solve_linear_system(A, b);

if isempty(x)
disp('The system of equations has no solution.');
else
disp(['The solution is: x = ', mat2str(x')]);
disp(['Solution type: ', sol_type]);
end
```

- **Sample Output**:

```
The solution is: x = 0 1 0
Solution type: Unique Solution
```

- **Extensions and Variations**:

  - Extend the function to handle systems of non-linear equations using numerical methods like Newton's method or fixed-point iteration.
  - Implement additional error handling and input validation to ensure the provided coefficient matrix and constant vector are valid and consistent.
  - Explore and implement techniques for finding the general solution for underdetermined systems, such as using the null space of the coefficient matrix.

5. **Creating a Simple Game**

- **Problem Statement and Background**: This problem involves designing and implementing a simple game using MATLAB's graphical capabilities (e.g., a number guessing game, a simple version of Tic-Tac-Toe or Hangman). The game should have a graphical user interface (GUI) for user input, displaying game state, and providing feedback. The game logic and rules should be implemented within MATLAB functions and callbacks.

- **Step-by-Step Approach**:

  a. Choose a simple game to implement, such as a number guessing game or a classic game like Tic-Tac-Toe or Hangman.
  b. Design the graphical user interface (GUI) using MATLAB's built-in tools (e.g., 'guide' or programmatic GUI creation).
  c. Create UI components for user input (e.g., text boxes, buttons), displaying game state (e.g., labels, static text), and providing feedback (e.g., message boxes, status indicators).
  d. Define a MATLAB function or script to Initialise the game state and set up the necessary variables and data structures.
  e. Implement the game logic and rules within separate MATLAB functions or callbacks associated with the GUI components.
  f. Handle user input and update the game state accordingly.

g. Update the GUI components to reflect the current game state and provide feedback to the user.

h. Implement win/lose conditions and end-game scenarios, displaying appropriate messages or prompts for restarting the game.

- **Sample Code (Number Guessing Game)**:

Listing A.5 Number Guessing Game

```
function number_guessing_game()
% Create the main figure
fig = figure('Name', 'Number Guessing Game', '
    MenuBar', 'none', 'Resize', 'off');

% Generate a random number between 1 and 100
answer = randi(100);

% Create UI components
prompt_text = uicontrol('Style', 'text', 'String',
    'Guess a number between 1 and 100:', 'Position'
    , [20 120 200 20]);
guess_edit = uicontrol('Style', 'edit', 'Position',
    [230 120 100 20]);
guess_button = uicontrol('Style', 'pushbutton', '
    String', 'Guess', 'Position', [340 120 80 20],
    'Callback', {@check_guess, answer});
result_text = uicontrol('Style', 'text', 'String',
    '', 'Position', [20 80 400 20]);

% Guess checking callback function
function check_guess(~, ~, answer)
    guess = str2double(get(guess_edit, 'String'));
    if isnan(guess) || guess < 1 || guess > 100
        set(result_text, 'String', 'Invalid input.
            Please enter a number between 1 and
            100.');
    elseif guess < answer
        set(result_text, 'String', 'Too low. Try
            again!');
    elseif guess > answer
        set(result_text, 'String', 'Too high. Try
            again!');
    else
        set(result_text, 'String', 'Congratulations
            ! You guessed the correct number!');
        answer = randi(100); % Generate a new
            random number for the next game
    end
end

end
```

- **Sample Output**: [A GUI window will be displayed with a text prompt, input field, "Guess" button, and a result text area. The user can enter their guess, and the program will provide feedback based on whether the guess is too low, too high, or correct.]

- **Extensions and Variations**:

  - Enhance the game with additional features like keeping track of the number of attempts, implementing a high score system, or allowing the user to set the difficulty level.
  - Implement more complex games like Tic-Tac-Toe or Hangman, with appropriate game boards, move validation, and win/lose conditions.
  - Explore advanced GUI techniques, such as creating custom UI components, adding animations or graphics, and incorporating sound effects or music.

A.2   Solutions to Chapter 2

1. **Problem Statement and Background**: The problem requires writing a MATLAB script that computes a given expression, $y = \frac{a^2 + b^3}{a - 2b}$, using two user inputs, a and b. It also requires handling the case where the denominator is zero and displaying an appropriate error message. This problem tests the understanding of **user input**, **arithmetic operations**, **conditional statements**, and **error handling** in MATLAB.

2. **Step-by-Step Approach**:

   a. Prompt the user to enter values for a and b.
   b. Compute the numerator: $a^2 + b^3$.
   c. Compute the denominator: $a - 2b$.
   d. Check if the denominator is zero:

      - If the denominator is zero, display an error message.
      - Otherwise, compute y by dividing the numerator by the denominator.

   e. Display the result.

3. **Sample Code**:

Listing A.6 Computing an expression with user input.

```
% Prompt the user for input
a = input('Enter the value of a: ');
b = input('Enter the value of b: ');

% Compute the numerator and denominator
numerator = a^2 + b^3;
denominator = a - 2*b;

% Check for zero denominator
if denominator == 0
```

```
disp('Error: Denominator cannot be zero.');
else
% Compute the expression
y = numerator / denominator;
disp(['The result of (a^2 + b^3) / (a - 2b) is: ',
    num2str(y)]);
end
```

4. **Sample Output**:

```
Enter the value of a: 2
Enter the value of b: 3
The result of (a^2 + b^3) / (a - 2b) is: -7
```

5. **Extensions and Variations**:

- Extend the script to handle complex numbers or matrices as input.
- Modify the script to perform additional operations on the computed result.
- Enhance the error handling to provide more informative messages or to handle other types of errors.

6. **Problem Statement and Background**: This problem requires creating a MATLAB script that generates a random 3x3 matrix with integer values between 1 and 10. The script should then find the maximum and minimum values in the matrix and display their indices. This problem tests the understanding of **matrix creation, random number generation, finding maximum and minimum values**, and **indexing** in MATLAB.

7. **Step-by-Step Approach**:

   a. Generate a random 3x3 matrix with integer values between 1 and 10 using the randi function.
   b. Find the maximum value in the matrix using the max function.
   c. Find the minimum value in the matrix using the min function.
   d. Find the indices of the maximum and minimum values using the find function.
   e. Display the matrix, the maximum and minimum values, and their indices.

8. **Sample Code**:

Listing A.7 Finding maximum and minimum values in a matrix.

```
% Generate a random 3x3 matrix
matrix = randi([1, 10], 3, 3);

% Find the maximum and minimum values
max_value = max(matrix(:));
min_value = min(matrix(:));

% Find the indices of the maximum and minimum values
[max_row, max_col] = find(matrix == max_value);
[min_row, min_col] = find(matrix == min_value);

% Display the results
```

```
disp('The matrix is:');
disp(matrix);
disp(['The maximum value is: ', num2str(max_value), '
      at indices (', num2str(max_row), ', ', num2str(
    max_col), ')']);
disp(['The minimum value is: ', num2str(min_value), '
      at indices (', num2str(min_row), ', ', num2str(
    min_col), ')']);
```

9. **Sample Output**:

```
The matrix is:
5  4  2
9  1  6
3  8  10

The maximum value is: 10 at indices (3, 3)
The minimum value is: 1 at indices (2, 2)
```

10. **Extensions and Variations**:

   - Extend the script to handle matrices of different sizes.
   - Modify the script to find the maximum and minimum values along specific dimensions (rows or columns) of the matrix.
   - Enhance the script to display the indices of all occurrences of the maximum and minimum values, in case they are not unique.

11. **Problem Statement and Background**: This problem requires writing a MAT-LAB function that takes two vectors as input and returns their dot product. The function should handle the case where the input vectors have different lengths and display an appropriate error message. This problem tests the understanding of **function creation**, **vector operations**, **error handling**, and **input validation** in MATLAB.

12. **Step-by-Step Approach**:

   a. Define a function that takes two input vectors.
   b. Check if the input vectors have the same length:

   - If the lengths are different, display an error message and return.
   - Otherwise, proceed to compute the dot product.

   c. Compute the dot product by multiplying the corresponding elements of the vectors and summing the products.
   d. Return the computed dot product.

13. **Sample Code**:

Listing A.8 Computing the dot product of two vectors.

```
function dot_product = vec_dot(vec1, vec2)
% Check if the input vectors have the same length
if length(vec1) ~= length(vec2)
```

```
disp('Error: Input vectors must have the same length.
    ');
dot_product = [];
return;
end

% Compute the dot product
dot_product = sum(vec1 .* vec2);

end
```

14. **Sample Usage**:

```
% Valid input
vec1 = [1, 2, 3];
vec2 = [4, 5, 6];
result = vec_dot(vec1, vec2);
disp(['The dot product is: ', num2str(result)]);

% Invalid input
vec3 = [1, 2];
vec4 = [3, 4, 5];
result = vec_dot(vec3, vec4);
```

15. **Sample Output**:

```
The dot product is: 32
Error: Input vectors must have the same length.
```

16. **Extensions and Variations**:

- Extend the function to handle matrix inputs and compute the matrix product.
- Modify the function to compute the dot product of complex vectors or matrices.
- Enhance the error handling to provide more informative messages or to handle other types of errors or edge cases.

17. **Problem Statement and Background**: This problem requires creating a MAT-LAB script that generates a random vector of length 10 with integer values between 1 and 20. The script should then count the number of occurrences of each value in the vector and display the results. This problem tests the understanding of **vector creation**, **random number generation**, **counting occurrences**, and **data analysis** in MATLAB.

18. **Step-by-Step Approach**:

   a. Generate a random vector of length 10 with integer values between 1 and 20 using the randi function.
   b. Create an array to store the counts of each value in the vector.
   c. Iterate through the vector and update the counts in the array.
   d. Display the vector, the unique values, and their corresponding counts.

19. **Sample Code**:

Listing A.9 Counting occurrences in a vector.

```
% Generate a random vector
vector = randi([1, 20], 1, 10);

% Initialise the counts array
counts = zeros(1, 20);

% Count the occurrences
for i = 1:length(vector)
counts(vector(i)) = counts(vector(i)) + 1;
end

% Display the results
disp('The vector is:');
disp(vector);
disp('The number of occurrences of each value is:');
for i = 1:20
if counts(i) > 0
disp([num2str(i), ' occurs ', num2str(counts(i)), '
    times.']);
end
end
```

20. **Sample Output**:

```
The vector is:
5 10 18 12 3 14 6 9 11 7

The number of occurrences of each value is:
3 occurs 1 times.
5 occurs 1 times.
6 occurs 1 times.
7 occurs 1 times.
9 occurs 1 times.
10 occurs 1 times.
11 occurs 1 times.
12 occurs 1 times.
14 occurs 1 times.
18 occurs 1 times.
```

21. **Extensions and Variations**:

- Extend the script to handle vectors of different lengths or different value ranges.
- Modify the script to display the results in a sorted or formatted manner.
- Enhance the script to perform additional statistical analysis on the data, such as computing the mean, median, or mode.

22. **Problem Statement and Background**: This problem requires writing a MAT-LAB script that prompts the user to enter a string, counts the number of vowels (a, e, i, o, u) and consonants in the string, and displays the results. This problem

tests the understanding of **string manipulation**, **pattern matching**, **user input**, and **character processing** in MATLAB.

23. **Step-by-Step Approach**:

   a. Prompt the user to enter a string.

   b. Convert the string to lowercase for case-insensitive matching.

   c. Initialise counters for vowels and consonants to zero.

   d. Iterate through each character in the string:

- If the character is a vowel, increment the vowel counter.
- If the character is a consonant, increment the consonant counter.

   e. Display the original string, the number of vowels, and the number of consonants.

24. **Sample Code**:

Listing A.10 Counting vowels and consonants in a string.

```
% Prompt the user for input
user_string = input('Enter a string: ', 's');

% Convert the string to lowercase
lowercase_string = lower(user_string);

% Initialise counters
vowel_count = 0;
consonant_count = 0;

% Count vowels and consonants
for i = 1:length(lowercase_string)
char = lowercase_string(i);
if ismember(char, 'aeiou')
vowel_count = vowel_count + 1;
elseif isstrprop(char, 'alpha')
consonant_count = consonant_count + 1;
end
end

% Display the results
disp(['The string is: ', user_string]);
disp(['The number of vowels is: ', num2str(
    vowel_count)]);
disp(['The number of consonants is: ', num2str(
    consonant_count)]);
```

25. **Sample Output**:

```
Enter a string: Hello World!
The string is: Hello World!
The number of vowels is: 3
The number of consonants is: 7
```

26. **Extensions and Variations**:

- Extend the script to handle other character types, such as digits or special characters.
- Modify the script to count the occurrences of specific vowels or consonants.
- Enhance the script to perform additional string operations, such as reversing the string or removing vowels/consonants.

A.3   Solutions to Chapter 3

1. **Problem Statement and Background**: Given a matrix A, create a new matrix B by replacing all negative elements in A with their absolute values.
2. **Step-by-Step Approach**:

   a. Create a logical matrix mask that identifies the negative elements in A.
   b. Use the logical mask to replace the negative elements in A with their absolute values, creating the new matrix B.

3. **Sample Code**:

Listing A.11 Replace Negative Elements with Absolute Values

```
A = [1 -2 3; -4 5 -6; 7 -8 9]

% Create a logical mask for negative elements
negative_mask = A < 0

% Replace negative elements with their absolute
   values
B = abs(A)
```

4. **Sample Output**:

```
A =   1        -2        3
     -4         5       -6
      7        -8        9

B =   1         2        3
      4         5        6
      7         8        9
```

5. **Extensions and Variations**:

- Instead of replacing negative elements with their absolute values, you could replace them with a specific value (e.g., 0 or a user-defined constant).
- You could also replace elements based on different conditions, such as replacing elements greater than a certain threshold or within a specific range.

1. **Problem Statement and Background**: Given a matrix A, find the row and column indices of the maximum element in the matrix.
2. **Step-by-Step Approach**:

   a. Find the maximum element in the matrix using the max function.
   b. Use the find function to get the linear indices of the maximum element(s).
   c. Convert the linear indices to row and column indices using the ind2sub function.

3. **Sample Code**:

   Listing A.12 Find Indices of Maximum Element

```
A = [1 4 2; 7 3 9; 5 6 8]

% Find the maximum element
max_element = max(A(:))

% Find the linear indices of the maximum element(s)
max_indices = find(A == max_element)

% Convert linear indices to row and column indices
[row_indices, col_indices] = ind2sub(size(A),
    max_indices)
```

4. **Sample Output**:

```
max_element =
9

row_indices =
3

col_indices =
3
```

5. **Extensions and Variations**:

   • If there are multiple occurrences of the maximum element, the find function will return all the corresponding linear indices. You can handle this case accordingly.
   • Instead of finding the maximum element, you could find the minimum element or elements satisfying a specific condition.

1. **Problem Statement and Background**: Given a matrix A, create a new matrix B by swapping the elements along the main diagonal with the elements along the secondary diagonal.
2. **Step-by-Step Approach**:

   a. Create a new matrix B by copying the elements of A.
   b. Extract the elements along the main diagonal of A using diag(A).

c. Extract the elements along the secondary diagonal of A using `diag(fliplr(A))`.

d. Swap the elements along the main diagonal of B with the elements along the secondary diagonal.

3. **Sample Code**:

Listing A.13 Swap Main and Secondary Diagonals

```
A = [1 2 3; 4 5 6; 7 8 9]

% Create a copy of A
B = A

% Extract the main diagonal elements of A
main_diag = diag(A)

% Extract the secondary diagonal elements of A
secondary_diag = diag(fliplr(A))

% Swap the main and secondary diagonal elements in B
B(1:size(B,1)+1:end) = secondary_diag
B(size(B,1):-1:1) = main_diag
```

4. **Sample Output**:

```
A = 1      2      3
    4      5      6
    7      8      9

B = 9      2      3
    4      5      8
    7      6      1
```

5. **Extensions and Variations**:

- Instead of swapping the main and secondary diagonals, you could swap other elements based on different conditions or patterns.
- You could also perform this operation on higher-dimensional arrays or tensors.

1. **Problem Statement and Background**: Given a matrix A, create a new matrix B by shifting each element in A one position to the right, wrapping around to the beginning of the row when reaching the end.

2. **Step-by-Step Approach**:

a. Create a new matrix B with the same dimensions as A.

b. Shift the elements of each row in A one position to the right using the `circshift` function.

c. Assign the shifted rows to the corresponding rows in B.

3. **Sample Code**:

Listing A.14 Shift Elements to the Right

```
A = [1 2 3 4; 5 6 7 8; 9 10 11 12]

% Create a new matrix B with the same dimensions as A
B = zeros(size(A))

% Shift each row of A one position to the right
for i = 1:size(A, 1)
B(i, :) = circshift(A(i, :), [0 1]);
end
```

4. **Sample Output**:

```
A = 1     2      3      4
    5     6      7      8
    9  10  11  12

B = 1
    2     4
    3     5      7
    6     8
    9
```

5. **Extensions and Variations**:

- Instead of shifting the elements to the right, you could shift them to the left by using a negative shift value in the `circshift` function.
- You could also shift the elements vertically (up or down) by applying the `circshift` function to the columns instead of the rows.
- You could combine horizontal and vertical shifts to create more complex patterns.

1. **Problem Statement and Background**: Given a matrix A, create a new matrix B by extracting the elements along the diagonals parallel to the main diagonal.
2. **Step-by-Step Approach**:

   a. Determine the number of diagonals parallel to the main diagonal, which is equal to the sum of the dimensions of the matrix minus one.
   b. Create an empty cell array to store the elements of each diagonal.
   c. For each diagonal, use the `diag` function to extract the elements along that diagonal.
   d. Concatenate the elements from all diagonals into a single matrix B.

3. **Sample Code**:

Listing A.15 Extract Diagonals Parallel to Main Diagonal

```
A = [1 2 3; 4 5 6; 7 8 9]

% Determine the number of diagonals
num_diagonals = size(A, 1) + size(A, 2) - 1

% Create a cell array to store the diagonals
diagonals = cell(1, num_diagonals)

% Extract each diagonal
for k = 1-size(A, 2):size(A, 1)-1
diagonals{k+size(A, 2)} = diag(A, k);
end

% Concatenate the diagonals into a single matrix
B = cell2mat(diagonals')
```

4. **Sample Output**:

```
A = 1       2       3
    4       5       6
    7       8       9

  B =
  1
  2       4
  3       5       7
  6       8
  9
```

5. **Extensions and Variations**:

- Instead of extracting diagonals parallel to the main diagonal, you could extract diagonals parallel to the secondary diagonal by modifying the range of the loop and the sign of the diagonal offset in the `diag` function.
- You could also extract specific diagonals or a subset of diagonals based on certain conditions or requirements.
- This approach can be extended to higher-dimensional arrays or tensors by modifying the loop structure and the way the diagonals are extracted.

A.4   Solutions to Chapter 4

1. **Problem Statement and Background**: This problem requires writing a MATLAB function that takes a number as input and returns the absolute value of that number using an **if** statement. The absolute value of a number is its distance from zero on the number line, ignoring the sign.

2. **Step-by-Step Approach**:

  a. Define a function that takes a number as input.

  b. Check if the number is positive or negative using an **if** statement.

  c. If the number is positive, return the number itself.

  d. If the number is negative, return the negative of the number (to make it positive).

3. **Sample Code**:

Listing A.16 Absolute value function using if statement.

```
function result = abs_value(num)
if num >= 0
result = num;
else
result = -num;
end
end
```

4. **Sample Output**:

```
>> abs_value(5)
ans =   5
>> abs_value(-3)
ans =   3
```

5. **Extensions and Variations**:

  • Extend the function to handle complex numbers or matrices.

  • Modify the function to use the built-in abs function instead of an **if** statement.

  • Enhance the function to handle specific edge cases or error conditions.

6. **Problem Statement and Background**: This problem requires implementing a MATLAB script that prompts the user to enter a character and determines whether it is a vowel or a consonant using a **switch** statement. A vowel is any of the five letters a, e, i, o, or u, and a consonant is any other letter in the alphabet.

7. **Step-by-Step Approach**:

  a. Prompt the user to enter a character.

  b. Convert the character to lowercase for case-insensitive matching.

  c. Use a **switch** statement to check if the character is a vowel.

  d. If the character is a vowel, display a message indicating it is a vowel.

  e. If the character is not a vowel, display a message indicating it is a consonant.

8. **Sample Code**:

Listing A.17 Vowel or consonant determination using switch statement.

```
% Prompt the user to enter a character
char = input('Enter a character: ', 's');

% Convert the character to lowercase
lowercase_char = lower(char);
```

```matlab
% Determine if it's a vowel or a consonant
switch lowercase_char
case 'a'
disp('The character is a vowel.');
case 'e'
disp('The character is a vowel.');
case 'i'
disp('The character is a vowel.');
case 'o'
disp('The character is a vowel.');
case 'u'
disp('The character is a vowel.');
otherwise
disp('The character is a consonant.');
end
```

9. **Sample Output**:

Enter a character: E The character is a vowel.

10. **Extensions and Variations**:

- Extend the script to handle non-alphabetic characters.
- Modify the script to display a message if the user enters an invalid input.
- Enhance the script to distinguish between uppercase and lowercase vowels/-consonants.

11. **Problem Statement and Background**: This problem requires creating a MAT-LAB function that takes three numbers as input and returns the maximum value among them using nested **if** statements. Finding the maximum value is a common operation in many applications, such as data analysis and optimization.

12. **Step-by-Step Approach**:

a. Define a function that takes three numbers as input.
b. Use nested **if** statements to compare the three numbers.
c. First, compare the first two numbers and store the maximum in a temporary variable.
d. Then, compare the temporary variable with the third number and update it if necessary.
e. Return the final maximum value stored in the temporary variable.

13. **Sample Code**:

Listing A.18 Maximum of three numbers using nested if statements.

```matlab
function max_value = find_max(num1, num2, num3)
% Compare the first two numbers
if num1 >= num2
temp_max = num1;
else
temp_max = num2;
end

% Compare the temporary maximum with the third number
```

```
if temp_max >= num3
    max_value = temp_max;
else
    max_value = num3;
end

end
```

14. **Sample Output**:

```
>> find_max(5, 8, 3)
ans =   8
>> find_max(-2, 10, -5)
ans =   10
```

15. **Extensions and Variations**:

- Extend the function to handle more than three numbers or to find the minimum value instead.
- Modify the function to use alternative methods, such as the built-in max function or a nested **if-elseif-else** statement.
- Enhance the function to handle specific edge cases or error conditions, such as NaN or Inf values.

16. **Problem Statement and Background**: This problem requires writing a MATLAB script that generates a random number between 1 and 10, and based on the value, displays a corresponding message using a **switch** statement. This can be useful in various applications where different actions need to be taken based on a specific condition or value.

17. **Step-by-Step Approach**:

a. Generate a random number between 1 and 10 using the randi function.
b. Use a **switch** statement to check the value of the random number.
c. For each possible value, include a case that displays a corresponding message.
d. Include a default case to handle any unexpected values.

18. **Sample Code**:

Listing A.19 Displaying messages based on random number using switch statement.

```
% Generate a random number between 1 and 10
random_num = randi([1, 10]);

% Display a message based on the random number
switch random_num
case 1
disp('The number is one.');
case 2
disp('The number is two.');
case 3
disp('The number is three.');
case 4
```

```
disp('The number is four.');
case 5
disp('The number is five.');
case 6
disp('The number is six.');
case 7
disp('The number is seven.');
```

19. **Problem Statement and Background**: This problem requires writing a MAT-LAB function that takes two numbers as input and returns their sum if both numbers are positive, their difference if one number is positive and the other is negative, or zero if both numbers are negative, using nested **if** statements. This type of conditional operation is common in many programming tasks.

20. **Step-by-Step Approach**:

   a. Define a function that takes two numbers as input.
   b. Use nested **if** statements to check the signs of the two numbers.
   c. If both numbers are positive, return their sum.
   d. If one number is positive and the other is negative, return their difference (positive - negative).
   e. If both numbers are negative, return zero.

21. **Sample Code**:

Listing A.20 Sum, difference, or zero based on number signs using nested if statements.

```
function result = sum_diff_zero(num1, num2)
if num1 >= 0 && num2 >= 0 % Both positive
result = num1 + num2;
elseif num1 >= 0 && num2 < 0 % One positive, one
    negative
result = num1 - num2;
elseif num1 < 0 && num2 >= 0 % One negative, one
    positive
result = num2 - num1;
else % Both negative
result = 0;
end
end
```

22. **Sample Output**:

```
>> sum_diff_zero(3, 5)
ans =   8
>> sum_diff_zero(-2, 7)
ans =   9
>> sum_diff_zero(-4, -6)
ans =   0
```

23. **Extensions and Variations**:

   • Extend the function to handle complex numbers or matrices.

- Modify the function to perform different operations based on the signs of the numbers.
- Enhance the function to handle specific edge cases or error conditions, such as zero values or NaN/Inf.

24. **Problem Statement and Background**: This problem requires implementing a MATLAB function that takes a year as input and determines whether it is a leap year or not using an **if-elseif-else** statement. A leap year is a year with 366 days instead of the usual 365 days, and it occurs every four years to keep the calendar in sync with the astronomical year.

25. **Step-by-Step Approach**:

   a. Define a function that takes a year as input.
   b. Use an **if-elseif-else** statement to check the conditions for a leap year.
   c. If the year is divisible by 4 and not divisible by 100, it is a leap year.
   d. If the year is divisible by 400, it is also a leap year.
   e. Otherwise, the year is not a leap year.
   f. Return a message indicating whether the year is a leap year or not.

26. **Sample Code**:

Listing A.21 Leap year determination using if-elseif-else statement.

```
function result = is_leap_year(year)
if mod(year, 4) == 0 && mod(year, 100) ~= 0 %
    Divisible by 4 and not divisible by 100
result = sprintf('%d is a leap year.', year);
elseif mod(year, 400) == 0 % Divisible by 400
result = sprintf('%d is a leap year.', year);
else
result = sprintf('%d is not a leap year.', year);
end
end
```

27. **Sample Output**:

```
>> is_leap_year(2024)
ans = "2024 is a leap year."
>> is_leap_year(2023)
ans = "2023 is not a leap year."
>> is_leap_year(2000)
ans = "2000 is a leap year."
```

28. **Extensions and Variations**:

- Extend the function to handle a range of years or an array of years.
- Modify the function to display additional information, such as the number of days in the leap year.
- Enhance the function to handle specific edge cases or error conditions, such as invalid year inputs.

29. **Problem Statement and Background**: This problem requires creating a MAT-LAB script that prompts the user to enter their age and displays a message indicating their age category (e.g., child, teenager, adult) using an **if-elseif-else** statement. This type of age categorization can be useful in various applications, such as targeted advertising or content filtering.

30. **Step-by-Step Approach**:

   a. Prompt the user to enter their age.
   b. Use an **if-elseif-else** statement to check the age range and determine the category.
   c. If the age is below a certain threshold (e.g., 13), categorize the user as a child.
   d. If the age is between the teenage range (e.g., 13-19), categorize the user as a teenager.
   e. If the age is above the adult threshold (e.g., 19), categorize the user as an adult.
   f. Display a message indicating the age category.

31. **Sample Code**:

Listing A.22 Age category determination using if-elseif-else statement.

```
% Prompt the user to enter their age
age = input('Enter your age: ');

% Determine the age category
if age < 13
disp('You are a child.');
elseif age >= 13 && age <= 19
disp('You are a teenager.');
else
disp('You are an adult.');
end
```

32. **Sample Output**:
   Enter your age: 25 You are an adult.

33. **Extensions and Variations**:

   • Extend the script to include additional age categories, such as senior citizen or infant.
   • Modify the script to handle invalid or non-numeric input.
   • Enhance the script to provide more detailed information or recommendations based on the age category.

34. **Problem Statement and Background**: This problem requires writing a MAT-LAB function that takes a character as input and determines whether it is a digit, an uppercase letter, a lowercase letter, or a special character using nested **if** statements. This type of character classification can be useful in various applications, such as data validation or text processing.

35. **Step-by-Step Approach**:

   a. Define a function that takes a character as input.
   b. Use nested **if** statements to check the character type.

c. First, check if the character is a digit using the isdigit function.

d. If not a digit, check if the character is an uppercase letter using the isletter and isupper functions.

e. If not an uppercase letter, check if the character is a lowercase letter using the isletter and islower functions.

f. If none of the above, categorize the character as a special character.

g. Return a message indicating the character type.

36. **Sample Code**:

Listing A.23 Character type determination using nested if statements.

```
function result = char_type(char_input)
if isdigit(char_input)
result = 'The character is a digit.';
elseif isletter(char_input) && isupper(char_input)
result = 'The character is an uppercase letter.';
elseif isletter(char_input) && islower(char_input)
result = 'The character is a lowercase letter.';
else
result = 'The character is a special character.';
end
end
```

37. **Sample Output**:

```
>> char_type('5')
ans = "The character is a digit."
>> char_type('A')
ans = "The character is an uppercase letter."
>> char_type('z')
ans = "The character is a lowercase letter."
>> char_type('?')
ans = "The character is a special character."
```

38. **Extensions and Variations**:

- Extend the function to handle multiple characters or strings.
- Modify the function to include additional character types, such as punctuation or whitespace.
- Enhance the function to handle specific edge cases or error conditions, such as empty or non-character input.

39. **Problem Statement and Background**: This problem requires implementing a MATLAB script that prompts the user to enter a number and displays whether it is positive, negative, or zero using an **if-elseif-else** statement. This type of number classification is a common task in many applications.

40. **Step-by-Step Approach**:

    a. Prompt the user to enter a number.
    b. Use an **if-elseif-else** statement to check the sign of the number.
    c. If the number is greater than zero, display a message indicating it is positive.
    d. If the number is less than zero, display a message indicating it is negative.
    e. If the number is zero, display a message indicating it is zero.

41. **Sample Code**:

    Listing A.24 Number sign determination using if-elseif-else statement.

    ```
    % Prompt the user to enter a number
    num = input('Enter a number: ');

    % Determine the sign of the number
    if num > 0
    disp('The number is positive.');
    elseif num < 0
    disp('The number is negative.');
    else
    disp('The number is zero.');
    end
    ```

42. **Sample Output**:
    Enter a number: −5 The number is negative.

43. **Extensions and Variations**:

    • Extend the script to handle complex numbers or matrices.
    • Modify the script to perform additional operations based on the sign of the number.
    • Enhance the script to handle specific edge cases or error conditions, such as non-numeric input.

44. **Problem Statement and Background**: This problem requires creating a MATLAB function that takes two numbers as input and returns their sum if both numbers are positive, their difference if one number is positive and the other is negative, or zero if both numbers are negative, using nested **if** statements. This type of conditional operation is common in many programming tasks.

45. **Step-by-Step Approach**:

    a. Define a function that takes two numbers as input.
    b. Use nested **if** statements to check the signs of the two numbers.
    c. If both numbers are positive, return their sum.
    d. If one number is positive and the other is negative, return their difference (positive - negative).
    e. If both numbers are negative, return zero.

46. **Sample Code**:

Listing A.25 Sum, difference, or zero based on number signs using nested if statements.

```
function result = sum_diff_zero(num1, num2)
if num1 >= 0 && num2 >= 0 % Both positive
result = num1 + num2;
elseif num1 >= 0 && num2 < 0 % One positive, one
    negative
result = num1 - num2;
elseif num1 < 0 && num2 >= 0 % One negative, one
    positive
result = num2 - num1;
else % Both negative
result = 0;
end
end
```

47. **Sample Output**:

```
>> sum_diff_zero(3, 5)
ans =   8
>> sum_diff_zero(-2, 7)
ans =   9
>> sum_diff_zero(-4, -6)
ans =   0
```

48. **Extensions and Variations**:

- Extend the function to handle complex numbers or matrices.
- Modify the function to perform different operations based on the signs of the numbers.
- Enhance the function to handle specific edge cases or error conditions, such as zero values or NaN/Inf.

49. **Problem Statement and Background**: This problem requires writing a MAT-LAB script that generates two random numbers between 1 and 6 (representing dice rolls) and displays a message indicating the outcome (e.g., "You rolled a double," "You rolled a high number," etc.) using a **switch** statement. This type of problem can be used to simulate simple games or random events.

50. **Step-by-Step Approach**:

   a. Generate two random numbers between 1 and 6 using the randi function.
   b. Use a **switch** statement to check the sum of the two random numbers.
   c. For each possible sum, include a case that displays a corresponding message.
   d. Include a default case to handle any unexpected sum values.

51. **Sample Code**:

Listing A.26 Dice roll outcome using switch statement.

```
% Generate two random numbers between 1 and 6
roll1 = randi([1, 6]);
roll2 = randi([1, 6]);
```

```
% Display the outcome based on the sum of the rolls
sum_rolls = roll1 + roll2;
switch sum_rolls
case 2
disp('You rolled a snake eyes!');
case 3
disp('You rolled a low number.');
case 7
disp('You rolled a high number.');
case 12
disp('You rolled a box cars!');
case {2, 12}
disp('You rolled a double!');
otherwise
disp('You rolled a regular combination.');
end
```

52. **Sample Output**:

    You rolled a high number.

53. **Extensions and Variations**:

    - Extend the script to handle more than two dice or different dice value ranges.
    - Modify the script to include additional outcomes or conditions, such as checking for specific combinations.
    - Enhance the script to perform additional operations or calculations based on the outcome.

    The basic syntax for various control statements is as follows:

Listing A.27 If statement syntax.

```
if condition
statements
end
```

Listing A.28 If-else statement syntax.

```
if condition
statements
else
statements
end
```

Listing A.29 If-elseif-else statement syntax.

```
if condition1
statements
elseif condition2
statements
else
statements
end
```

Listing A.30 Switch statement syntax.

```
switch expression
case value1
statements
case value2
statements
otherwise
statements
end
```

These control statements allow you to execute different blocks of code based on certain conditions or expressions, enabling you to create more complex and intelligent programs in MATLAB.

A.5   Solutions to Chapter 5

1. **Problem Statement and Background**: This problem requires writing a MATLAB script that creates a vector of random integers between 1 and 20 with a length of 10. A **for** loop is then used to iterate over the vector and print all elements that are even. This type of problem is useful for understanding how to generate random data and iterate over vectors using loops.

2. **Step-by-Step Approach**:

   a. Create a vector of 10 random integers between 1 and 20 using the randi function.
   b. Initialize a **for** loop to iterate over the vector from the first index to the last index.
   c. Within the loop, check if the current element is even using the mod function (or the rem function for remainders).
   d. If the element is even, print it using the disp function.

3. **Sample Code**:

Listing A.31 Printing even elements of a vector using a for loop.

```
% Create a vector of 10 random integers between 1 and
    20
vec = randi([1, 20], 1, 10);

% Print the even elements
disp('Even elements in the vector:');
for i = 1:length(vec)
if mod(vec(i), 2) == 0
disp(vec(i));
end
end
```

4. **Sample Output**:

```
Even elements in the vector:
6
20
14
2
```

5. **Extensions and Variations**:

- Modify the script to print the indices of the even elements instead of their values.
- Extend the script to count the number of even elements in the vector and print the count.
- Enhance the script to allow the user to specify the vector length and value range.

6. **Problem Statement and Background**: This problem requires writing a MAT-LAB function that takes a vector as input and returns a new vector containing only the positive elements. A **for** loop is used to iterate over the input vector and select the positive elements. This type of problem is useful for understanding how to filter or extract specific elements from a vector based on a condition.

7. **Step-by-Step Approach**:

   a. Define a function that takes a vector as input.
   b. Initialize an empty vector to store the positive elements.
   c. Use a **for** loop to iterate over the input vector.
   d. Within the loop, check if the current element is positive.
   e. If the element is positive, append it to the new vector.
   f. After the loop, return the new vector containing only the positive elements.

8. **Sample Code**:

Listing A.32 Extracting positive elements from a vector using a for loop.

```
function pos_vec = extract_positive(vec)
pos_vec = []; % Initialize empty vector
for i = 1:length(vec)
if vec(i) > 0
pos_vec = [pos_vec, vec(i)]; % Append positive
    element
end
end
end
```

9. **Sample Output**:

```
>> vec = [-3, 0, 5, -2, 7, -1];
>> pos_vec = extract_positive(vec)
pos_vec =
      5      7
```

10. **Extensions and Variations**:

   - Modify the function to handle matrices or higher-dimensional arrays.
   - Extend the function to allow the user to specify a different condition for element selection (e.g., negative elements, elements within a range).
   - Enhance the function to handle specific edge cases or error conditions, such as an empty input vector.

11. **Problem Statement and Background**: This problem requires writing a MAT-LAB script that creates a 3x3 matrix with random values between 1 and 10. Nested **for** loops are then used to iterate over the matrix and print the elements in reverse row order. This type of problem is useful for understanding how to manipulate and access elements in matrices using nested loops.

12. **Step-by-Step Approach**:

   a. Create a 3x3 matrix with random values between 1 and 10 using the randi function.
   b. Initialize an outer **for** loop to iterate over the rows of the matrix from the last row to the first row.
   c. Within the outer loop, initialize an inner **for** loop to iterate over the columns of the matrix from the first column to the last column.
   d. Within the inner loop, print the current element using the disp function.

13. **Sample Code**:

Listing A.33 Printing matrix elements in reverse row order using nested for loops.

```
% Create a 3x3 matrix with random values between 1
    and 10
mat = randi([1, 10], 3, 3);

% Print the elements in reverse row order
disp('Matrix elements in reverse row order:');
for i = 3:-1:1 % Iterate over rows in reverse order
for j = 1:3 % Iterate over columns
disp(mat(i, j));
end
end
```

14. **Sample Output**:

```
Matrix elements in reverse row order:
7
5
1
3
9
6
4
2
8
```

15. **Extensions and Variations**:

   - Modify the script to print the elements in reverse column order instead of reverse row order.
   - Extend the script to handle matrices of different sizes or higher dimensions.
   - Enhance the script to allow the user to specify the value range for the random matrix elements.

16. **Problem Statement and Background**: This problem requires writing a MAT-LAB function that takes a scalar value and a vector as input. A **for** loop is used to iterate over the vector and multiply each element by the scalar value. The modified vector is then returned. This type of problem is useful for understanding how to perform element-wise operations on vectors using loops.

17. **Step-by-Step Approach**:

   a. Define a function that takes a scalar value and a vector as input.
   b. Initialize an empty vector to store the modified elements.
   c. Use a **for** loop to iterate over the input vector.
   d. Within the loop, multiply the current element by the scalar value and append the result to the new vector.
   e. After the loop, return the new vector containing the modified elements.

18. **Sample Code**:

Listing A.34 Multiplying vector elements by a scalar using a for loop.

```
function new_vec = scalar_multiply(scalar, vec)
new_vec = []; % Initialize empty vector
for i = 1:length(vec)
new_vec = [new_vec, scalar * vec(i)]; % Append
    modified element
end
end
```

19. **Sample Output**:

```
>> vec = [1, 2, 3, 4, 5];
>> new_vec = scalar_multiply(2, vec)
new_vec =
    2    4    6    8    10
```

20. **Extensions and Variations**:

   - Modify the function to handle matrices or higher-dimensional arrays.
   - Extend the function to perform different element-wise operations (e.g., addition, subtraction, division).
   - Enhance the function to handle specific edge cases or error conditions, such as an empty input vector or a scalar value of zero.

21. **Problem Statement and Background**: This problem requires writing a MAT-LAB script that creates a vector of random integers between 1 and 100 with a

length of 20. A **while** loop is then used to iterate over the vector and print all elements that are divisible by 3 or 5. This type of problem is useful for understanding how to use while loops and iterate over vectors based on a condition.

22. **Step-by-Step Approach**:

    a. Create a vector of 20 random integers between 1 and 100 using the randi function.

    b. Initialize a counter variable i to 1.

    c. Initialize a **while** loop that continues as long as i is less than or equal to the length of the vector.

    d. Within the loop, check if the current element is divisible by 3 or 5 using the mod function (or the rem function for remainders).

    e. If the element is divisible by 3 or 5, print it using the disp function.

    f. Increment the counter variable i by 1.

23. **Sample Code**:

Listing A.35 Printing elements divisible by 3 or 5 using a while loop.

```
% Create a vector of 20 random integers between 1 and
    100
vec = randi([1, 100], 1, 20);

% Print elements divisible by 3 or 5
disp('Elements divisible by 3 or 5:');
i = 1;
while i <= length(vec)
if mod(vec(i), 3) == 0 || mod(vec(i), 5) == 0
disp(vec(i));
end
i = i + 1;
end
```

24. **Sample Output**:

```
Elements divisible by 3 or 5:
15
30
45
60
75
90
```

25. **Extensions and Variations**:

- Modify the script to print the indices of the divisible elements instead of their values.
- Extend the script to count the number of elements divisible by 3 or 5 and print the count.
- Enhance the script to allow the user to specify the vector length and value range, as well as the divisibility conditions.

26. **Problem Statement and Background**: This problem requires writing a MAT-LAB function that takes a matrix as input and computes the sum of all elements in the matrix using nested **for** loops. This type of problem is useful for understanding how to iterate over matrices and perform calculations on their elements.

27. **Step-by-Step Approach**:

    a. Define a function that takes a matrix as input.
    b. Initialize a variable sum to 0 to store the cumulative sum.
    c. Use a nested **for** loop structure to iterate over the rows and columns of the matrix.
    d. Within the inner loop, add the current element to the sum variable.
    e. After the nested loops, return the final value of sum.

28. **Sample Code**:

Listing A.36 Computing the sum of matrix elements using nested for loops.

```
function total_sum = matrix_sum(mat)
total_sum = 0; % Initialize sum to 0
for i = 1:size(mat, 1) % Iterate over rows
for j = 1:size(mat, 2) % Iterate over columns
total_sum = total_sum + mat(i, j); % Add current
    element to sum
end
end
end
```

29. **Sample Output**:

```
>> mat = [1 2 3; 4 5 6; 7 8 9];
>> total_sum = matrix_sum(mat)
total_sum =
    45
```

30. **Extensions and Variations**:

    • Modify the function to handle higher-dimensional arrays.
    • Extend the function to compute other statistical measures, such as the mean, variance, or standard deviation of the matrix elements.
    • Enhance the function to handle specific edge cases or error conditions, such as an empty input matrix.

31. **Problem Statement and Background**: This problem requires writing a MAT-LAB script that creates two vectors, A and B, of random integers between 1 and 10 with a length of 5. Vectorisation is then used to compute the element-wise sum, difference, and product of the two vectors. This type of problem is useful for understanding how to perform element-wise operations on vectors efficiently using vectorisation techniques.

32. **Step-by-Step Approach**:

   a. Create two vectors, A and B, of length 5 with random integers between 1 and 10 using the randi function.
   b. Compute the element-wise sum of A and B using the + operator and vectorisation.
   c. Compute the element-wise difference of A and B using the - operator and vectorisation.
   d. Compute the element-wise product of A and B using the .* operator and vectorisation.
   e. Display the results.

33. **Sample Code**:

Listing A.37 Vectorised element-wise operations on vectors.

```
% Create two vectors with random integers between 1
    and 10
A = randi([1, 10], 1, 5);
B = randi([1, 10], 1, 5);

% Compute element-wise operations using vectorisation
sum_AB = A + B;
diff_AB = A - B;
prod_AB = A .* B;

% Display the results
disp('Vector A:');
disp(A);
disp('Vector B:');
disp(B);
disp('Element-wise sum:');
disp(sum_AB);
disp('Element-wise difference:');
disp(diff_AB);
disp('Element-wise product:');
disp(prod_AB);
```

34. **Sample Output**:

```
Vector A:
     3      8      1      5      6
Vector B:
     2      4      9      7     10
Element-wise sum:
     5     12     10     12     16
Element-wise difference:
     1      4     -8     -2     -4
Element-wise product:
     6     32      9     35     60
```

35. **Extensions and Variations**:

- Modify the script to handle matrices or higher-dimensional arrays.
- Extend the script to perform additional element-wise operations, such as division or exponentiation.
- Enhance the script to allow the user to specify the vector length and value range.

36. **Problem Statement and Background**: This problem requires writing a MATLAB function that takes a vector as input and returns the maximum and minimum values in the vector using a **for** loop. This type of problem is useful for understanding how to find the extrema (maximum and minimum values) in a dataset using loops.

37. **Step-by-Step Approach**:

a. Define a function that takes a vector as input.
b. Initialize variables max_val and min_val to the first element of the vector.
c. Use a **for** loop to iterate over the remaining elements of the vector, starting from the second element.
d. Within the loop, update the max_val variable if the current element is greater than the current maximum value.
e. Within the loop, update the min_val variable if the current element is less than the current minimum value.
f. After the loop, return both max_val and min_val.

38. **Sample Code**:

Listing A.38 Finding maximum and minimum values in a vector using a for loop.

```
function [max_val, min_val] = find_extrema(vec)
max_val = vec(1); % Initialize max_val to first
    element
min_val = vec(1); % Initialize min_val to first
    element
for i = 2:length(vec) % Start from second element
if vec(i) > max_val
max_val = vec(i); % Update max_val
end
if vec(i) < min_val
min_val = vec(i); % Update min_val
end
end
end
```

39. **Sample Output**:

```
>> vec = [5, 3, 8, 1, 6];
>> [max_val, min_val] = find_extrema(vec)
max_val =
      8
min_val =
      1
```

40. **Extensions and Variations**:

- Modify the function to handle matrices or higher-dimensional arrays.
- Extend the function to find the maximum and minimum values along specific dimensions (e.g., rows or columns) of a matrix.
- Enhance the function to handle specific edge cases or error conditions, such as an empty input vector.

41. **Problem Statement and Background**: This problem requires writing a MAT-LAB script that creates a vector of random integers between 1 and 20 with a length of 15. A **for** loop is then used to iterate over the vector and replace all occurrences of the number 3 with the value −1. This type of problem is useful for understanding how to modify elements in a vector based on a condition using loops.

42. **Step-by-Step Approach**:

a. Create a vector of 15 random integers between 1 and 20 using the randi function.

b. Use a **for** loop to iterate over the vector.

c. Within the loop, check if the current element is equal to 3.

d. If the element is equal to 3, replace it with −1.

e. After the loop, display the modified vector.

43. **Sample Code**:

Listing A.39 Replacing occurrences of a value in a vector using a for loop.

```
% Create a vector of 15 random integers between 1 and
    20
vec = randi([1, 20], 1, 15);

% Replace occurrences of 3 with -1
for i = 1:length(vec)
if vec(i) == 3
vec(i) = -1; % Replace with -1
end
end

% Display the modified vector
disp('Modified vector:');
disp(vec);
```

44. **Sample Output**:

```
Modified vector:
    17    -1     9     6     5     1    14    18    -1     3    11    -1
    16    20     7
```

45. **Extensions and Variations**:

- Modify the script to replace occurrences of multiple values with different replacement values.

- Extend the script to handle matrices or higher-dimensional arrays.
- Enhance the script to allow the user to specify the vector length, value range, and the value(s) to be replaced.

46. **Problem Statement and Background**: This problem requires writing a MAT-LAB function that takes a matrix as input and computes the sum of the diagonal elements using a single **for** loop. This type of problem is useful for understanding how to access and manipulate diagonal elements in a matrix using loops.

47. **Step-by-Step Approach**:

   a. Define a function that takes a matrix as input.
   b. Initialize a variable $diag\_sum$ to 0 to store the sum of the diagonal elements.
   c. Use a single **for** loop to iterate over the diagonal elements of the matrix.
   d. Within the loop, add the current diagonal element to the $diag\_sum$ variable.
   e. After the loop, return the final value of $diag\_sum$.

48. **Sample Code**:

Listing A.40 Computing the sum of diagonal elements using a for loop.

```
function diag_sum = diagonal_sum(mat)
diag_sum = 0; % Initialize sum to 0
for i = 1:min(size(mat)) % Iterate over diagonal
    elements
diag_sum = diag_sum + mat(i, i); % Add diagonal
    element to sum
end
end
```

49. **Sample Output**:

```
>> mat = [1 2 3; 4 5 6; 7 8 9];
>> diag_sum = diagonal_sum(mat)
diag_sum =
     15
```

50. **Extensions and Variations**:

- Modify the function to compute the sum of the off-diagonal elements or the secondary diagonal elements.
- Extend the function to handle matrices of different sizes or higher dimensions.
- Enhance the function to handle specific edge cases or error conditions, such as an empty input matrix or non-square matrices.

The basic syntax for various loop statements is as follows:

Listing A.41 For loop syntax.

```
for variable = expression
% Statement(s)
end
```

Listing A.42 While loop syntax.

```
while condition
% Statement(s)
end
```

Listing A.43 Nested loops syntax.

```
for outer_variable = outer_expression
for inner_variable = inner_expression
% Statement(s)
end
end
```

In summary, this chapter covered various examples and problem statements related to loops in MATLAB, including:

- Using for loops to perform operations on vector elements
- Employing while loops to iterate based on conditions
- Utilizing nested for loops to iterate over matrices
- Applying vectorization techniques for efficient element-wise operations
- Finding maximum and minimum values in a vector
- Modifying vector elements based on conditions
- Computing the sum of diagonal elements in a matrix

These examples aimed to provide a solid understanding of how to use different loop constructs and apply them to solve practical problems in MATLAB. Additionally, extensions and variations were suggested to further enhance the learning experience and problem-solving skills.

It is important to note that while loops are powerful tools for iterating and performing operations, MATLAB's vectorized operations are often more efficient and preferred for many computations involving arrays and matrices. However, loops remain essential for certain tasks, such as conditional operations or iterating over specific elements based on criteria.

By mastering the concepts covered in this chapter, you will be well-equipped to tackle a wide range of programming challenges that involve iterative processes and data manipulation in MATLAB.

These problems cover various aspects of using loops (**for** and **while**) in MATLAB, including iterating over vectors and matrices, performing element-wise operations, and implementing basic algorithms. The solutions provide step-by-step approaches, sample code, and sample outputs, along with suggestions for extensions and variations to further enhance your understanding and skills.

A.6 Solutions to Chapter 6

1. **Problem Statement and Background**: Write a MATLAB script that prompts the user to enter their name and age, and then displays a personalised greeting.

2. **Step-by-Step Approach**:

   a. Prompt the user to enter their name using the input function.
   b. Prompt the user to enter their age using the input function.
   c. Construct a personalized greeting message using the entered name and age.
   d. Display the greeting message using the fprintf function.

3. **Sample Code**:

Listing A.44 Personalized greeting script.

```
% Prompt the user to enter their name
name = input('Enter your name: ', 's');

% Prompt the user to enter their age
age = input('Enter your age: ');

% Construct the greeting message
greeting = sprintf('Hello, %s! You are %d years old.'
    , name, age);

% Display the greeting message
fprintf('%s\n', greeting);
```

4. **Sample Output**: If the user enters "John" as their name and "25" as their age, the output will be:

```
Enter your name: John
Enter your age: 25
Hello, John! You are 25 years old.
```

5. **Extensions and Variations**:

   - Modify the script to handle invalid input (e.g., non-numeric age or empty name).
   - Add additional information to the greeting message, such as a personalized message based on the user's age range.

6. **Problem Statement and Background**: Create a MATLAB function that takes two numbers as input and returns their sum, difference, product, and quotient (if applicable).

7. **Step-by-Step Approach**:

   a. Define a function named 'arithmetic' that takes two input arguments 'a' and 'b'.
   b. Inside the function, calculate the sum, difference, product, and quotient (if applicable) of 'a' and 'b'.
   c. Return the calculated values as output arguments from the function.

8. **Sample Code**:

Listing A.45 Arithmetic operations function.

```
function [sum, diff, prod, quot] = arithmetic(a, b)
% ARITHMETIC Performs arithmetic operations on two
    numbers
% [sum, diff, prod, quot] = arithmetic(a, b) returns
    the
% sum, difference, product, and quotient (if
    applicable)
% of a and b

sum  = a + b;
diff = a - b;
prod = a * b;
if b ~= 0
quot = a / b;
else
quot = NaN; % Undefined for division by zero
end
end
```

9. **Sample Output**: To call the function and perform arithmetic operations on 5 and 3, use the following code:

```
[s, d, p, q] = arithmetic(5, 3)
s =
8
d =
2
p =
15
q =
1.6667
```

10. **Extensions and Variations**:

   - Modify the function to handle other arithmetic operations, such as modulus or exponentiation.
   - Add error handling for non-numeric input or other edge cases.
   - Create a script that prompts the user to enter two numbers and calls the arithmetic function to display the results.

11. **Problem Statement and Background**: Write a MATLAB script that generates a random vector of 10 integers between 1 and 100, and then calculates the mean, median, and standard deviation of the vector.

12. **Step-by-Step Approach**:

   a. Generate a random vector of 10 integers between 1 and 100 using the 'randi' function.

b. Calculate the mean of the vector using the 'mean' function.

c. Calculate the median of the vector using the 'median' function.

d. Calculate the standard deviation of the vector using the 'std' function.

e. Display the generated vector, mean, median, and standard deviation.

13. **Sample Code**:

Listing A.46 Vector statistics script.

```
% Generate a random vector of 10 integers between 1
    and 100
vec = randi([1, 100], 1, 10);

% Calculate the mean, median, and standard deviation
mean_val = mean(vec);
median_val = median(vec);
std_val = std(vec);

% Display the vector and statistics
fprintf('Random vector: ');
disp(vec);
fprintf('Mean: %.2f\n', mean_val);
fprintf('Median: %.2f\n', median_val);
fprintf('Standard deviation: %.2f\n', std_val);
```

14. **Sample Output**: The output will be a randomly generated vector and its mean, median, and standard deviation, for example:

```
Random vector: 8 82 48 99 31 77 63 12 57 14
Mean: 49.10
Median: 54.50
Standard deviation: 31.56
```

15. **Extensions and Variations**:

- Modify the script to generate a vector of a different size or within a different range of values.
- Add functionality to calculate additional statistical measures, such as quartiles or mode.
- Create a function that takes a vector as input and returns the calculated statistics.

16. **Problem Statement and Background**: Create a MATLAB function that takes a string as input and returns the number of vowels (a, e, i, o, u) in the string.

17. **Step-by-Step Approach**:

a. Define a function named countVowels that takes a string str as input.

b. Initialise a counter variable vowel_count to 0.

c. Convert the input string to lowercase using the lower function.

d. Iterate through each character in the string using a loop.

e. Inside the loop, check if the current character is a vowel (a, e, i, o, u) using a conditional statement.

f. If the character is a vowel, increment the vowel_count by 1.

g. After the loop finishes, return vowel_count as the output of the function.

18. **Sample Code**:

Listing A.47 Count vowels function.

```
function vowel_count = countVowels(str)
% COUNTVOWELS Counts the number of vowels in a string
% vowel_count = countVowels(str) returns the number
    of
% vowels (a, e, i, o, u) in the input string str

vowel_count = 0;
str = lower(str); % Convert to lowercase
for i = 1:length(str)
char = str(i);
if char == 'a' || char == 'e' || char == 'i' || char
    == 'o' || char == 'u'
vowel_count = vowel_count + 1;
end
end
end
```

19. **Sample Output**: To call the function and count the vowels in the string "Hello, World!", use the following code:

```
countVowels('Hello, World!')
ans =
3
```

20. **Extensions and Variations**:

- Modify the function to count vowels in a case-insensitive manner (i.e., treat uppercase and lowercase vowels the same).
- Add functionality to count specific vowels only (e.g., count only the occurrences of 'a' and 'e').
- Create a script that prompts the user to enter a string and calls the countVowels function to display the number of vowels.

21. **Problem Statement and Background**: Write a MATLAB script that prompts the user to enter the coefficients of a quadratic equation (

$$ax^2 + bx + c = 0$$

) and calculates its roots using the quadratic formula.

22. **Step-by-Step Approach**:

a. Prompt the user to enter the coefficients a, b, and c using the input function.

b. Calculate the discriminant (

$$b^2 - 4ac$$

) using the entered coefficients.

c. Check if the discriminant is positive, zero, or negative to determine the nature of the roots.

d. If the discriminant is positive, calculate the two distinct real roots using the quadratic formula.

e. If the discriminant is zero, calculate the single real root using the quadratic formula.

f. If the discriminant is negative, output a message indicating that the roots are complex conjugates.

g. Display the calculated roots (if any) using the fprintf function.

23. **Sample Code**:

Listing A.48 Quadratic equation solver script.

```
% Prompt the user to enter the coefficients
a = input('Enter the coefficient a: ');
b = input('Enter the coefficient b: ');
c = input('Enter the coefficient c: ');

% Calculate the discriminant
discriminant = b^2 - 4ac;

% Check the nature of the roots based on the
    discriminant
if discriminant > 0
% Two distinct real roots
root1 = (-b + sqrt(discriminant)) / (2a);
root2 = (-b - sqrt(discriminant)) / (2a);
fprintf('The roots of the equation are: %.2f and %.2f
    \n', root1, root2);
elseif discriminant == 0
% Single real root
root = -b / (2*a);
fprintf('The root of the equation is: %.2f\n', root);
else
% Complex conjugate roots
fprintf('The roots of the equation are complex
    conjugates.\n');
end
```

24. **Sample Output**: If the user enters coefficients a = 1, b = −3, and c = 2, the output will be:

```
Enter the coefficient a: 1
Enter the coefficient b: -3
Enter the coefficient c: 2
The roots of the equation are: 2.00 and 1.00
```

25. **Extensions and Variations**:

- Modify the script to handle invalid input (e.g., non-numeric coefficients or a = 0, which makes it a linear equation).

- Add functionality to display the roots in a more user-friendly format (e.g., using complex number notation for complex roots).
- Create a function that takes the coefficients as input and returns the calculated roots as output.

26. **Problem Statement and Background**: Create a MATLAB function that takes a vector of numbers as input and returns a new vector containing only the unique elements.

27. **Step-by-Step Approach**:

   a. Define a function named uniqueElements that takes a vector vec as input.
   b. Initialise an empty vector unique_vec to store the unique elements.
   c. Iterate through each element in the input vector vec using a loop.
   d. Inside the loop, check if the current element is already present in the unique_vec vector using the ismember function.
   e. If the current element is not present in unique_vec, append it to the end of unique_vec.
   f. After the loop finishes, unique_vec will contain all the unique elements from the input vector.
   g. Return unique_vec as the output of the function.

28. **Sample Code**:

Listing A.49 Unique elements function.

```
function unique_vec = uniqueElements(vec)
% UNIQUEELEMENTS Finds the unique elements in a
    vector
% unique_vec = uniqueElements(vec) returns a vector
    containing
% the unique elements from the input vector vec

unique_vec = [];
for i = 1:length(vec)
if ~ismember(vec(i), unique_vec)
unique_vec = [unique_vec, vec(i)];
end
end
end
```

29. **Sample Output**: To call the function and find the unique elements in the vector [1, 2, 3, 2, 4, 1, 5], use the following code:

```
vec = [1, 2, 3, 2, 4, 1, 5];
unique_vec = uniqueElements(vec)
unique_vec =
1 2 3 4 5
```

30. **Extensions and Variations**:

- Modify the function to handle different data types (e.g., strings or cell arrays) instead of numeric vectors.
- Add functionality to sort the output vector in ascending or descending order.
- Create a script that prompts the user to enter the elements of the vector and calls the uniqueElements function to display the unique elements.

31. **Problem Statement and Background**: Write a MATLAB script that generates a random 3x3 matrix and calculates its determinant, trace, and inverse (if applicable).

32. **Step-by-Step Approach**:

   a. Generate a random 3x3 matrix using the randi function.
   b. Calculate the determinant of the matrix using the det function.
   c. Calculate the trace of the matrix by summing the diagonal elements.
   d. Check if the determinant is non-zero to determine if the matrix is invertible.
   e. If the matrix is invertible, calculate its inverse using the inv function.
   f. Display the generated matrix, determinant, trace, and inverse (if applicable).

33. **Sample Code**:

Listing A.50 Matrix operations script.

```matlab
% Generate a random 3x3 matrix
A = randi([1, 10], 3, 3);

% Calculate the determinant
det_A = det(A);

% Calculate the trace
trace_A = sum(diag(A));

% Check if the matrix is invertible
if det_A ~= 0
% Calculate the inverse
inv_A = inv(A);
fprintf('The inverse of the matrix is:\n');
disp(inv_A);
else
fprintf('The matrix is not invertible (determinant is
    zero).\n');
end

% Display the matrix, determinant, and trace
fprintf('The matrix is:\n');
disp(A);
fprintf('Determinant: %d\n', det_A);
fprintf('Trace: %d\n', trace_A);
```

34. **Sample Output**: The output will depend on the randomly generated matrix, but it may look something like this:

```
The matrix is:
9 3 7
1 5 8
6 4 2
Determinant: -33
Trace: 16
The inverse of the matrix is:
0.0606 0.0303 -0.1818
0.1818 -0.0303 -0.0606
-0.2424 0.1818 0.2121
```

35. **Extensions and Variations**:

  • Modify the script to generate a matrix of a different size (e.g., 4x4 or NxN).
  • Add functionality to calculate additional matrix properties, such as the rank or eigenvalues/eigenvectors.
  • Create a function that takes a matrix as input and returns its determinant, trace, and inverse (if applicable).

36. **Problem Statement and Background**: Create a MATLAB function that takes a positive integer as input and returns the sum of its digits.

37. **Step-by-Step Approach**:

  a. Define a function named sumDigits that takes a positive integer num as input.
  b. Initialise a variable sum to 0 to store the sum of digits.
  c. Convert the input number to a string using the num2str function.
  d. Iterate through each character in the string using a loop.
  e. Inside the loop, convert the current character to a numeric value using the str2double function and add it to the sum variable.
  f. After the loop finishes, return sum as the output of the function.

38. **Sample Code**:

Listing A.51 Sum of digits function.

```
function sum = sumDigits(num)
% SUMDIGITS Calculates the sum of digits of a number
% sum = sumDigits(num) returns the sum of digits
% of the input positive integer num

sum = 0;
str_num = num2str(num);
for i = 1:length(str_num)
digit = str2double(str_num(i));
sum = sum + digit;
end
end
```

39. **Sample Output**: To call the function and calculate the sum of digits for the number 12345, use the following code:

```
sumDigits(12345)
ans =
15
```

40. **Extensions and Variations**:

   • Modify the function to handle negative integers by taking the absolute value of the input number.
   • Add functionality to handle non-integer input or other edge cases.
   • Create a script that prompts the user to enter a number and calls the sumDigits function to display the sum of its digits.

41. **Problem Statement and Background**: Write a MATLAB script that prompts the user to enter the side lengths of a triangle and determines whether the triangle is equilateral, isosceles, or scalene.

42. **Step-by-Step Approach**:

   a. Prompt the user to enter the three side lengths of the triangle using the input function.
   b. Check if all three side lengths are equal. If so, the triangle is equilateral.
   c. Otherwise, check if any two side lengths are equal. If so, the triangle is isosceles.
   d. If none of the side lengths are equal, the triangle is scalene.
   e. Display the classification of the triangle (equilateral, isosceles, or scalene) using the fprintf function.

43. **Sample Code**:

Listing A.52 Triangle classification script.

```
% Prompt the user to enter the side lengths
side1 = input('Enter the length of side 1: ');
side2 = input('Enter the length of side 2: ');
side3 = input('Enter the length of side 3: ');

% Check the triangle type
if side1 == side2 && side2 == side3
fprintf('The triangle is equilateral.\n');
elseif side1 == side2 || side1 == side3 || side2 ==
    side3
fprintf('The triangle is isosceles.\n');
else
fprintf('The triangle is scalene.\n');
end
```

44. **Sample Output**: If the user enters side lengths 5, 5, and 7, the output will be:

```
Enter the length of side 1: 5
Enter the length of side 2: 5
Enter the length of side 3: 7
The triangle is isosceles.
```

45. **Extensions and Variations**:

- Modify the script to handle invalid input (e.g., non-numeric side lengths or side lengths that do not satisfy the triangle inequality).
- Add functionality to calculate additional properties of the triangle, such as its perimeter or area.
- Create a function that takes the side lengths as input and returns the classification of the triangle.

46. **Problem Statement and Background**: Create a MATLAB function that takes a vector of numbers as input and returns a new vector containing the cumulative sum of the elements.

47. **Step-by-Step Approach**:

   a. Define a function named cumulativeSum that takes a vector vec as input.
   b. Initialise an empty vector cumsum_vec to store the cumulative sum.
   c. Iterate through each element in the input vector vec using a loop.
   d. Inside the loop, calculate the cumulative sum by adding the current element to the previous cumulative sum value.
   e. Append the calculated cumulative sum to the cumsum_vec vector.
   f. After the loop finishes, return cumsum_vec as the output of the function.

48. **Sample Code**:

Listing A.53 Cumulative sum function.

```
function cumsum_vec = cumulativeSum(vec)
% CUMULATIVESUM Calculates the cumulative sum of a
    vector
% cumsum_vec = cumulativeSum(vec) returns a vector
    containing
% the cumulative sum of the elements in the input
    vector vec

cumsum_vec = zeros(size(vec));
cumsum = 0;
for i = 1:length(vec)
cumsum = cumsum + vec(i);
cumsum_vec(i) = cumsum;
end
end
```

49. **Sample Output**: To call the function and calculate the cumulative sum for the vector [1, 2, 3, 4, 5], use the following code:

```
vec = [1, 2, 3, 4, 5];
cumsum_vec = cumulativeSum(vec)
cumsum_vec =
1  3  6  10  15
```

50. **Extensions and Variations**:

- Modify the function to handle different data types (e.g., strings or cell arrays) instead of numeric vectors.
- Add functionality to calculate the cumulative product or other cumulative operations instead of the cumulative sum.
- Create a script that prompts the user to enter the elements of the vector and calls the cumulativeSum function to display the cumulative sum.

51. Create a MATLAB function to calculate 'the coefficient of determination or R^2' [1] with the function name 'SECF_assess_R2.m', in which, the first line is

```
[R2] = SECF\_assess\_R2(y\_test, y\_calculation)
```

a. **Problem Statement and Background**:

- Create a MATLAB function to calculate **the coefficient of determination or R^2** with the function name `SECF_assess_R2.m`.
- The function should take two input arguments, `y_test` and `y_calculation`, and output the R^2 value.
- The coefficient of determination (R^2) measures the proportion of the variance in the dependent variable that is predictable from the independent variable(s).

b. **Step-by-Step Approach**:

- **Define Function Signature**: Start by defining the function signature in MATLAB.
- **Compute Deviations**: Calculate deviations from the mean for the test data.
- **Compute Total Variation**: Compute the total variation to be accounted for (SStot).
- **Compute Residuals**: Calculate the residuals (differences between the test data and the calculated data).
- **Compute Residual Variation**: Compute the variation not accounted for (SSerr).
- **Calculate R^2**: Calculate the R^2 value using the formula $R^2 = 1 - \frac{SSerr}{SStot}$.

c. **Sample Code**: The basic syntax is:

Listing A.54 Function to calculate R^2.

```
% SECF_assess_R2 Begin

function [R2] = SECF_assess_R2( y_test,
    y_calculation )

    % deviations - measure of spread
    sample_dev = y_test - mean(y_test);

    % total variation to be accounted for
    SStot = sum(sample_dev.^2);
```

```
% residuals - measure of mismatch
resid = y_test - y_calculation;

% variation NOT accounted for
SSerr = sum(resid.^2);

% residual norm - the 2-norm of the vector
    of the residuals for the fit.
% One common goodness of fit involves a
    least-squares approximation.
% This describes the distance of the entire
    set of data points from the
% fitted curve. The normalization of the
    residual error minimizing the
% square of the sum of squares of all
    residual errors.

normr = sqrt(SSerr);

% R2 Error (percent of error explained)
% The coefficient of determination (also
    referred to as the R2 value) for
% the fit indicates the percent of the
    variation in the data that is
% explained by the model.

R2 = 1 - SSerr/SStot;

% SECF_assess_R2 End
```

d. **Sample Output**:

- Given a test data set `y_test` and calculated data set `y_calculat ion`, the function will output the R^2 value.
- Example output:

```
>> y_test = [3, 5, 7, 9];
>> y_calculation = [2.8, 5.1, 6.9, 9.2];
>> R2 = SECF_assess_R2(y_test, y_calculation)
R2 = 0.9984
```

e. **Extensions and Variations**:

- **Weighted** R^2: One can extend the function to calculate a weighted R^2 by incorporating weights for each data point.
- **Adjusted** R^2: For models with multiple predictors, the adjusted R^2 can be calculated to account for the number of predictors.
- **Different Norms**: Instead of using the 2-norm, other norms like the 1-norm or infinity norm could be used to calculate residuals.

- **Other Goodness-of-Fit Metrics**: Metrics such as Mean Squared Error (MSE), Root Mean Squared Error (RMSE), and Mean Absolute Error (MAE) can be added to the function for a more comprehensive assessment.

52. Create two MATLAB functions to calculate 'the moving mean of the average precision (mmAP)' and 'the moving mean of standard derivation (mmSTD)' [4].

   a. **Problem Statement and Background**:

   - Create two MATLAB functions to calculate **the moving mean of the average precision (mmAP)** and **the moving mean of standard deviation (mmSTD)**.
   - The definitions of the two trend indices, mmAP and mmSTD, are provided in Eqs. (A.1) and (A.2), respectively.
   - These indices are used to mitigate short-term fluctuations by capturing the longer-term trend across the evolutionary process.
   - The indices are defined as follows:

$$mmAP\left(f_j\right) = \frac{1}{p}\sum_{i=1}^{p}\left(\frac{1}{i}\sum_{j=1}^{i}MEAN\left(f_j\right)\right) \qquad (A.1)$$

$$mmSTD\left(f_j\right) = \frac{1}{p}\sum_{i=1}^{p}\left(\frac{1}{i}\sum_{j=1}^{i}STD\left(f_j\right)\right) \qquad (A.2)$$

   b. **Step-by-Step Approach**:

   - **Define Function Signature**: Start by defining the function signature in MATLAB for both mmAP and mmSTD.
   - **Initialize Parameters**: Initialize necessary parameters such as the length of the input vector.
   - **Calculate Moving Mean of Mean (mmAP)**:

      - Iterate through each element of the vector.
      - Calculate the mean up to the current index.
      - Compute the cumulative moving mean.

   - **Calculate Moving Mean of Standard Deviation (mmSTD)**:

      - Iterate through each element of the vector.
      - Calculate the standard deviation up to the current index.
      - Compute the cumulative moving standard deviation.

   c. **Sample Code**: The basic syntax for calculating mmAP and mmSTD is provided below.

Listing A.55 Function to calculate mmAP.

```
% Function to calculate mmAP
function mmAP = calculate_mmAP(f)
p = length(f);
mmAP = 0;
for i = 1:p
current_mean = mean(f(1:i));
mmAP = mmAP + (1/i) * current_mean;
end
mmAP = mmAP / p;
end
```

Listing A.56 Function to calculate mmSTD.

```
% Function to calculate mmSTD
function mmSTD = calculate_mmSTD(f)
p = length(f);
mmSTD = 0;
for i = 1:p
current_std = std(f(1:i));
mmSTD = mmSTD + (1/i) * current_std;
end
mmSTD = mmSTD / p;
end
```

d. **Sample Output**: The sample output for the functions can be illustrated with an example vector.

Listing A.57 Sample Output for mmAP and mmSTD Functions.

```
% Sample input vector
f = [1, 2, 3, 4, 5, 6, 7, 8, 9, 10];

% Calculate mmAP
mmAP_value = calculate_mmAP(f);
fprintf('The moving mean of average
    precision (mmAP) is: %.4f\n',
    mmAP_value);

% Calculate mmSTD
mmSTD_value = calculate_mmSTD(f);
fprintf('The moving mean of standard
    deviation (mmSTD) is: %.4f\n',
    mmSTD_value);
```

The expected output for the given vector 'f' would be:

```
The moving mean of average precision (mmAP) is: 5.5000
The moving mean of standard deviation (mmSTD) is: 2.8723
```

e. **Extensions and Variations**:

- **Weighted Moving Mean**: Extend the functions to calculate weighted moving means, where more recent values are given higher weights.

- **Different Window Sizes**: Implement functionality to use different window sizes for the moving average, allowing for more flexibility in capturing trends.
- **Robust Statistics**: Incorporate robust statistical measures such as median and interquartile range (IQR) instead of mean and standard deviation to handle outliers.
- **Real-time Data Processing**: Modify the functions to process streaming data in real-time, updating the moving mean and standard deviation as new data comes in.
- **Visualisation**: Add Visualisation features to plot the moving mean and standard deviation over time, providing a graphical representation of the trends.

These problems cover various aspects of working with scripts and functions in MATLAB, including input/output operations, control structures, arithmetic operations, string manipulation, matrix operations, and vector operations. The solutions provide step-by-step approaches, sample code, sample output, and suggestions for extensions and variations to enhance learning and understanding.

A.7   Solutions to Chapter 7

1. **Problem Statement and Background**: Write a MATLAB function that takes a text file as input and counts the occurrences of each unique word in the file. The function should return a struct or a cell array containing the unique words and their corresponding counts.
2. **Step-by-Step Approach**:
   a. Define a function named 'wordCountsInFile' that takes a filename as input.
   b. Open the file using the 'fopen' function.
   c. Read the contents of the file into a string using the 'fscanf' function.
   d. Close the file using the 'fclose' function.
   e. Split the file contents into individual words using the 'strsplit' function and appropriate delimiters (e.g., whitespace characters).
   f. Initialise a struct or a cell array to store the unique words and their counts.
   g. Iterate through the words and update the counts in the struct or cell array.
   h. Return the struct or cell array containing the unique words and their counts.
3. **Sample Code**:

Listing A.58 Word counts in a file.

```
function wordCounts = wordCountsInFile(filename)
% WORDCOUNTSINFILE Counts the occurrences of each
    unique word in a file
% wordCounts = wordCountsInFile(filename)
```

```
% returns a struct or cell array containing unique
    words and counts

% Open the file
fileID = fopen(filename, 'r');

% Read the file contents
fileContents = fscanf(fileID, '%c');

% Close the file
fclose(fileID);

% Split the file contents into words
words = strsplit(strtrim(fileContents), '\s+');

% Initialise a struct to store word counts
wordCounts = struct();

% Count the occurrences of each word
for i = 1:length(words)
word = words{i};
if isfield(wordCounts, word)
wordCounts.(word) = wordCounts.(word) + 1;
else
wordCounts.(word) = 1;
end
end
end
```

4. **Sample Output**:

```
wordCounts = wordCountsInFile('example.txt')
wordCounts =
struct with fields:
the: 2
quick: 1
brown: 1
fox: 1
jumps: 1
over: 1
lazy: 1
dog: 1
```

5. **Extensions and Variations**:

- Modify the function to handle case-insensitive word counting.
- Add functionality to exclude common words (e.g., "the," "a," "and") from the word count.
- Allow the function to take multiple filenames as input and combine the word counts from all files.
- Implement alternative data structures (e.g., containers.Map) to store the word counts.

6. **Problem Statement and Background**: Create a MATLAB script that reads data from a CSV (Comma-Separated Values) file and performs basic data analysis tasks, such as calculating the mean, median, and standard deviation for each column of numerical data.

7. **Step-by-Step Approach**:

   a. Prompt the user to enter the filename of the CSV file using the 'input' function.
   b. Use the 'readtable' function to read the contents of the CSV file into a table. The basic syntax is:

Listing A.59 Reading a CSV file.

```
data = readtable(filename);
```

   c. Identify the numerical columns in the table using the isnumeric function.
   d. For each numerical column, calculate the mean, median, and standard deviation using the mean, median, and std functions, respectively.
   e. Display the results using the fprintf function or create a new table with the statistical measures.

8. **Sample Code**:

Listing A.60 Data analysis on a CSV file.

```
% Prompt the user for the filename
filename = input('Enter the filename of the CSV file: ', 's');

% Read the CSV file into a table
data = readtable(filename);

% Get the variable names (column names)
varNames = data.Properties.VariableNames;

% Initialise a table to store the statistical
    measures
statsTable = table();

% Iterate over the columns
for i = 1:length(varNames)
colName = varNames{i};

% Check if the column is numerical
if isnumeric(data.(colName))
    % Calculate statistical measures
    mean_value = mean(data.(colName));
    median_value = median(data.(colName));
    std_value = std(data.(colName));

    % Store the measures in the statsTable
    statsTable.(colName) = [mean_value; median_value;
        std_value];
end
```

```
end

% Display the statistical measures
disp(statsTable)
```

9. **Sample Output**:

```
Enter the filename of the CSV file: data.csv

statsTable =
7*3 table
Age  Height  Weight

_____  _____  _____
35              172         65.3
33              171         68.4
32.5            169         62.1
```

10. **Extensions and Variations**: ·

- Add error handling to check if the specified CSV file exists and can be read.
- Implement additional statistical measures, such as quartiles, mode, or variance.
- Allow the user to specify which statistical measures to calculate and display.
- Provide an option to save the statistical measures to a separate file or generate a report.

11. **Problem Statement and Background**: Implement a MATLAB function that takes a text file as input and removes all occurrences of a specified string from the file. The function should create a new file with the modified contents.

12. **Step-by-Step Approach**:

   a. Define a function named removeStringFromFile that takes two inputs: the filename and the string to be removed.
   b. Open the input file using the fopen function.
   c. Read the contents of the file into a string using the fscanf function.
   d. Close the input file using the fclose function.
   e. Use the strrep function to replace all occurrences of the specified string with an empty string in the file contents.
   f. Create a new output file using the fopen function with write mode ('w').
   g. Write the modified contents to the output file using the fprintf function.
   h. Close the output file using the fclose function.

13. **Sample Code**:

Listing A.61 Removing a string from a file.

```
function removeStringFromFile(filename,
    stringToRemove)
% REMOVESTRINGFROMFILE Removes all occurrences of a
    string from a file
% removeStringFromFile(filename, stringToRemove)
% creates a new file with the modified contents
```

```
% Open the input file
fileID = fopen(filename, 'r');

% Read the file contents
fileContents = fscanf(fileID, '%c');

% Close the input file
fclose(fileID);

% Remove the specified string from the file contents
modifiedContents = strrep(fileContents,
    stringToRemove, '');

% Create a new output file
outputFilename = [filename '_modified.txt'];
outputFileID = fopen(outputFilename, 'w');

% Write the modified contents to the output file
fprintf(outputFileID, '%s', modifiedContents);

% Close the output file
fclose(outputFileID);

fprintf('File "%s" has been created with the modified
    contents.\n', outputFilename);
end
```

14. **Sample Output**:

```
removeStringFromFile('example.txt', 'the')
File "example_modified.txt" has been created with the
    modified contents.
```

15. **Extensions and Variations**:

- Modify the function to handle case-insensitive string removal.
- Implement a regular expression-based approach for removing patterns or substrings rather than exact string matches.
- Add functionality to remove multiple strings from the file in a single operation.
- Provide an option to overwrite the original file or create a new file with the modified contents.

16. **Problem Statement and Background**: Write a MATLAB script that prompts the user to enter a series of file names and then concatenates the contents of all the specified files into a single output file.

17. **Step-by-Step Approach**:

a. Initialise an empty cell array to store the file contents.
b. Use a loop (e.g., while loop) to repeatedly prompt the user to enter a filename using the input function.
c. Open the specified file using the fopen function.

d. Read the contents of the file into a string using the fscanf function.

e. Close the file using the fclose function.

f. Append the file contents to the cell array.

g. Provide an option for the user to stop entering filenames (e.g., enter an empty string or a specific value like 'done').

h. Prompt the user to enter an output filename.

i. Create a new output file using the fopen function with write mode ('w').

j. Concatenate the contents of the cell array into a single string.

k. Write the concatenated string to the output file using the fprintf function.

l. Close the output file using the fclose function.

18. **Sample Code**:

Listing A.62 Concatenating multiple files.

```
% Initialise a cell array to store file contents
fileContents = {};

% Prompt the user for filenames
while true
filename = input('Enter a filename (or "done" to
    finish): ', 's');
if strcmp(filename, 'done')
break;
end

% Open the file
fileID = fopen(filename, 'r');

% Read the file contents
contents = fscanf(fileID, '%c');

% Close the file
fclose(fileID);

% Append the contents to the cell array
fileContents{end+1} = contents;

end

% Prompt the user for the output filename
outputFilename = input('Enter the output filename: ',
    's');

% Open the output file
outputFileID = fopen(outputFilename, 'w');

% Concatenate the file contents
concatenatedContents = cat(2, fileContents{:});

% Write the concatenated contents to the output file
fprintf(outputFileID, '%s', concatenatedContents);
```

```
% Close the output file
fclose(outputFileID);

fprintf('Contents have been concatenated and written
    to "%s".\n', outputFilename);
```

19. **Sample Output**:

```
Enter a filename (or "done" to finish): file1.txt
Enter a filename (or "done" to finish): file2.txt
Enter a filename (or "done" to finish): file3.txt
Enter a filename (or "done" to finish): done
Enter the output filename: combined.txt
Contents have been concatenated and written to "
    combined.txt".
```

20. **Extensions and Variations**:

- Add error handling to check if the specified input files exist and can be read-/written.
- Implement an option to include separators (e.g., newlines, tabs, or custom separators) between the values when writing the modified contents to the output file.
- Provide a way to specify the encoding (e.g., UTF-8, ASCII) for reading and writing the files.
- Allow the user to choose whether to overwrite the original file or create a new file with the modified contents.
- Implement support for handling binary files in addition to text files.
- Extend the functionality to perform other text processing operations, such as removing or replacing specific characters or patterns, converting between different character encodings, or inserting/deleting lines or sections of text.
- Develop a graphical user interface (GUI) to provide a more user-friendly experience for selecting files, specifying replacement words/phrases, and displaying the results.
- Enhance the script to handle large files efficiently by reading and processing the files in chunks or using memory-mapped files.
- Implement multithreading or parallel processing techniques to improve performance when processing multiple files or large files.
- Integrate the script with version control systems (e.g., Git) to track changes made to the files and enable collaboration.
- Develop a command-line interface (CLI) for running the script from the terminal, allowing users to specify input files, replacement words/phrases, and other options through command-line arguments.
- Implement logging and error reporting mechanisms to capture and display informative error messages and debugging information.

- Integrate the script with other text processing tools or libraries, such as regular expression engines or natural language processing libraries, to provide additional functionality or enhance the existing features.

21. **Problem Statement and Background**: Create a MATLAB function that takes a matrix as input and saves it to a binary file. The function should also include the ability to load the matrix from the binary file at a later time.

22. **Step-by-Step Approach**:

   a. Define a function named saveMatrixToBinary that takes two inputs: the matrix and the filename.
   b. Open a new binary file for writing using the fopen function with the 'wb' mode.
   c. Write the matrix to the binary file using the fwrite function.
   d. Close the binary file using the fclose function.
   e. Define another function named loadMatrixFromBinary that takes the filename as input.
   f. Open the binary file for reading using the fopen function with the 'rb' mode.
   g. Read the matrix from the binary file using the fread function, specifying the appropriate data type and size.
   h. Close the binary file using the fclose function.
   i. Return the loaded matrix.

23. **Sample Code**:

Listing A.63 Saving and loading a matrix to/from a binary file.

```
function saveMatrixToBinary(matrix, filename)
% SAVEMATRIXTOBINARY Saves a matrix to a binary file
% saveMatrixToBinary(matrix, filename)

% Open the binary file for writing
fileID = fopen(filename, 'wb');

% Write the matrix to the binary file
fwrite(fileID, matrix, 'double');

% Close the binary file
fclose(fileID);

fprintf('Matrix saved to "%s".\n', filename);
end

function loadedMatrix = loadMatrixFromBinary(filename
    )
% LOADMATRIXFROMBINARY Loads a matrix from a binary
    file
% loadedMatrix = loadMatrixFromBinary(filename)

% Open the binary file for reading
fileID = fopen(filename, 'rb');
```

```
% Get the size of the matrix
fileInfo = dir(filename);
fileSize = fileInfo.bytes;
numElements = fileSize / 8; % Assuming double
    precision

% Read the matrix from the binary file
loadedMatrix = fread(fileID, numElements, 'double');

% Reshape the matrix to its original dimensions
loadedMatrix = reshape(loadedMatrix, sqrt(numElements
    ), sqrt(numElements));

% Close the binary file
fclose(fileID);
end
```

24. **Sample Output**:

```
A = magic(5);
saveMatrixToBinary(A, 'matrix.bin');
Matrix saved to "matrix.bin".
B = loadMatrixFromBinary('matrix.bin');
isequal(A, B)
ans =
1
```

25. **Extensions and Variations**:

- Add error handling to check if the specified binary file exists and can be read/written.
- Implement support for saving and loading matrices of different data types (e.g., single precision, integers).
- Allow the user to specify additional metadata to be saved along with the matrix, such as dimensions, data type, or variable name.
- Implement compression or encoding techniques to reduce the file size for large matrices.

26. **Problem Statement and Background**: Develop a MATLAB script that reads data from an Excel file and creates a bar plot or a histogram to visualize the data distribution.

27. **Step-by-Step Approach**:

    a. Prompt the user to enter the filename of the Excel file using the input function.
    b. Use the readmatrix function to read the contents of the Excel file into a matrix. The basic syntax is:

Listing A.64 Reading an Excel file.

```
data = readmatrix(filename);
```

c. Determine the type of Visualisation (bar plot or histogram) based on user input or the characteristics of the data.

d. For a bar plot:

- Use the bar function to create a bar plot of the data.
- Customize the plot with labels, title, and other desired formatting options.

e. For a histogram:

- Use the histogram function to create a histogram of the data.
- Customize the histogram with bin settings, labels, title, and other desired formatting options.

28. **Sample Code**:

Listing A.65 Visualizing data from an Excel file.

```
% Prompt the user for the filename
filename = input('Enter the filename of the Excel
    file: ', 's');

% Read the Excel file into a matrix
data = readmatrix(filename);

% Determine the type of Visualisation
visualType = input('Enter "bar" for a bar plot or "
    hist" for a histogram: ', 's');

% Create the Visualisation
figure;
if strcmp(visualType, 'bar')
bar(data);
title('Bar Plot');
xlabel('Category');
ylabel('Value');
elseif strcmp(visualType, 'hist')
histogram(data);
title('Histogram');
xlabel('Value');
ylabel('Frequency');
else
fprintf('Invalid Visualisation type specified.\n');
end
```

29. **Sample Output**:

```
Enter the filename of the Excel file: data.xlsx
Enter "bar" for a bar plot or "hist" for a histogram:
    hist
```

30. **Extensions and Variations**:

- Add error handling to check if the specified Excel file exists and can be read.
- Implement additional Visualisation options, such as line plots, scatter plots, or box plots.

- Allow the user to specify which columns or rows of the Excel file to visualize.
- Provide options for customizing the appearance of the Visualisation (e.g., colors, labels, legends).
- Implement support for reading data from different sheet names or ranges within the Excel file.

31. **Problem Statement and Background**: Implement a MATLAB function that takes a text file as input and performs basic text processing tasks, such as counting the number of lines, words, and characters, as well as identifying the most frequently occurring word in the file.

32. **Step-by-Step Approach**:

   a. Define a function named textFileAnalysis that takes a filename as input.
   b. Open the file using the fopen function.
   c. Read the contents of the file into a string using the fscanf function.
   d. Close the file using the fclose function.
   e. Count the number of lines in the file using the sum function with the condition fileContents == newline.
   f. Split the file contents into individual words using the strsplit function and appropriate delimiters (e.g., whitespace characters).
   g. Count the number of words using the length function.
   h. Count the number of characters in the file using the length function on the file contents string.
   i. Initialise a struct or a cell array to store the word counts.
   j. Iterate through the words and update the counts in the struct or cell array.
   k. Identify the most frequently occurring word by finding the maximum value in the word counts.
   l. Return a struct or a cell array containing the text analysis results.

33. **Sample Code**:

Listing A.66 Text file analysis.

```
function analysisResults = textFileAnalysis(filename)
% TEXTFILEANALYSIS Performs basic text analysis on a
    file
% analysisResults = textFileAnalysis(filename)
% returns a struct containing analysis results

% Open the file
fileID = fopen(filename, 'r');

% Read the file contents
fileContents = fscanf(fileID, '%c');

% Close the file
fclose(fileID);

% Count the number of lines
```

```
numLines = sum(fileContents == newline);

% Split the file contents into words
words = strsplit(strtrim(fileContents), '\s+');

% Count the number of words
numWords = length(words);

% Count the number of characters
numChars = length(fileContents);

% Initialise a struct to store word counts
wordCounts = struct();

% Count the occurrences of each word
for i = 1:length(words)
word = words{i};
if isfield(wordCounts, word)
wordCounts.(word) = wordCounts.(word) + 1;
else
wordCounts.(word) = 1;
end
end

% Find the most frequent word
maxCount = 0;
mostFrequentWord = '';
fieldNames = fields(wordCounts);
for i = 1:length(fieldNames)
field = fieldNames{i};
if wordCounts.(field) > maxCount
maxCount = wordCounts.(field);
mostFrequentWord = field;
end
end

% Store the analysis results in a struct
analysisResults = struct();
analysisResults.numLines = numLines;
analysisResults.numWords = numWords;
analysisResults.numChars = numChars;
analysisResults.wordCounts = wordCounts;
analysisResults.mostFrequentWord = mostFrequentWord;
end
```

34. **Sample Output**:

```
results = textFileAnalysis('example.txt')
results =
struct with fields:
numLines: 3
numWords: 9
numChars: 36
```

```
wordCounts: [1*8 struct]
mostFrequentWord: 'the'
```

35. **Extensions and Variations**:

- Modify the function to handle case-insensitive word counting.
- Add functionality to exclude common words (e.g., "the," "a," "and") from the word count and most frequent word calculation.
- Implement additional text analysis tasks, such as counting the number of unique words, calculating word length statistics, or identifying the longest/shortest word.
- Provide an option to save the analysis results to a file or generate a report.

36. **Problem Statement and Background**: Write a MATLAB script that prompts the user to enter a directory path and then lists all the files in that directory, along with their sizes and modification dates.

37. **Step-by-Step Approach**:

   a. Prompt the user to enter a directory path using the input function.
   b. Use the dir function to get a structure array containing information about all the files and folders in the specified directory.
   c. Iterate through the structure array and extract the file names, sizes, and modification dates.
   d. Display the file information using the fprintf function or create a table to display the data in a formatted manner.

38. **Sample Code**:

Listing A.67 Listing files in a directory.

```
% Prompt the user for the directory path
dirPath = input('Enter the directory path: ', 's');

% Get information about files and folders in the
    directory
fileInfo = dir(dirPath);

% Create a table to store the file information
fileTable = table('Size', [0], 'Name', {''}, 'Date',
    {''}, 'VariableNames', {'Size (bytes)', 'File Name
    ', 'Modification Date'});

% Iterate through the file information
for i = 1:length(fileInfo)
entry = fileInfo(i);
% Check if the entry is a file
if ~entry.isdir
    % Extract file information
    fileName = entry.name;
    fileSize = entry.bytes;
    fileDate = datestr(entry.datenum);
```

```
    % Add the file information to the table
    fileTable = [fileTable; {fileSize, fileName,
        fileDate}];
end
end

% Display the file information
disp(fileTable);
```

39. **Sample Output**:

```
Enter the directory path: /path/to/directory
Size (bytes)    File Name                Modification Date
            ---------------    ----------
            ------------------
                1024          file1.txt      28-May-2023
                  10:15:32
                2048          file2.dat      15-Apr-2023
                  14:22:18
                512           file3.csv      01-Jun-2023
                  09:45:01
```

40. **Extensions and Variations**:

- Add error handling to check if the specified directory exists and is accessible.
- Implement additional filtering options, such as displaying only files with specific extensions or excluding hidden files.
- Provide an option to recursively list files in subdirectories.
- Allow the user to sort the file information based on file size, name, or modification date.
- Implement support for displaying additional file attributes, such as file permissions or owner information.

41. **Problem Statement and Background**: Create a MATLAB function that takes a text file as input and replaces all occurrences of a specified word or phrase with a new word or phrase. The function should create a new file with the modified contents.

42. **Step-by-Step Approach**:

a. Define a function named replaceWordInFile that takes three inputs: the filename, the word/phrase to be replaced, and the new word/phrase.
b. Open the input file using the fopen function.
c. Read the contents of the file into a string using the fscanf function.
d. Close the input file using the fclose function.
e. Use the strrep function to replace all occurrences of the specified word/phrase with the new word/phrase in the file contents.
f. Create a new output file using the fopen function with write mode ('w').
g. Write the modified contents to the output file using the fprintf function.
h. Close the output file using the fclose function.

43. **Sample Code**:

Listing A.68 Replacing a word/phrase in a file.

```
function replaceWordInFile(filename, oldWord, newWord
    )
% REPLACEWORDIBNFILE Replaces a word/phrase in a file
    with a new word/phrase
% replaceWordInFile(filename, oldWord, newWord)
% creates a new file with the modified contents

% Open the input file
fileID = fopen(filename, 'r');

% Read the file contents
fileContents = fscanf(fileID, '%c');

% Close the input file
fclose(fileID);

% Replace the specified word/phrase with the new word
    /phrase
modifiedContents = strrep(fileContents, oldWord,
    newWord);

% Create a new output file
outputFilename = [filename '_modified.txt'];
outputFileID = fopen(outputFilename, 'w');

% Write the modified contents to the output file
fprintf(outputFileID, '%s', modifiedContents);

% Close the output file
fclose(outputFileID);

fprintf('File "%s" has been created with the modified
    contents.\n', outputFilename);
end
```

44. **Sample Output**:

```
replaceWordInFile('example.txt', 'hello', 'goodbye')
File "example_modified.txt" has been created with the
    modified contents.
```

45. **Extensions and Variations**:

- Modify the function to handle case-insensitive word/phrase replacement.
- Implement support for replacing multiple words/phrases in a single operation.
- Add functionality to replace words/phrases based on regular expressions or patterns.
- Provide an option to overwrite the original file or create a new file with the modified contents.

- Implement error handling to check if the specified input file exists and can be read/written.

46. **Problem Statement and Background**: Write a MATLAB script that reads data from a text file, where each line represents a data point with multiple values separated by commas. The script should calculate basic statistical measures (e.g., mean, median, standard deviation) for each column of data.

47. **Step-by-Step Approach**:

   a. Prompt the user to enter the filename of the text file using the input function.
   b. Open the file using the fopen function.
   c. Read the contents of the file into a cell array of strings using the textscan function with the '%s' format specifier and delimiter '\n'.
   d. Close the file using the fclose function.
   e. Split each line (string) in the cell array into individual values using the strsplit function with the delimiter ','.
   f. Convert the cell array of strings into a numeric matrix using the str2double function.
   g. Calculate the statistical measures (mean, median, and standard deviation) for each column of the matrix using the mean, median, and std functions, respectively.
   h. Display the results in a formatted table or matrix.

48. **Sample Code**:

Listing A.69 Calculating statistical measures from a text file.

```
% Prompt the user for the filename
filename = input('Enter the filename of the text file
   : ', 's');

% Open the file
fileID = fopen(filename, 'r');

% Read the file contents into a cell array
fileContents = textscan(fileID, '%s', 'Delimiter', '\
   n');
fileContents = fileContents{1};

% Close the file
fclose(fileID);

% Split each line into individual values
data = cellfun(@(x) str2double(strsplit(x, ',')),
   fileContents, 'UniformOutput', false);

% Convert the cell array into a numeric matrix
data = cell2mat(data);

% Calculate statistical measures for each column
means = mean(data, 1);
```

```
medians = median(data, 1);
stds = std(data, 1);

% Display the results
numColumns = size(data, 2);
fprintf('Statistical Measures:\n');
for i = 1:numColumns
fprintf('Column %d:\n', i);
fprintf(' Mean: %.2f\n', means(i));
fprintf(' Median: %.2f\n', medians(i));
fprintf(' Std Dev: %.2f\n', stds(i));
end
```

49. **Sample Output**:

```
Enter the filename of the text file: data.txt

Statistical Measures:
Column 1:
Mean: 25.67
Median: 27.00
Std Dev: 10.21
Column 2:
Mean: 15.33
Median: 16.00
Std Dev: 5.13
Column 3:
Mean: 92.67
Median: 90.00
Std Dev: 12.66
```

50. **Extensions and Variations**:

- Implement error handling to check if the specified input file exists and can be read.
- Provide an option to calculate additional statistical measures, such as quartiles, range, or variance.
- Allow the user to specify which columns or rows to include in the statistical analysis.
- Implement support for handling missing data or invalid values in the input file.
- Provide an option to save the statistical measures to a file or generate a report.

A.8   Solutions to Chapter 8

1. **Problem Statement and Background**: Create a line plot of the function $y = x^2 - 2x + 1$ over the range $-2 \leq x \leq 4$. Annotate the plot with a title, x-label, y-label, and a legend.
   Step-by-Step Approach:

   a. Define the x range: x = -2:0.1:4;

b. Calculate y values:
$$y = x.2 - 2 * x + 1$$

;

c. Plot x vs y: **plot(x,y)**
d. Add title: **title('Quadratic Function Plot')**
e. Label x-axis: **xlabel('x')**
f. Label y-axis: **ylabel('**
$$y = x2 - 2x + 1$$

')
g. Add legend: **legend('**
$$y = x2 - 2x + 1$$

')

Sample Code:

Listing A.70 Line plot with annotations.

```
x = -2:0.1:4;
y = x.^2 - 2*x + 1;
plot(x,y)
title('Quadratic Function Plot')
xlabel('x')
ylabel('y = x^2 - 2x + 1')
legend('y = x^2 - 2x + 1')
```

Sample Output: (A line plot with the specified annotations would be shown)
Extensions and Variations:

- Plot multiple functions on the same axes for comparison
- Use different **line colors** and **styles**
- Add **grid lines** to the plot

2. **Problem Statement and Background**: Plot the vectorized sine function $y = \sin(x)$ over one period $0 \leq x \leq 2\pi$. Customize the line color and style.
Step-by-Step Approach:

a. Define x range: **x = 0:0.1:2*pi;**
b. Calculate sine values: **y = sin(x);**
c. Plot x vs y: **plot(x,y)**
d. Set line color: **plot(x,y,'Color','r')**
e. Set line style: **plot(x,y,'-')**

Sample Code:

Listing A.71 Sine plot with custom line.

```
x = 0:0.1:2*pi;
y = sin(x);
plot(x,y,'Color','r','LineStyle','--')
```

Sample Output: (A red dashed sine wave plot would be shown)
Extensions and Variations:

- Plot **multiple cycles** by adjusting x range
- Add **axis labels** and **title**
- Use **line specification** shortcut syntax

3. **Problem Statement and Background**: Generate a scatter plot of random (x,y) points within the range $0 \leq x, y \leq 10$. Add gridlines and use different marker styles and colors to distinguish clusters.
   Step-by-Step Approach:

   a. Generate random x,y data: $x = 10\text{rand}(50,1); y = 10\text{rand}(50,1);$
   b. Create scatter plot: **scatter(x,y)**
   c. Add grid: **grid on**
   d. Customize markers: **scatter(x,y,36,'filled','MarkerEdgeColor','r')**
   e. Add 2nd cluster:

   ```
   x2 = 6 + 2rand(30,1);
   y2 = 6 + 2rand(30,1);
   hold on;
   scatter(x2, y2, '^', 'MarkerEdgeColor', 'g');
   ```

Sample Code:

Listing A.72 Scatter plot with clustered data.

```
x = 10rand(50,1); y = 10rand(50,1);
scatter(x,y,36,'filled','MarkerEdgeColor','r')
hold on
x2 = 6 + 2rand(30,1); y2 = 6 + 2rand(30,1);
scatter(x2,y2,'^','MarkerEdgeColor','g')
grid on
```

Sample Output: (A scatter plot with red filled circles and green triangles clustered)
Extensions and Variations:

- Add a **legend** to identify clusters
- Use different **color maps** to color-code data
- Apply **clustering algorithms** for grouping

4. **Problem Statement and Background**: Create a bar chart to visualize student scores in five different courses. Customize the bar colors, add data labels, and rotate the x-tick labels for better readability.

   Step-by-Step Approach:

   a. Define course names and scores: courses = 'Math','Physics',' Chemistry','Biology','CS'; scores = [82 76 89 92 85];
   b. Plot bar chart: **bar(scores)**
   c. Set x-ticks: **set(gca,'XTickLabel',courses)**
   d. Rotate x-labels: **xtickangle(45)**
   e. Customize colors: **bar(scores,'FaceColor','flat')**
   f. Add data labels: **bar(scores,'FaceColor','flat','EdgeColor ','k'); text(1:5,scores,num2str(scores'),'vert',' bottom','horiz','center');**

   Sample Code:

Listing A.73 Bar chart with customizations.

```
courses = {'Math','Physics','Chemistry','Biology','CS
    '};
scores = [82 76 89 92 85];
bar(scores,'FaceColor','flat','EdgeColor','k')
set(gca,'XTickLabel',courses,'XTickLabelRotation',45)
text(1:5,scores,num2str(scores'),'vert','bottom','
    horiz','center');
```

   Sample Output: (A bar chart with flat color bars, rotated x-ticks, and data labels)

   Extensions and Variations:

   - Group bars using **bar()** options
   - Create **stacked** or **clustered** bar plots
   - Add y-axis gridlines and title

5. **Problem Statement and Background**: Load a image dataset (e.g., from MAT-LAB's sample data) and display the image. Adjust the colormap and add a descriptive title.

   Step-by-Step Approach:

   a. Load image: img = imread('saturn.png');
   b. Display image: **imshow(img)**
   c. Apply colormap: **colormap(pink)**
   d. Add title: **title('Saturn Image')**

   Sample Code:

Listing A.74 Display image with colormap.

```
img = imread('saturn.png');
imshow(img)
colormap(pink)
title('Saturn Image')
```

Sample Output: (The Saturn image loaded and displayed with a pink colormap)
Extensions and Variations:

- Try different **colormaps** like `hot`, `cool`, etc.
- Adjust **brightness** and **contrast**
- Add **colorbar** to show mapping

6. **Problem Statement and Background**: Generate a 3D surface plot of the function $z = x^2 + y^2$ over the ranges $-2 \leq x, y \leq 2$. Add appropriate view angles, lighting, and color the surface by z-values.
   Step-by-Step Approach:

   a. Define x,y ranges: `[x,y] = meshgrid(-2:0.1:2);`
   b. Calculate z:

   $$z = x.2 + y.2$$

   ;
   c. Create surface: **`surf(x,y,z)`**
   d. Set view angles: **`view(30,45)`**
   e. Add lighting: **`camlight`**
   f. Color by z-values: **`colormap(hot)`**

   Sample Code:

Listing A.75 3D surface plot customizations.

```
[x,y] = meshgrid(-2:0.1:2);
z = x.^2 + y.^2;
surf(x,y,z); view(30,45); camlight; colormap(hot)
```

Sample Output: (A 3D colored surface plot of the function from specified view angles)
Extensions and Variations:

- Add **contour lines** using `hold on; contour3(x,y,z)`
- Adjust **color limits** using `caxis`
- Create an **animation** by rotating view angles

7. **Problem Statement and Background**: Plot multiple line graphs on the same axes to compare growth trends for different populations over time. Use legends and line styles to differentiate the plots.
   Step-by-Step Approach:

   a. Define data: `years = 1950:10:2020; popA = [500 632 801 965 1200]; popB = [300 420 580 750 950];`
   b. Create plot: **`plot(years,popA,'-o',years,popB,'-s')`**
   c. Add legend: **`legend('Population A','Population B')`**
   d. Add labels: **`xlabel('Year'); ylabel('Population');`**
   e. Add title: **`title('Population Growth Trends')`**

Sample Code:

Listing A.76 Multiple line plots with legend.

```
years = 1950:10:2020;
popA = [500 632 801 965 1200];
popB = [300 420 580 750 950];
plot(years,popA,'-o',years,popB,'--s')
legend('Population A','Population B')
xlabel('Year'); ylabel('Population');
title('Population Growth Trends')
```

Sample Output: (Two line plots showing increasing population trends distinguished by legends)

Extensions and Variations:

- Add more population data with different line styles
- Use **subplots** to visualize additional growth factors
- Compute and plot **growth rates** instead of populations

8. **Problem Statement and Background**: Load a real-world dataset (e.g., from a CSV file) and create a histogram to visualize the distribution of a chosen variable. Experiment with different bin widths and overlay a kernel density estimate.

   Step-by-Step Approach:

   a. Load data: `data = readtable('housing.csv');`
   b. Extract variable: `prices = data.Price;`
   c. Plot histogram: **histogram(prices)**
   d. Adjust bins: **histogram(prices,20)**
   e. Overlay KDE: **hold on; kernel = ksdensity(prices); plot (kernel.Support,kernel.Density)**

   Sample Code:

Listing A.77 Histogram with kernel density.

```
data = readtable('housing.csv');
prices = data.Price;
histogram(prices,20)
hold on;
kernel = ksdensity(prices);
plot(kernel.Support,kernel.Density,'LineWidth',2)
```

Sample Output: (A histogram with adjusted bins and an overlaid kernel density curve)

Extensions and Variations:

- Explore **normality** tests and **outlier** detection
- Try different **kernel functions** for density estimate
- Plot **conditional distributions** based on other variables

9. **Problem Statement and Background**: Create a filled contour plot of a 2D Gaussian function. Add a colorbar and label the contour levels.
   Step-by-Step Approach:

   a. Define x,y ranges: `[x,y] = meshgrid(-3:0.1:3);`
   b. Calculate z:

   $$z = exp(-(x.2 + y.2))$$

   ;
   c. Create contour: **contourf(x,y,z)**
   d. Add colorbar: **colorbar**
   e. Label levels: **clabel(c,c.LevelList)**

   Sample Code:

Listing A.78 Filled contour plot.

```
[x,y] = meshgrid(-3:0.1:3);
z = exp(-(x.^2 + y.^2));
contourf(x,y,z)
colorbar
c = contour(x,y,z);
clabel(c,c.LevelList)
```

Sample Output: (A filled contour plot of a 2D Gaussian with labeled contour levels would be displayed)
This code generates a filled contour plot of a 2D Gaussian function using the following steps:

-
```
[x,y] = meshgrid(-3:0.1:3);
```

creates 2D grid matrices x and y ranging from -3 to 3 with a step size of 0.1.

-
```
z = exp(-(x.^2 + y.^2));
```

calculates the z-values for the 2D Gaussian function using the formula $z = e^{-(x^2+y^2)}$.

-
```
contourf(x,y,z)
```

creates a filled contour plot of the z-values over the x-y grid.

-
```
colorbar
```

adds a colorbar to the plot, which shows the mapping of colors to the contour levels.

-
```
c = contour(x,y,z);
```

generates contour lines for the same z-values.

```
clabel(c,c.LevelList)
```

labels each contour line with its corresponding contour level value.

10. **Problem Statement and Background**: Animate a bouncing ball simulation by plotting the ball's position at each time step. Control the animation speed and add a trail effect.

Step-by-Step Approach:

a. Define initial conditions: y0 = 0; v0 = 5; g = 9.8; dt = 0.1; t = 0;

b. Create figure: figure; axis([0 10 -1 6]); hold on;

c. Loop for animation:

```
while y0 >= 0
y = y0 + v0t - 0.5gt^2;
v = v0 - gt;
plot(t,y,'ro','MarkerSize',10);
trail_x(end+1) = t; trail_y(end+1) = y;
plot(trail_x,trail_y,'r-');
t = t + dt; y0 = y; v0 = v;
pause(0.05);
end
```

Sample Code:

Listing A.79 Bouncing ball animation.

```
y0 = 0; v0 = 5; g = 9.8; dt = 0.1; t = 0;
figure; axis([0 10 -1 6]); hold on;
trail_x = []; trail_y = [];
while y0 >= 0
y = y0 + v0t - 0.5gt^2;
v = v0 - gt;
plot(t,y,'ro','MarkerSize',10);
trail_x(end+1) = t; trail_y(end+1) = y;
plot(trail_x,trail_y,'r-');
t = t + dt; y0 = y; v0 = v;
pause(0.05);
end
```

Sample Output: (An animated plot showing a bouncing ball with a trail effect)

Extensions and Variations:

- Add **ground line** and **bouncing effect**
- Simulate **drag force** for realistic motion
- Create an **interactive control** for changing parameters

The basic syntax for animations and dynamic plotting is:

Listing A.80 Animation syntax.

```
% Create a figure
figure;

% Loop for updating plot
while condition
% Update data
% Plot current state
plot(...);

% Control animation speed
pause(seconds);

end
```

A.9   Solutions to Chapter 9

1. **Problem Statement and Background**: Write a MATLAB script that prompts the user to enter a number and determines whether it is positive, negative, or zero using **if-else** statements.
2. **Step-by-Step Approach**:

   a. Prompt the user to enter a number using the **input** function.
   b. Use an **if-elseif-else** statement to check if the number is greater than 0, less than 0, or equal to 0.
   c. If the number is greater than 0, display a message indicating that it is positive.
   d. If the number is less than 0, display a message indicating that it is negative.
   e. If the number is equal to 0, display a message indicating that it is zero.

3. **Sample Code**:

Listing A.81 Determine sign of a number.

```
num = input('Enter a number: ');

if num > 0
disp('The number is positive.')
elseif num < 0
disp('The number is negative.')
else
disp('The number is zero.')
end
```

4. **Sample Output**:

```
Enter a number: 5
The number is positive.
```

```
Enter a number: -3
The number is negative.
```

```
Enter a number: 0
The number is zero.
```

5. **Extensions and Variations**:

   - Add input validation to ensure that the user enters a valid numeric value.
   - Modify the script to handle different number ranges or categories (e.g., positive, negative, zero, small, large, etc.).
   - Extend the script to perform additional operations or calculations based on the sign of the number.

6. **Problem Statement and Background**: Implement a MATLAB function that takes two numbers as input and returns the maximum value using **if-else** statements.

7. **Step-by-Step Approach**:

   a. Define a function that takes two input arguments (the two numbers).
   b. Use an **if-else** statement to compare the two numbers.
   c. If the first number is greater than the second, return the first number as the maximum value.
   d. Otherwise, return the second number as the maximum value.

8. **Sample Code**:

Listing A.82 Find maximum of two numbers.

```
function max_value = findMax(num1, num2)
if num1 > num2
max_value = num1;
else
max_value = num2;
end
end
```

9. **Sample Output**:

```
max_value = findMax(5, 8)
max_value = 8
```

```
max_value = findMax(10, 3)
max_value = 10
```

10. **Extensions and Variations**:

   - Modify the function to handle cases where the two numbers are equal.
   - Extend the function to find the maximum of an arbitrary number of input values, not just two.
   - Incorporate input validation to ensure that the input values are valid numbers.

11. **Problem Statement and Background**: Write a MATLAB script that prompts the user to enter a character and determines whether it is a vowel or a consonant using a **switch** statement.

12. **Step-by-Step Approach**:

   a. Prompt the user to enter a character using the **input** function.
   b. Convert the input character to lowercase using the **lower** function.
   c. Use a **switch** statement to check if the character is a vowel ('a', 'e', 'i', 'o', 'u').
   d. If the character is a vowel, display a message indicating that it is a vowel.
   e. Otherwise, display a message indicating that it is a consonant.

13. **Sample Code**:

Listing A.83 Determine vowel or consonant.

```
char = input('Enter a character: ', 's');
char = lower(char);

switch char
case 'a'
disp('The character is a vowel.')
case 'e'
disp('The character is a vowel.')
case 'i'
disp('The character is a vowel.')
case 'o'
disp('The character is a vowel.')
case 'u'
disp('The character is a vowel.')
otherwise
disp('The character is a consonant.')
end
```

14. **Sample Output**:

```
Enter a character: A
The character is a vowel.
```

```
Enter a character: b
The character is a consonant.
```

15. **Extensions and Variations**:

- Modify the script to handle uppercase and lowercase characters without explicitly converting the input.
- Extend the script to handle non-alphabetic characters and provide appropriate messages for them.
- Incorporate input validation to ensure that the user enters a valid character.

16. **Problem Statement and Background**: Implement a MATLAB function that takes a year as input and determines whether it is a leap year or not using **if-else** statements. A year is considered a leap year if it is divisible by 4, except for years divisible by 100, which are not leap years unless they are also divisible by 400.

17. **Step-by-Step Approach**:

a. Define a function that takes a year as input.
b. Use **if-else** statements to check the conditions for a leap year.
c. If the year is divisible by 4 and not divisible by 100, or if it is divisible by 400, it is a leap year.
d. Return a boolean value (true or false) indicating whether the year is a leap year or not.

18. **Sample Code**:

Listing A.84 Determine leap year.

```
function is_leap_year = isLeapYear(year)
is_leap_year = false;
if (mod(year, 4) == 0 && mod(year, 100) ~= 0) || mod(
    year, 400) == 0
is_leap_year = true;
end
end
```

19. **Sample Output**:

```
is_leap_year = isLeapYear(2024)
is_leap_year = 1 % true
```

```
is_leap_year = isLeapYear(2100)
is_leap_year = 0 % false
```

20. **Extensions and Variations**:

- Modify the function to handle different calendars or systems for determining leap years.
- Incorporate input validation to ensure that the input year is a valid positive integer.

- Extend the function to return additional information, such as the number of days in the given year or the next leap year.

21. **Problem Statement and Background**: Write a MATLAB script that prompts the user to enter a letter grade (A, B, C, D, or F) and calculates the corresponding grade point average (GPA) using **if-elseif-else** statements.

22. **Step-by-Step Approach**:

   a. Prompt the user to enter a letter grade using the **input** function.
   b. Convert the input to uppercase using the **upper** function.
   c. Use an **if-elseif-else** statement to check the letter grade and assign the corresponding GPA.
   d. Display the calculated GPA.

23. **Sample Code**:

Listing A.85 Calculate GPA from letter grade.

```
grade = input('Enter a letter grade (A, B, C, D, or F
   ): ', 's');
grade = upper(grade);

switch grade
case 'A'
gpa = 4.0;
case 'B'
gpa = 3.0;
case 'C'
gpa = 2.0;
case 'D'
gpa = 1.0;
case 'F'
gpa = 0.0;
otherwise
disp('Invalid grade entered.');
return
end

disp(['The GPA for the grade ', grade, ' is ',
   num2str(gpa)])
```

24. **Sample Output**:

```
Enter a letter grade (A, B, C, D, or F): b
The GPA for the grade B is 3
```

```
Enter a letter grade (A, B, C, D, or F): X
Invalid grade entered.
```

25. **Extensions and Variations**:

   - Modify the script to handle additional grade scales or grading systems.
   - Incorporate input validation to ensure that the user enters a valid letter grade.
   - Extend the script to calculate the overall GPA based on multiple courses and their respective grades.

26. **Problem Statement and Background**: Implement a MATLAB function that takes three numbers as input and returns the median value (the middle value when the numbers are arranged in ascending or descending order) using **if-elseif-else** statements.

27. **Step-by-Step Approach**:

   a. Define a function that takes three numbers as input.
   b. Use **if-elseif-else** statements to compare the numbers and determine the median value.
   c. Return the median value.

28. **Sample Code**:

Listing A.86 Find median of three numbers.

```
function median_value = findMedian(num1, num2, num3)
if (num1 >= num2 && num1 <= num3) || (num1 >= num3 &&
     num1 <= num2)
median_value = num1;
elseif (num2 >= num1 && num2 <= num3) || (num2 >=
     num3 && num2 <= num1)
median_value = num2;
else
median_value = num3;
end
end
```

29. **Sample Output**:

```
median_value = findMedian(5, 8, 3)
median_value = 5
```

```
median_value = findMedian(10, 3, 7)
median_value = 7
```

30. **Extensions and Variations**:

   - Modify the function to handle cases where two or more numbers are equal.
   - Extend the function to find the median of an arbitrary number of input values, not just three.
   - Incorporate input validation to ensure that the input values are valid numbers.

31. **Problem Statement and Background**: Write a MATLAB script that generates a random integer between 1 and 10, prompts the user to guess the number, and provides feedback using **if-else** statements. The script should continue prompting the user until the correct guess is made.

32. **Step-by-Step Approach**:

   a. Generate a random integer between 1 and 10 using the **randi** function.
   b. Prompt the user to enter their guess using the **input** function.
   c. Use an **if-else** statement to compare the user's guess with the random number.
   d. If the guess is correct, display a message indicating that the user guessed correctly.
   e. If the guess is incorrect, display a message indicating whether the guess was too high or too low.
   f. Repeat steps 2–5 until the user guesses the correct number.

33. **Sample Code**:

Listing A.87 Guess the number game.

```
secret_number = randi([1, 10]);
guess = -1;

while guess ~= secret_number
guess = input('Guess the number between 1 and 10: ';

if guess == secret_number
    disp('Congratulations! You guessed the number
        correctly.')
elseif guess < secret_number
    disp('Your guess is too low. Try again.')
else
    disp('Your guess is too high. Try again.')
end

end
```

34. **Sample Output**:

```
Guess the number between 1 and 10: 5
Your guess is too high. Try again.
Guess the number between 1 and 10: 3
Your guess is too low. Try again.
Guess the number between 1 and 10: 4
Congratulations! You guessed the number correctly.
```

35. **Extensions and Variations**:

   • Modify the script to keep track of the number of attempts made by the user.
   • Incorporate input validation to ensure that the user enters a valid integer within the specified range.
   • Extend the script to allow the user to play multiple rounds and keep score.

36. **Problem Statement and Background**: Implement a MATLAB function that takes a character as input and determines whether it is an uppercase letter, lowercase letter, or neither using a **switch** statement.

37. **Step-by-Step Approach**:

   a. Define a function that takes a character as input.
   b. Use a **switch** statement to check if the character is an uppercase letter, lowercase letter, or neither.
   c. For uppercase letters, return a specific value (e.g., 1).
   d. For lowercase letters, return a different value (e.g., 2).
   e. For non-alphabetic characters, return a third value (e.g., 0).

38. **Sample Code**:

Listing A.88 Determine character type.

```
function char_type = determineCharType(char)
char_type = 0; % Default value for non-alphabetic
    characters

switch char
    case 'A':'Z'
        char_type = 1; % Uppercase letter
    case 'a':'z'
        char_type = 2; % Lowercase letter
    otherwise
        % Non-alphabetic character, keep default
            value
end

end
```

39. **Sample Output**:

```
char_type = determineCharType('A')
char_type = 1 % Uppercase letter
```

```
char_type = determineCharType('b')
char_type = 2 % Lowercase letter
```

```
char_type = determineCharType('$')
char_type = 0 % Non-alphabetic character
```

40. **Extensions and Variations**:

   • Modify the function to handle additional character types, such as digits or special characters.
   • Incorporate input validation to ensure that the input is a valid character.

- Extend the function to return a descriptive string instead of numeric values for better readability.

41. **Problem Statement and Background**: Write a MATLAB script that prompts the user to enter their age and displays a message indicating their eligibility to vote based on the following criteria: - If the age is 18 or above, the user is eligible to vote. - If the age is below 18, the user is not eligible to vote.

42. **Step-by-Step Approach**:

   a. Prompt the user to enter their age using the **input** function.
   b. Use an **if-else** statement to check if the age is greater than or equal to 18.
   c. If the age is greater than or equal to 18, display a message indicating that the user is eligible to vote.
   d. If the age is less than 18, display a message indicating that the user is not eligible to vote.

43. **Sample Code**:

Listing A.89 Voting eligibility.

```
age = input('Enter your age: ');

if age >= 18
disp('You are eligible to vote.')
else
disp('You are not eligible to vote.')
end
```

44. **Sample Output**:

```
Enter your age: 21
You are eligible to vote.
```

```
Enter your age: 16
You are not eligible to vote.
```

45. **Extensions and Variations**:

- Modify the script to handle different voting age requirements based on location or jurisdiction.
- Incorporate input validation to ensure that the user enters a valid positive integer for age.
- Extend the script to provide additional information, such as the number of years remaining until the user becomes eligible to vote.

46. **Problem Statement and Background**: Implement a MATLAB function that takes a string as input and determines whether it is a palindrome (a word, phrase,

number, or other sequence of characters that reads the same backward as forward) using **if** statements.

47. **Step-by-Step Approach**:

    a. Define a function that takes a string as input.
    b. Remove any non-alphabetic characters and convert the string to lowercase.
    c. Reverse the string using the **flip** function.
    d. Use an **if** statement to compare the original string with the reversed string.
    e. If the strings are equal, return true (indicating that the input is a palindrome).
    f. Otherwise, return false (indicating that the input is not a palindrome).

48. **Sample Code**:

Listing A.90 Determine if a string is a palindrome.

```
function is_palindrome = isPalindrome(str)
% Remove non-alphabetic characters and convert to
    lowercase
str = regexprep(lower(str), '[^a-z]', '');
% Reverse the string
reversed_str = flip(str);

% Compare the original string with the reversed
    string
if strcmp(str, reversed_str)
    is_palindrome = true;
else
    is_palindrome = false;
end

end
```

49. **Sample Output**:

```
is_palindrome = isPalindrome('racecar')
is_palindrome = 1 % true
```

```
is_palindrome = isPalindrome('hello')
is_palindrome = 0 % false
```

50. **Extensions and Variations**:

    • Modify the function to handle palindromes that include spaces and punctuation.
    • Incorporate input validation to ensure that the input is a valid string.
    • Extend the function to provide additional information, such as the length of the palindrome or the longest palindromic substring within the input string.

51. Utilise Computational Intelligence Aided Design (CIAD) techniques [2–4] to create a digital twin of a Micro Aerial Vehicle (MAV) robotic system for precision agriculture applications in future farming scenarios.

a. **Problem Statement and Background**:

- Design a MATLAB program to create a digital twin of an MAV used in future farming. The MAV will monitor crop health and provide data for analysis.
- Use CI methods to optimise the MAV's flight path for efficient coverage of the farm.
- Integrate real-time data collection and analysis, simulating the MAV's operations and decision-making process.

b. **Step-by-Step Approach**:

    i. **Define Variables and Data Structures**:
- Define the farm's grid layout and initialise the MAV's parameters (e.g., battery life, speed, sensor range).
- Use arrays or matrices to represent the farm and the MAV's state.

    ii. **Implement CI Algorithms**:
- Implement a simple genetic algorithm (GA) to optimise the MAV's flight path.
- Define the fitness function to maximise coverage and minimise energy consumption.

    iii. **Simulate MAV Operations**:
- Write functions to simulate the MAV's movement, data collection, and decision-making processes.
- Use loops and conditional statements to model the MAV's behaviour over time.

    iv. **Visualise Results**:
- Plot the MAV's flight path and coverage area using MATLAB's plotting functions.
- Display real-time data collected by the MAV in a dynamic plot.

c. **Sample Code**:

Listing A.91 Sample code for MAV digital twin simulation.

```
    % Define farm grid and MAV parameters
farmSize = [100, 100];
mavPosition = [1, 1];
batteryLife = 100;
sensorRange = 10;
coverageMap = zeros(farmSize);

% Genetic algorithm parameters
populationSize = 50;
numGenerations = 100;
mutationRate = 0.01;

% Fitness function
fitnessFunction = @(path) evaluatePath(path,
    coverageMap, mavPosition, sensorRange);
```

```matlab
% Generate initial population
population = initializePopulation(populationSize,
    farmSize);

% Run genetic algorithm
for generation = 1:numGenerations
fitnessValues = arrayfun(fitnessFunction,
    population);
population = evolvePopulation(population,
    fitnessValues, mutationRate);
end

% Simulate MAV operations
for t = 1:batteryLife
mavPosition = updatePosition(mavPosition,
    population(1, :));
coverageMap = updateCoverage(coverageMap,
    mavPosition, sensorRange);
plotFarm(farmSize, coverageMap, mavPosition);
pause(0.1); % Simulate real-time operation
end

function fitness = evaluatePath(path, coverageMap,
    startPos, sensorRange)
% Evaluate the fitness of a given path
mavPos = startPos;
for step = 1:length(path)
mavPos = updatePosition(mavPos, path(step));
coverageMap = updateCoverage(coverageMap, mavPos,
    sensorRange);
end
fitness = sum(coverageMap(:));
end

function newPopulation = evolvePopulation(
    population, fitnessValues, mutationRate)
% Implement selection, crossover, and mutation to
    evolve the population
% ...
end

function mavPos = updatePosition(currentPos, step)
% Update the MAV's position based on the step
% Assuming step is a vector [dx, dy]
mavPos = currentPos + step;
end

function coverageMap = updateCoverage(coverageMap,
    mavPos, sensorRange)
% Update the coverage map based on the MAV's
    current position
```

```
[x, y] = meshgrid(mavPos(1)-sensorRange:mavPos(1)+
    sensorRange, mavPos(2)-sensorRange:mavPos(2)+
    sensorRange);
validIdx = (x > 0 & x <= size(coverageMap, 1) & y >
    0 & y <= size(coverageMap, 2));
coverageMap(sub2ind(size(coverageMap), x(validIdx),
    y(validIdx))) = 1;
end

function plotFarm(farmSize, coverageMap, mavPos)
% Plot the farm grid, MAV position, and coverage
    map
imagesc(coverageMap);
hold on;
plot(mavPos(2), mavPos(1), 'ro', 'MarkerSize', 10,
    'LineWidth', 2);
title('MAV Coverage Map');
xlabel('X Position');
ylabel('Y Position');
hold off;
end
```

d. **Sample Output**:

- The output will be a series of plots showing the MAV's coverage of the farm over time. Each plot will display the areas covered by the MAV's sensors and its current position.
- The final plot should show the entire farm grid, highlighting the areas that have been covered by the MAV.

e. **Extensions and Variations**:

- Multi-MAV Coordination: Extend the problem to include multiple MAVs coordinating to cover the farm more efficiently. Implement communication and coordination algorithms to optimise their collective paths.
- Dynamic Obstacles: Introduce dynamic obstacles (e.g., weather conditions, moving animals) that the MAV must detect and avoid in real-time.
- Data Analysis: Integrate data analysis tools to process the data collected by the MAV and provide insights into crop health, soil conditions, etc.
- Energy Management: Implement more sophisticated energy management strategies to maximise the MAV's operational time and efficiency.

f. **Explanation of Important Concepts:**

- **Digital Twin**: A virtual model designed to accurately reflect a physical object. In this problem, the MAV's operations and environment are simulated digitally.
- **Computational Intelligence (CI)**: Techniques such as genetic algorithms are used to optimise the MAV's flight path, demonstrating how CI can solve complex, non-linear problems.

- **Flight Path Optimisation**: Using CI methods to determine the most efficient path for the MAV to cover the entire farm.
- **Real-Time Simulation**: The MATLAB program simulates the MAV's operations in real-time, providing immediate feedback and visualisation of its performance.

This problem and its solution not only cover key MATLAB programming concepts but also illustrate the practical application of CI in digital twinning and autonomous systems for agriculture.

52. Please collaborate with ChatGPT or a similar AI tool to create a MATLAB function that can generate plots similar to the example figure provided in Fig. 9.25 [5].

   a. **Problem Statement and Background**:

   - We need to create a MATLAB function named 'plot_errorbands' that can generate plots with shaded error bands similar to a given example figure.
   - This function should take input data, time points, error values, categories for the legend, and axis labels.
   - **Error bands** are used to visualize the variability or uncertainty in the data.

   b. **Step-by-Step Approach**:

   - **Define the function signature**: The function will take five inputs: 'timepoints', 'data', 'error', 'categories', and 'labels'.
   - **Create a figure and hold on**: Use 'figure' and 'hold on' to prepare the plot for multiple lines and shaded areas.
   - **Define colors and line styles**: Use 'lines' to generate a color scheme and define different line styles for differentiation.
   - **Plot shaded error bands**: Use 'fill' to create shaded areas representing the error bands.
   - **Plot data lines**: Use 'plot' to draw the actual data lines over the error bands.
   - **Add labels and legend**: Use 'xlabel', 'ylabel', and 'legend' to add descriptive labels and a legend to the plot.

   c. **Sample Code**:
   The basic syntax is:

Listing A.92 plot_errorbands function syntax.

```
function plot_errorbands(timepoints, data,
    error, categories, labels)
  % plot_errorbands - Plots data with shaded
    error bands
  %
  % Syntax: plot_errorbands(timepoints, data,
      error, categories, labels)
  %
  % timepoints - Vector of time points
```

```matlab
% data - Matrix of data values (each row is
     a different line)
% error - Matrix of error values (same size
     as data)
% categories - Cell array of category names
     (for legend)
% labels - Cell array with two elements: {'
    xlabel', 'ylabel'}

figure;
hold on;

% Define colors and line styles
colors = lines(size(data, 1));
linestyles = {'-', '--'};

% Plot each data line with error bands
for i = 1:size(data, 1)
    % Plot the error bands
    fill([timepoints, fliplr(timepoints)],
        ...
        [data(i, :) + error(i, :), fliplr(
            data(i, :) - error(i, :))], ...
        colors(i, :), 'FaceAlpha', 0.2, '
            EdgeColor', 'none');

    % Plot the data line
    plot(timepoints, data(i, :), 'Color',
        colors(i, :), ...
        'LineStyle', linestyles{mod(i-1,
            length(linestyles))+1}, ...
        'LineWidth', 1.5);
end

% Add labels and legend
xlabel(labels{1});
ylabel(labels{2});
legend(categories, 'Location', 'Best');

hold off;
end
```

d. **Sample Output**:

- When the function 'plot_errorbands' is called with appropriate input data, it generates a plot with multiple lines and shaded error bands.
- The plot includes labels for the x-axis and y-axis and a legend for different categories.
- The visual representation helps in understanding the data trends and the associated uncertainties.

As shown in Fig. A.1, the example of *plot_errorbands* function, all codes are in a file named *plot_errorbands_example*01.*m*.

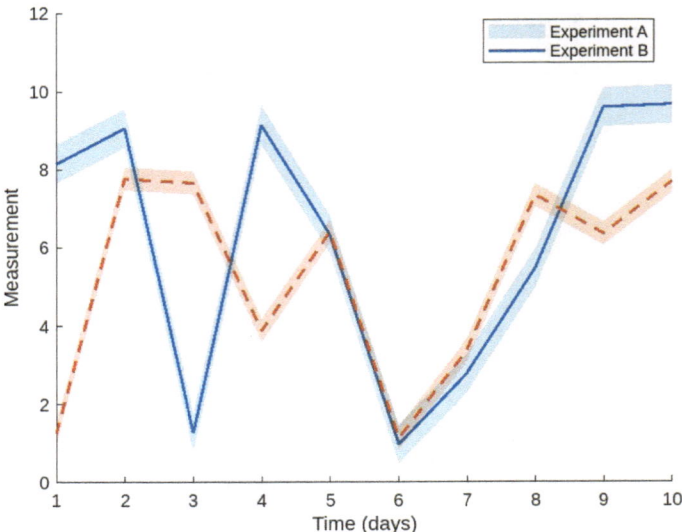

Fig. A.1 Example: line plot with error band using ChatGPT

Listing A.93 An example of plot_errorbands function

```
% Generate some data
timepoints = 1:10;
data = [10*rand(1,10); 8*rand(1,10)];
error = [0.5*ones(1,10); 0.3*ones(1,10)];
categories = {'Experiment A', 'Experiment B'};
labels = {'Time (days)', 'Measurement'};

% Plot the error bands
plot_errorbands(timepoints, data, error, categories
    , labels);
```

e. **Extensions and Variations**:

- **Customization**: The function can be extended to allow customization of colors, line styles, and transparency of the error bands.
- **Additional Features**: Adding options to save the plot as an image file or to include grid lines and titles can enhance the function.
- **Different Plot Types**: The function can be modified to support different types of plots (e.g., bar plots, scatterplots) while still incorporating error bands.

A.10   Solutions to Chapter 10

1. **Bank Account Management**

   a. **Problem Statement and Background**: Develop an object-oriented system to manage various types of bank accounts, including features like deposits, withdrawals, balance checking, and account-specific functionalities (e.g., interest calculation, overdraft fees).

   b. **Step-by-Step Approach**:
      i. Define the `BankAccount` class with properties for account number, account holder's name, and balance.
      ii. Implement methods for depositing, withdrawing, and checking the balance.
      iii. Incorporate error handling to prevent negative balances and invalid transactions.
      iv. Create subclasses for different account types (e.g., `SavingsAccount`, `CheckingAccount`) and implement their respective functionalities.
      v. Test the system by creating instances of different account types and performing various operations.

   c. **Sample Code**:

Listing A.94 Bank Account Management.

```
    classdef BankAccount
properties
accountNumber
accountHolderName
balance
end

methods
    function obj = BankAccount(accountNumber,
        accountHolderName, initialBalance)
        obj.accountNumber = accountNumber;
        obj.accountHolderName = accountHolderName;
        obj.balance = initialBalance;
    end

    function deposit(obj, amount)
        if amount > 0
            obj.balance = obj.balance + amount;
        else
            error('Invalid deposit amount.');
        end
    end

    function withdraw(obj, amount)
        if amount > 0 && amount <= obj.balance
            obj.balance = obj.balance - amount;
        else
```

```matlab
                error('Insufficient funds or invalid
                    withdrawal amount.');
            end
        end

        function currentBalance = getBalance(obj)
            currentBalance = obj.balance;
        end
    end
end

classdef SavingsAccount < BankAccount
properties
interestRate
end

methods
    function obj = SavingsAccount(accountNumber,
        accountHolderName, initialBalance,
        interestRate)
        obj = obj@BankAccount(accountNumber,
            accountHolderName, initialBalance);
        obj.interestRate = interestRate;
    end

    function calculateInterest(obj)
        interest = obj.balance * obj.interestRate;
        obj.deposit(interest);
    end
end
end
```

d. **Sample Output**:

```matlab
    % Create a savings account
savingsAccount = SavingsAccount('SA001', 'John Doe'
    , 5000, 0.03);

% Deposit some money
savingsAccount.deposit(2000);
currentBalance = savingsAccount.getBalance(); %
    Output: 7000

% Calculate and apply interest
savingsAccount.calculateInterest();
currentBalance = savingsAccount.getBalance(); %
    Output: 7150

% Withdraw money
savingsAccount.withdraw(1500);
currentBalance = savingsAccount.getBalance(); %
    Output: 5650
```

e. **Extensions and Variations**:

- Implement additional account types like `CheckingAccount` with over-draft fees or `CreditCardAccount` with interest calculations and credit limits.
- Add functionality for transferring funds between accounts, either within the same bank or across different banks.
- Incorporate account statements or transaction history tracking.
- Implement account closures and account number validation.
- Explore database integration for persistent storage of account information.

2. **Geometric Shapes**

a. **Problem Statement and Background**: Create an object-oriented system to represent and calculate properties of various geometric shapes, such as area, perimeter, and volume (for 3D shapes).

b. **Step-by-Step Approach**:

i. Define an abstract class called `Shape` with a method for calculating the area.

ii. Derive concrete subclasses for different shapes (e.g., `Circle`, `Rectangle`, `Triangle`) and implement their respective area calculation methods.

iii. Create a function that takes an array of `Shape` objects and calculates the total area of all shapes.

iv. Optionally, extend the problem by adding methods for calculating perimeter, volume (for 3D shapes), or other relevant properties.

v. Test the system by creating instances of different shape objects and calculating their properties.

c. **Sample Code**:

Listing A.95 Geometric Shapes.

```
    classdef Shape
properties
end

methods (Abstract)
    area = calculateArea(obj)
end

end

classdef Circle < Shape
properties
radius
end

methods
    function obj = Circle(radius)
        obj.radius = radius;
```

```matlab
    end

    function area = calculateArea(obj)
        area = pi * obj.radius^2;
    end
end

end

classdef Rectangle < Shape
properties
length
width
end

methods
    function obj = Rectangle(length, width)
        obj.length = length;
        obj.width = width;
    end

    function area = calculateArea(obj)
        area = obj.length * obj.width;
    end
end

end

function totalArea = calculateTotalArea(shapes)
totalArea = 0;
for i = 1:length(shapes)
totalArea = totalArea + shapes(i).calculateArea();
end
end
```

d. **Sample Output**:

```matlab
    % Create shape objects
circle1 = Circle(5);
rectangle1 = Rectangle(3, 4);
rectangle2 = Rectangle(6, 2);

% Create an array of shapes
shapes = [circle1, rectangle1, rectangle2];

% Calculate total area
totalArea = calculateTotalArea(shapes); % Output:
   78.5398
```

e. **Extensions and Variations**:

- Implement additional shape classes like Triangle, Sphere, Cube, or other 3D shapes.

- Add methods for calculating perimeter, volume, or other relevant properties for each shape class.
- Implement methods for drawing or visualizing the shapes using MATLAB graphics.
- Explore inheritance hierarchies and polymorphism for more complex shape representations (e.g., regular polygons, composite shapes).
- Incorporate shape transformations such as scaling, rotation, or translation.

3. **Student Record Management**

a. **Problem Statement and Background**: Develop an object-oriented system to manage student records, including features like adding/updating grades, calculating grade point averages (GPA), generating report cards, and handling different types of students (e.g., undergraduate, graduate).

b. **Step-by-Step Approach**:
   i. Define the `Student` class with properties for name, ID, and grades.
   ii. Implement methods for adding/updating grades, calculating the GPA, and generating a report card.
   iii. Create a subclass called `GraduateStudent` that inherits from `Student` and includes additional properties and methods specific to graduate students (e.g., thesis, research projects).
   iv. Design a system to manage a collection of students, perform operations like sorting or filtering based on specific
   v. Criteria, and generate statistical reports.
   vi. Test the system by creating instances of different student types, adding grades, and performing various operations.

c. **Sample Code**:

Listing A.96 Student Record Management.

```
classdef Student
properties
name
id
grades = []
end

methods
    function obj = Student(name, id)
        obj.name = name;
        obj.id = id;
    end

    function addGrade(obj, grade)
        obj.grades = [obj.grades, grade];
    end

    function gpa = calculateGPA(obj)
        gpa = mean(obj.grades);
    end
```

```matlab
    function generateReportCard(obj)
        disp(['Name: ', obj.name]);
        disp(['ID: ', obj.id]);
        disp('Grades:');
        disp(obj.grades);
        disp(['GPA: ', num2str(obj.calculateGPA())
            ]);
    end
end

end

classdef GraduateStudent < Student
properties
thesis
researchProjects = {}
end

methods
    function obj = GraduateStudent(name, id, thesis
        )
        obj = obj@Student(name, id);
        obj.thesis = thesis;
    end

    function addResearchProject(obj, project)
        obj.researchProjects{end+1} = project;
    end
end

end
```

d. **Sample Output**:

```matlab
% Create undergraduate student
undergrad = Student('John Doe', '12345');
undergrad.addGrade(85);
undergrad.addGrade(92);
undergrad.addGrade(78);
undergrad.generateReportCard();

% Create graduate student
grad = GraduateStudent('Jane Smith', '67890', '
    Machine Learning Algorithms');
grad.addGrade(90);
grad.addGrade(88);
grad.addGrade(94);
grad.addResearchProject('Deep Learning for Image
    Recognition');
grad.generateReportCard();
```

e. **Extensions and Variations**:

- Implement additional student types like `PartTimeStudent`, `InternationalStudent`, or `OnlineStudent` with their respective rules and requirements.
- Add functionality for managing course registrations, schedules, and pre-requisites.
- Incorporate features like academic probation, Dean's List, or honors recognition based on GPA thresholds.
- Explore database integration for persistent storage of student records and transcripts.
- Implement user authentication and authorization for secure access to student records.

4. **Employee Payroll System**

a. **Problem Statement and Background**: Create an object-oriented system to manage employee payroll, including different types of employees (e.g., hourly, salaried, contract), calculation of monthly salaries, overtime pay, bonuses, and deductions.

b. **Step-by-Step Approach**:
   i. Define an abstract class called `Employee` with properties for name, ID, and a method for calculating the monthly salary.
   ii. Derive concrete subclasses for different employee types (e.g., `Hourly Employee`, `SalariedEmployee`, `ContractEmployee`) and implement their respective salary calculation methods.
   iii. Incorporate features like overtime pay, bonuses, or deductions based on employee type or performance.
   iv. Design a system to manage a company's payroll, generate paychecks, and produce reports for HR or financial analysis.
   v. Test the system by creating instances of different employee types, calculating their monthly salaries, and generating payroll reports.

c. **Sample Code**:

Listing A.97 Employee Payroll System.

```
classdef Employee
properties
name
id
end

methods (Abstract)
    monthlySalary = calculateMonthlySalary(obj)
end

end

classdef HourlyEmployee < Employee
```

```matlab
properties
hourlyRate
hoursWorked
end

methods
    function obj = HourlyEmployee(name, id,
        hourlyRate)
        obj.name = name;
        obj.id = id;
        obj.hourlyRate = hourlyRate;
    end

    function monthlySalary = calculateMonthlySalary
        (obj)
        regularHours = min(obj.hoursWorked, 160);
        overtimeHours = max(0, obj.hoursWorked -
            160);
        monthlySalary = (regularHours * obj.
            hourlyRate) + (overtimeHours * 1.5 * obj
            .hourlyRate);
    end
end

end

classdef SalariedEmployee < Employee
properties
monthlySalary
end

methods
    function obj = SalariedEmployee(name, id,
        monthlySalary)
        obj.name = name;
        obj.id = id;
        obj.monthlySalary = monthlySalary;
    end

    function salary = calculateMonthlySalary(obj)
        salary = obj.monthlySalary;
    end
end

end

% ... (additional employee types like
    ContractEmployee)

function payroll = generatePayroll(employees)
payroll = [];
for i = 1:length(employees)
employee = employees(i);
```

```matlab
paycheck.name = employee.name;
paycheck.id = employee.id;
paycheck.monthlySalary = employee.
    calculateMonthlySalary();
payroll = [payroll, paycheck];
end
end
```

d. **Sample Output**:

```matlab
% Create employee instances
hourlyEmployee = HourlyEmployee('John Doe', '12345'
    , 25);
hourlyEmployee.hoursWorked = 180;
salariedEmployee = SalariedEmployee('Jane Smith', '
    67890', 5000);

% Generate payroll
employees = [hourlyEmployee, salariedEmployee];
payroll = generatePayroll(employees);
```

e. **Extensions and Variations**:

- Implement additional employee types like CommissionEmployee or PartTimeEmployee with their respective salary calculation rules.
- Add features for managing employee benefits, tax deductions, and other payroll-related calculations.
- Incorporate features for tracking employee attendance, leave balances, and time-off requests.
- Explore database integration for persistent storage of employee records and payroll history.
- Implement user authentication and authorization for secure access to payroll information.

5. **Library Management System**

a. **Problem Statement and Background**: Develop an object-oriented system for managing a library, including features like book cataloging, checkout/return operations, reservations, and user accounts.

b. **Step-by-Step Approach**:
   i. Define a class called Book with properties for title, author, publication date, and availability status.
   ii. Implement methods for checking out, returning, and reserving books.
   iii. Create a class called Library that manages a collection of Book
   iv. Objects and provides functionality for searching, sorting, and filtering books based on various criteria.
   v. Incorporate features like user accounts, late fees, and notifications for overdue books or reserved items.

vi. Test the system by creating instances of books, users, and performing various library operations.
c. **Sample Code**:

Listing A.98 Library Management System.

```
classdef Book
properties
title
author
publicationDate
available = true
end

methods
    function obj = Book(title, author,
        publicationDate)
        obj.title = title;
        obj.author = author;
        obj.publicationDate = publicationDate;
    end

    function checkout(obj)
        if obj.available
            obj.available = false;
        else
            error('Book is not available for
                checkout.');
        end
    end

    function returnBook(obj)
        obj.available = true;
    end

    function reserve(obj)
        if ~obj.available
            disp(['Book "', obj.title, '" has been
                reserved.']);
        else
            disp(['Book "', obj.title, '" is
                currently available.']);
        end
    end
end

end

classdef Library
properties
books = Book.empty();
end
```

```
methods
    function addBook(obj, book)
        obj.books(end+1) = book;
    end

    function searchBooks(obj, query)
        matches = [];
        for i = 1:length(obj.books)
            book = obj.books(i);
            if contains(book.title, query) ||
                contains(book.author, query)
                matches(end+1) = book;
            end
        end
        disp('Search Results:');
        for i = 1:length(matches)
            book = matches(i);
            disp(['Title: ', book.title, ', Author:
                ', book.author]);
        end
    end
end

end
```

d. **Sample Output**:

```
% Create books
book1 = Book('To Kill a Mockingbird', 'Harper Lee',
    '1960');
book2 = Book('1984', 'George Orwell', '1949');
book3 = Book('Pride and Prejudice', 'Jane Austen',
    '1813');

% Create a library and add books
library = Library();
library.addBook(book1);
library.addBook(book2);
library.addBook(book3);

% Search for books
library.searchBooks('Mockingbird');

% Check out a book
book1.checkout();

% Reserve a book
book2.reserve();
```

e. **Extensions and Variations**:

- Implement user accounts with borrowing history, fines, and notifications.
- Add features for managing book acquisitions, donations, and removals from the library.

- Incorporate features for managing library staff, roles, and permissions.
- Explore database integration for persistent storage of library catalogs and user records.
- Implement user authentication and authorization for secure access to library resources.

6. **Vehicle Rental System**

   a. **Problem Statement and Background**: Create an object-oriented system for a vehicle rental company, including different types of vehicles (e.g., cars, trucks, motorcycles), rental rate calculations, availability tracking, and customer management.

   b. **Step-by-Step Approach**:
      i. Define an abstract class called Vehicle with properties for make, model, year, and a method for calculating the daily rental rate.
      ii. Derive concrete subclasses for different vehicle types (e.g., Car, Truck, Motorcycle) and implement their respective rental rate calculation methods.
      iii. Create a class called RentalCompany that manages a fleet of Vehicle objects and provides functionality for renting, returning, and tracking vehicle availability.
      iv. Incorporate features like customer accounts, discounts for long-term rentals, and additional services like insurance or GPS navigation.
      v. Test the system by creating instances of different vehicle types, renting and returning vehicles, and generating rental reports.

   c. **Sample Code**:

Listing A.99 Vehicle Rental System.

```
classdef Vehicle
properties
make
model
year
end

methods (Abstract)
    dailyRate = calculateDailyRate(obj)
end

end

classdef Car < Vehicle
methods
function dailyRate = calculateDailyRate(obj)
dailyRate = 50; % Default daily rate for cars
end
end
end
```

```
classdef Truck < Vehicle
properties
payload
end

methods
    function dailyRate = calculateDailyRate(obj)
        dailyRate = 75 + obj.payload * 0.1; % Daily
            rate based on payload
    end
end

end

classdef RentalCompany
properties
vehicles = Vehicle.empty();
end

methods
    function addVehicle(obj, vehicle)
        obj.vehicles(end+1) = vehicle;
    end

    function rentVehicle(obj, vehicle)
        if any(obj.vehicles == vehicle)
            disp(['Renting ', vehicle.make, ' ',
                vehicle.model]);
        else
            error('Vehicle not found in the rental
                fleet.');
        end
    end

    function returnVehicle(obj, vehicle)
        if any(obj.vehicles == vehicle)
            disp(['Returning ', vehicle.make, ' ',
                vehicle.model]);
        else
            error('Vehicle not found in the rental
                fleet.');
        end
    end
end
end
end
```

d. **Sample Output**:

```
% Create vehicles
car1 = Car('Toyota', 'Camry', 2020);
truck1 = Truck('Ford', 'F-150', 2018, 1500);

% Create a rental company and add vehicles
rentalCompany = RentalCompany();
```

```
rentalCompany.addVehicle(car1);
rentalCompany.addVehicle(truck1);

% Rent a vehicle
rentalCompany.rentVehicle(car1);

% Return a vehicle
rentalCompany.returnVehicle(truck1);
```

e. **Extensions and Variations**:

- Implement additional vehicle types like SUV, Van, or RV with their respective rental rate calculations.
- Add features for managing customer accounts, reservations, and payment processing.
- Incorporate features like vehicle maintenance schedules, fuel tracking, and insurance claims.
- Explore database integration for persistent storage of vehicle fleets, customer records, and rental history.
- Implement user authentication and authorization for secure access to rental system features.

7. **Social Media Platform**

a. **Problem Statement and Background**: Develop an object-oriented system for a social media platform, including user profiles, posts, comments, and friend connections.

b. **Step-by-Step Approach**:
   i. Define a class called User with properties for username, profile information (name, bio, etc.), and friend connections.
   ii. Create a class called Post with properties for content, author (User), timestamp, and comments.
   iii. Implement a Comment class with properties for content, author (User), timestamp, and the post it belongs to.
   iv. Create a SocialMediaPlatform class that manages users, posts, comments, and friend connections, and provides functionality for creating posts, commenting, and managing friend connections.
   v. Test the system by creating users, making friend connections, posting content, commenting, and simulating social media interactions.

c. **Sample Code**:

Listing A.100 Social Media Platform.

```
classdef User
properties
username
name
bio
friends = User.empty();
```

```matlab
    end

    methods
        function obj = User(username, name, bio)
            obj.username = username;
            obj.name = name;
            obj.bio = bio;
        end

        function addFriend(obj, friend)
            obj.friends(end+1) = friend;
        end
    end

end

classdef Post
properties
content
author
timestamp
comments = Comment.empty();
end

methods
    function obj = Post(content, author)
        obj.content = content;
        obj.author = author;
        obj.timestamp = datetime('now');
    end

    function addComment(obj, comment)
        obj.comments(end+1) = comment;
    end
end

end

classdef Comment
properties
content
author
timestamp
post
end

methods
    function obj = Comment(content, author, post)
        obj.content = content;
        obj.author = author;
        obj.timestamp = datetime('now');
        obj.post = post;
        post.addComment(obj);
```

```matlab
        end
    end

    end

classdef SocialMediaPlatform
properties
users = User.empty();
posts = Post.empty();
end

methods
    function addUser(obj, user)
        obj.users(end+1) = user;
    end

    function post = createPost(obj, content, author
        )
        post = Post(content, author);
        obj.posts(end+1) = post;
    end

    function comment = addComment(obj, post,
        content, author)
        comment = Comment(content, author, post);
    end

    function makeFriends(obj, user1, user2)
        user1.addFriend(user2);
        user2.addFriend(user1);
    end
end

end
```

d. **Sample Output**:

```matlab
% Create users
user1 = User('john_doe', 'John Doe', 'Software
    Engineer');
user2 = User('jane_smith', 'Jane Smith', 'Artist');

% Create a social media platform
platform = SocialMediaPlatform();
platform.addUser(user1);
platform.addUser(user2);

% Make friend connection
platform.makeFriends(user1, user2);

% Create a post
post1 = platform.createPost('Had a great day today!
    ', user1);
```

```
% Add a comment
comment1 = platform.addComment(post1, 'Awesome!',
    user2);
```

e. **Extensions and Variations**:

- Implement features for news feeds, post sharing, and content discovery.
- Add features for post reactions (likes, dislikes, etc.), hashtags, and mentions.
- Incorporate features like group management, event creation, and messaging.
- Explore database integration for persistent storage of user profiles, posts, comments, and connections.
- Implement user authentication, privacy settings, and content moderation features.

8. **Online Shopping Cart**

a. **Problem Statement and Background**: Develop an object-oriented system for an online shopping cart, including product catalog management, shopping cart functionality, order processing, and customer accounts.

b. **Step-by-Step Approach**:
   i. Define a class called `Product` with properties for name, description, price, and stock quantity.
   ii. Create a class called `ShoppingCart` that manages a collection of `Product` objects and provides functionality for adding, removing, and updating quantities in the cart.
   iii. Implement an `Order` class that represents a customer's order, including the ordered products, total cost, and shipping information.
   iv. Incorporate features like customer accounts, order history, and payment processing.
   v. Test the system by creating product catalogs, simulating shopping cart operations, and processing orders.

c. **Sample Code**:

Listing A.101 Online Shopping Cart.

```
classdef Product
properties
name
description
price
stockQuantity
end

methods
    function obj = Product(name, description, price
        , stockQuantity)
        obj.name = name;
        obj.description = description;
        obj.price = price;
```

```
                obj.stockQuantity = stockQuantity;
        end

    end

classdef ShoppingCart
properties
items = Product.empty();
quantities = [];
end

methods
    function addItem(obj, product, quantity)
        idx = find([obj.items.name] == product.name
            );
        if isempty(idx)
            obj.items(end+1) = product;
            obj.quantities(end+1) = quantity;
        else
            obj.quantities(idx) = obj.quantities(
                idx) + quantity;
        end
    end

    function removeItem(obj, product)
        idx = find([obj.items.name] == product.name
            );
        if ~isempty(idx)
            obj.items(idx) = [];
            obj.quantities(idx) = [];
        end
    end

    function updateQuantity(obj, product,
        newQuantity)
        idx = find([obj.items.name] == product.name
            );
        if ~isempty(idx)
            obj.quantities(idx) = newQuantity;
        end
    end

    function totalCost = calculateTotalCost(obj)
        totalCost = 0;
        for i = 1:length(obj.items)
            totalCost = totalCost + obj.items(i).
                price * obj.quantities(i);
        end
    end
end

end
```

```matlab
classdef Order
properties
customer
items
quantities
totalCost
shippingAddress
end

methods
    function obj = Order(customer, shoppingCart)
        obj.customer = customer;
        obj.items = shoppingCart.items;
        obj.quantities = shoppingCart.quantities;
        obj.totalCost = shoppingCart.
            calculateTotalCost();
        % Prompt user for shipping address
    end

    function processOrder(obj)
        % Process payment
        disp('Order processed successfully!');
    end
end
end

end
```

d. **Sample Output**:

```matlab
% Create products
product1 = Product('T-Shirt', 'Cotton T-Shirt',
    19.99, 100);
product2 = Product('Jeans', 'Denim Jeans', 49.99,
    50);

% Create a shopping cart
cart = ShoppingCart();

% Add items to the cart
cart.addItem(product1, 2);
cart.addItem(product2, 1);

% Calculate total cost
totalCost = cart.calculateTotalCost(); % Output:
    89.97

% Create and process an order
order = Order('John Doe', cart);
order.processOrder();
```

e. **Extensions and Variations**:

- Implement customer accounts with order history and personalized recommendations.
- Add features for product reviews, ratings, and wish lists.
- Incorporate features like discounts, promotions, and gift cards.
- Explore database integration for persistent storage of product catalogs, customer information, and order history.
- Implement user authentication and authorization for secure access to customer accounts and order information.

9. **Flight Reservation System**

a. **Problem Statement and Background**: Develop an object-oriented system for airline reservations, including flight management, seat reservations, passenger management, and ticket booking.

b. **Step-by-Step Approach**:

i. Define a class called `Flight` with properties for flight number, origin, destination, departure time, and available seats.

ii. Create a class called `Passenger` with properties for name, contact information, and frequent flyer status.

iii. Implement a `Reservation` class that represents a passenger's reservation, including the flight, passenger details, and seat assignment.

iv. Create an `AirlineReservationSystem` class that manages flights, passengers, and reservations, and provides functionality for booking, canceling, and modifying reservations.

v. Test the system by creating flights, managing passenger information, and simulating reservation scenarios.

c. **Sample Code**:

Listing A.102 Airline Reservation System.

```
classdef Flight
properties
flightNumber
origin
destination
departureTime
availableSeats
end

methods
    function obj = Flight(flightNumber, origin,
        destination, departureTime, totalSeats)
        obj.flightNumber = flightNumber;
        obj.origin = origin;
        obj.destination = destination;
        obj.departureTime = departureTime;
        obj.availableSeats = totalSeats;
```

```
        end
    end

    end

classdef Passenger
properties
name
contact
frequentFlyerStatus
end

methods
    function obj = Passenger(name, contact,
        frequentFlyerStatus)
        obj.name = name;
        obj.contact = contact;
        obj.frequentFlyerStatus =
            frequentFlyerStatus;
    end
end

end

classdef Reservation
properties
flight
passenger
seatNumber
end

methods
    function obj = Reservation(flight, passenger,
        seatNumber)
        obj.flight = flight;
        obj.passenger = passenger;
        obj.seatNumber = seatNumber;
        flight.availableSeats = flight.
            availableSeats - 1;
    end
end

end

classdef AirlineReservationSystem
properties
flights
passengers
reservations
end

methods
    function obj = AirlineReservationSystem()
```

```matlab
            obj.flights = Flight.empty();
            obj.passengers = Passenger.empty();
            obj.reservations = Reservation.empty();
        end

        function addFlight(obj, flight)
            obj.flights(end+1) = flight;
        end

        function addPassenger(obj, passenger)
            obj.passengers(end+1) = passenger;
        end

        function reservation = bookReservation(obj,
            flightNumber, passenger)
            flight = obj.getFlight(flightNumber);
            if ~isempty(flight) && flight.
                availableSeats > 0
                seatNumber = flight.availableSeats;
                reservation = Reservation(flight,
                    passenger, seatNumber);
                obj.reservations(end+1) = reservation;
            else
                reservation = [];
                warning('Flight not found or no
                    available seats.');
            end
        end

        function cancelReservation(obj, reservation)
            idx = find([obj.reservations.seatNumber] ==
                reservation.seatNumber);
            if ~isempty(idx)
                reservation.flight.availableSeats =
                    reservation.flight.availableSeats +
                    1;
                obj.reservations(idx) = [];
            end
        end

        function flight = getFlight(obj, flightNumber)
            idx = find([obj.flights.flightNumber] ==
                flightNumber);
            if ~isempty(idx)
                flight = obj.flights(idx);
            else
                flight = [];
            end
        end
    end
end
```

d. **Sample Output**:

```
% Create flights
flight1 = Flight('AA123', 'New York', 'Los Angeles'
    , datetime(2023, 6, 1, 9, 0, 0), 150);
flight2 = Flight('UA456', 'Chicago', 'San Francisco
    ', datetime(2023, 6, 2, 11, 30, 0), 120);

% Create passengers
passenger1 = Passenger('John Doe', '123-456-7890',
    'Gold');
passenger2 = Passenger('Jane Smith', '987-654-321C'
    , 'Silver');

% Create an airline reservation system
reservationSystem = AirlineReservationSystem();
reservationSystem.addFlight(flight1);
reservationSystem.addFlight(flight2);
reservationSystem.addPassenger(passenger1);
reservationSystem.addPassenger(passenger2);

% Book a reservation
reservation1 = reservationSystem.bookReservation('
    AA123', passenger1);

% Cancel a reservation
reservationSystem.cancelReservation(reservation1);
```

e. **Extensions and Variations**:

- Implement features for flight schedules, connections, and multi-leg journeys.
- Add features for seat selection, meal preferences, and special assistance.
- Incorporate features like baggage tracking, check-in, and boarding procedures.
- Explore database integration for persistent storage of flight schedules. passenger information, and reservation records.
- Implement user authentication and authorization for secure access to reservation features.

10. **Weather Monitoring System**

a. **Problem Statement and Background**: Develop an object-oriented system for monitoring weather conditions, including temperature, humidity, precipitation, and wind speed/direction. The system should support multiple weather stations and provide data Visualisation and analysis capabilities.

b. **Step-by-Step Approach**:

- Define a class called `WeatherData` with properties for temperature, humidity, precipitation, wind speed, and wind direction.
- Create a class called `WeatherStation` with properties for location, station ID, and a collection of `WeatherData` objects.

- Implement a `WeatherMonitoringSystem` class that manages weather stations, collects weather data, and provides functionality for data Visualisation and analysis.
- Create helper classes or functions for data Visualisation (e.g., plotting temperature, humidity, or precipitation over time) and data analysis (e.g., calculating averages, extremes, or trends).
- Test the system by simulating weather data collection from multiple stations, visualizing the data, and performing analysis on the collected data.

c. **Sample Code**:

Listing A.103 Weather Monitoring System.

```
classdef WeatherData
properties
temperature
humidity
precipitation
windSpeed
windDirection
end

methods
    function obj = WeatherData(temperature,
        humidity, precipitation, windSpeed,
        windDirection)
        obj.temperature = temperature;
        obj.humidity = humidity;
        obj.precipitation = precipitation;
        obj.windSpeed = windSpeed;
        obj.windDirection = windDirection;
    end
end

end

classdef WeatherStation
properties
location
stationID
weatherData = WeatherData.empty();
end

methods
    function obj = WeatherStation(location,
        stationID)
        obj.location = location;
        obj.stationID = stationID;
    end

    function addWeatherData(obj, weatherData)
        obj.weatherData(end+1) = weatherData;
    end
```

```matlab
end

end

classdef WeatherMonitoringSystem
properties
weatherStations = WeatherStation.empty();
end

methods
    function addWeatherStation(obj, weatherStation)
        obj.weatherStations(end+1) = weatherStation
            ;
    end

    function visualizeData(obj, stationID, dataType
        )
        station = obj.getWeatherStation(stationID);
        if ~isempty(station)
            switch dataType
                case 'temperature'
                    data = [station.weatherData.
                        temperature];
                    plot(data);
                    title(['Temperature at Station
                        ', station.stationID]);
                    xlabel('Time');
                    ylabel('Temperature (\celsius)'
                        );
                case 'humidity'
                    data = [station.weatherData.
                        humidity];
                    plot(data);
                    title(['Humidity at Station ',
                        station.stationID]);
                    xlabel('Time');
                    ylabel('Humidity (%)');
                % Add more cases for other data
                    types
            end
        else
            error('Weather station not found.');
        end
    end

    function averageData = analyzeData(obj,
        stationID, dataType)
        station = obj.getWeatherStation(stationID);
        if ~isempty(station)
            switch dataType
                case 'temperature'
                    data = [station.weatherData.
                        temperature];
```

```matlab
                        averageData = mean(data);
                    case 'humidity'
                        data = [station.weatherData.
                            humidity];
                        averageData = mean(data);
                    % Add more cases for other data
                        types
                end
            else
                error('Weather station not found.');
            end
        end

        function station = getWeatherStation(obj,
            stationID)
            idx = find([obj.weatherStations.stationID]
                == stationID);
            if ~isempty(idx)
                station = obj.weatherStations(idx);
            else
                station = [];
            end
        end
    end
end
```

d. **Sample Output**:

```matlab
% Create weather stations
station1 = WeatherStation('New York', 'NY001');
station2 = WeatherStation('Los Angeles', 'LA001');

% Add weather data
station1.addWeatherData(WeatherData(25, 70, 0, 10,
    'N'));
station1.addWeatherData(WeatherData(28, 65, 0.2,
    15, 'NW'));
station2.addWeatherData(WeatherData(30, 40, 0, 5, '
    W'));
station2.addWeatherData(WeatherData(32, 35, 0.1, 8,
    'SW'));

% Create a weather monitoring system
weatherSystem = WeatherMonitoringSystem();
weatherSystem.addWeatherStation(station1);
weatherSystem.addWeatherStation(station2);

% Visualize temperature data
weatherSystem.visualizeData('NY001', 'temperature')
    ;

% Analyze humidity data
avgHumidity = weatherSystem.analyzeData('LA001', '
    humidity');
```

```
disp(['Average humidity at LA001: ', num2str(
    avgHumidity), '%']);
```

e. **Extensions and Variations**:

- Implement features for weather forecasting and alert systems based on historical data and weather patterns.
- Add support for additional weather data types, such as atmospheric pressure, cloud cover, and visibility.
- Incorporate real-time data acquisition from weather APIs or sensor networks.
- Explore database integration for persistent storage of weather data and station information.
- Implement user authentication and authorization for secure access to weather monitoring features.

Appendix B
Frequently Asked Questions (FAQs)

B.1 Prerequisites

While no strict prerequisites are required, some background in the following areas will help readers get the most out of this textbook:

- Programming experience—Prior experience in any programming language like C, Python, Java, etc. will be helpful to understand basic programming constructs and data structures.
- Mathematics—Foundational knowledge in mathematics including calculus, linear algebra, probability and statistics will enable better understanding of examples and applications.
- Engineering and Science basics—Some familiarity with basic engineering or science concepts will provide context for many of the examples. However, the book covers fundamentals as well.
- Computer skills—Basic computer skills including proficiency with an operating system, file management, office productivity tools etc. will be useful.

That said, the book is designed in a modular fashion allowing even beginners with no prior experience to pick up MATLAB skills systematically. The programming aspects are built up gradually with abundant examples. Necessary mathematical and scientific context is provided along the way.

Readers with some amount of prior experience in programming, mathematics or an engineering and science discipline will likely be able to progress through the material more quickly. However, the book can be used even by complete beginners starting from first principles.

The terms **computing** and **calculating** are often used interchangeably, but they can have distinct meanings, especially in the context of **technology** and **mathematics**.

Calculating generally refers to the process of finding a **numerical answer** to a problem. It involves **arithmetic operations** such as addition, subtraction, multiplication, and division. Calculating can be done by hand, with a calculator, or by a computer as part of a larger set of operations.

© The Editor(s) (if applicable) and The Author(s) 2025
Y. Chen and L. Huang, *MATLAB Roadmap to Applications*,
https://doi.org/10.1007/978-981-97-8788-3

Examples of calculating include:

- Working out the tip at a restaurant.
- Balancing a checkbook.
- Determining the area of a rectangle by multiplying its width by its height.

Computing, on the other hand, is a broader term that encompasses calculating but also includes other types of operations and processes. It can refer to any type of **information processing** that can be done by an **algorithm** or a **computer**. This includes not only arithmetic but also **data processing, algorithmic decision-making, simulation**, and more complex mathematical operations.

Examples of computing include:

- Running a complex simulation of weather patterns using a supercomputer.
- Processing large datasets to find patterns and insights (data mining).
- Operating an autonomous vehicle that has to make real-time navigational decisions.

In summary, while **calculating** is specifically about **numerical computation**, **computing** is a more inclusive term that covers a wide range of **information processing tasks**, which can be numerical, symbolic, or based on data operations, often performed by **computers**.

B.2   Story of MATLAB Logo

The MATLAB logo has an interesting story behind its design and symbolism. The logo features a distinct wave-like shape with a colorful gradient, representing the diverse applications and versatility of MATLAB across various fields.

The origins of the MATLAB logo can be traced back to the late 1970s when Cleve Moler, the co-founder of MathWorks (the company behind MATLAB), was working on the first version of the software. At that time, Cleve was working on developing a program that could solve matrix problems efficiently.

During the development process, Cleve encountered a matrix that was particularly challenging to work with, and he spent a significant amount of time trying to find a solution. After many unsuccessful attempts, he eventually succeeded in solving the problem, and this breakthrough became a pivotal moment in the creation of MATLAB.

To commemorate this achievement, Cleve decided to create a visual representation of the solution, which would later become the MATLAB logo. He plotted the values of the matrix as a series of lines, resulting in a distinctive wave-like pattern.

The shape of the logo was inspired by the concept of waves and oscillations, which are fundamental in many areas of science and engineering, such as signal processing, control systems, and physics. The colorful gradient used in the logo was chosen to represent the wide range of applications and disciplines that MATLAB supports.

Over the years, the MATLAB logo has undergone minor updates and refinements, but its core design and symbolism have remained largely unchanged. The logo has become an iconic representation of MATLAB's capabilities and its role as a powerful computational tool used by researchers, engineers, and scientists worldwide.

Today, the MATLAB logo is instantly recognizable and serves as a visual reminder of the software's rich history, its origins in solving complex matrix problems, and its versatility in tackling a diverse array of computational challenges across various domains (Fig. B.1).

Listing B.1 MATLAB code to display the MATLAB logo [6]

```
% Initialize the data for the MATLAB logo
L = 160*membrane(1,100);

% Create a figure and an axes to display the logo
f = figure;
ax = axes;

% Create the surface for the logo using the membrane
    data and adjust properties
s = surface(L);
s.EdgeColor = 'none';
view(3)

% Adjust the axes limits
ax.XLim = [1 201];
ax.YLim = [1 201];
ax.ZLim = [-53.4 160];

% Adjust the camera properties
ax.CameraPosition = [-145.5 -229.7 283.6];
ax.CameraTarget = [77.4 60.2 63.9];
ax.CameraUpVector = [0 0 1];
ax.CameraViewAngle = 36.7;

% Change the axes position and aspect ratio
ax.Position = [0 0 1 1];
ax.DataAspectRatio = [1 1 .9];

% Create and position lights for the logo
l1 = light;
l1.Position = [160 400 80];
l1.Style = 'local';
l1.Color = [0 0.8 0.8];

l2 = light;
l2.Position = [.5 -1 .4];
l2.Color = [0.8 0.8 0];

% Change the logo color
s.FaceColor = [0.9 0.2 0.2];

% Set lighting and specular properties for the surface
```

```
s.FaceLighting = 'gouraud';
s.AmbientStrength = 0.3;
s.DiffuseStrength = 0.6;
s.BackFaceLighting = 'lit';
s.SpecularStrength = 1;
s.SpecularColorReflectance = 1;
s.SpecularExponent = 7;

% Final adjustments and display options
axis off
f.Color = 'white';
```

B.3 Tips: Programming Style and Best Practices

Following these guidelines will make your code much easier to read and under- stand, and therefore easier to work with, mainten and modify.

B.3.1 Variables and Names

- Use mnemonic variable names (names that make sense; for example, radius instead of xyz)
- Although variables named result and RESULT are different, avoid this as it would be confusing
- Do not use names of built-in functions as variable names
- Store results in named variables (rather than using ans) if they are to be used later
- Do not use ans in expressions
- Make sure variable names have fewer characters than namelengthmax
- If different sets of random numbers are desired, set the seed for the random functions using mg

Fig. B.1 MATLAB Logo
generated by the codes in [6]

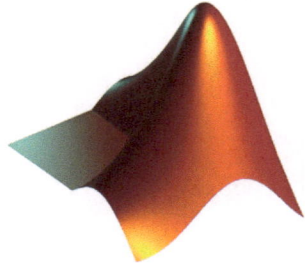

B.3.2 Setting up MATLAB .m Files in a Good Readable Style

- **Follow a Logical Order**: To initialize, compute, process results, plot/output. Split code into cells or sections for readability. Keep a consistent style for braces, spacing, capitalization across all .m files in a project.
- **Use Clear Variable Names**: Use descriptive variable names that indicate meaning and units. Avoid short or cryptic names like a, b, c. Avoid single-letter names except for simple loop counters, such as i, j, k.
- **Comment Generously**: Make sure to provide comments that explain the purpose of the code and any complex logic. Start comments with % for single lines or use %{ and %} for block comments.
- **Indentation and Spacing**: Use proper indentation to make code blocks easily identifiable. Also, use spacing between operators and operands for better readability.
- **Section Your Code**: Use sections within your script to separate different parts of your code logically. You can create a section by using %% followed by a section title. Logically group related variables, functions, etc. Use consistent indentation (e.g. 2 or 4 spaces) for nested statements and functions. Indentation visually indicates control flow.
- **Consistent Naming Conventions**: Be consistent with your naming conventions throughout the script. For example, use camelCase or underscores to separate words in variable names.
- **Avoid Hardcoding**: Avoid hardcoding values that might change, such as file paths or parameters. Instead, define them at the top of your script. Avoid global variables. Pass variables between functions using input arguments and return values.
- **Function Usage**: If your script is lengthy or contains repeated code, consider breaking it down into functions. Name functions clearly according to their usage. Start with a verb like calculate_stats(), plot_data() etc. Put main script code in a separate section after all function definitions. Start main script with initializing inputs/variables.
- **Vectorization**: Where possible, use vectorized operations instead of loops to improve performance and conciseness. Use **built-in MATLAB functions** and vectorization to express complex operations concisely. Avoid for or while loops if vectorized alternative exists.
- **Error Checking**: Implement error checking for operations that might fail, such as file I/O, and provide informative error messages. Check for errors and edge cases using **try/catch**, **input validation**, **asserts** etc. Print/log errors meaningfully.

B.3.3   Writing Readable and Maintainable Conditional Statements

To enhance code readability and maintainability, it's essential to follow best practices when writing conditional statements:

- Use descriptive variable and function names that convey their purpose.
- Indent code blocks consistently to improve visual clarity.
- Add comments to explain the purpose and logic of complex conditional statements.
- Break down complex conditions into smaller, more readable expressions using logical operators (AND (&&), OR (||), and NOT (~)).
- Consider using **switch** statements for scenarios with multiple distinct cases, as they can be more concise and readable than nested **if-elseif** statements.

B.3.4   Optimizing Code Efficiency and Performance

While MATLAB is designed to optimize code execution, there are certain practices that can further enhance the efficiency and performance of conditional statements:

- Avoid unnecessary computations or comparisons within conditional statements by performing calculations outside the conditional blocks whenever possible.
- Use vectorized operations instead of loops when working with arrays or matrices, as they are generally more efficient in MATLAB.
- Consider precomputing or caching frequently used values to reduce redundant computations.
- Utilize **MATLAB's built-in functions** and optimized algorithms whenever possible, as they are typically more efficient than custom implementations.

Here's an example of optimizing code performance by using vectorized operations:

Listing B.2 Vectorized code for computing square roots.

```
% Define a large vector
vec = magic(1000);

% Compute square roots using a loop (inefficient)
tic
for i = 1:numel(vec)
sqrt_vec(i) = sqrt(vec(i));
end
toc

% Compute square roots using vectorized operation (
    efficient)
tic
sqrt_vec_vectorized = sqrt(vec);
toc
```

B.3.5   Debugging Techniques for Conditional Statements

Debugging conditional statements can be challenging, especially when dealing with complex conditions or nested statements. Here are some techniques that can help:

- Use **MATLAB's built-in debugger** to step through the code and inspect variable values at each conditional branch. This can help identify logical errors or unexpected behavior.
- Add **disp()** or **fprintf()** statements to print out relevant information, such as variable values or condition evaluations, to better understand the program flow.
- Simplify complex conditions by breaking them down into smaller, more manageable pieces and testing each piece individually.
- Employ temporary logging or tracing mechanisms to record the execution path and variable states during program execution, which can aid in identifying issues.
- Consider writing unit tests or test cases specifically designed to validate the behavior of conditional statements under various scenarios.

Here's an example of using **disp()** statements to debug a conditional statement:

Listing B.3 Debugging conditional statements with disp().

```
x = 10;
y = 5;

disp('Before if statement:')
disp(['x = ' num2str(x), ', y = ' num2str(y)])

if x > y
disp('x is greater than y')
else
disp('x is not greater than y')
end
```

B.3.6   Guidelines for Choosing the Appropriate Conditional Statement

When deciding between using an **if-elseif-else** statement or a **switch** statement, consider the following guidelines:

- Use **if-elseif-else** statements when dealing with ranges, inequalities, or complex conditions involving logical operations.
- Prefer **switch** statements when evaluating equality against a fixed set of discrete values or expressions.
- If the conditions involve string comparisons or logical operations on strings, **if-elseif-else** statements may be more suitable due to MATLAB's handling of string comparisons in **switch** statements.

- For simple cases with a small number of conditions, either statement type can be used, and the choice may depend on personal preference or code readability.

It's important to note that while these guidelines provide a general starting point, the decision should also consider factors such as code readability, maintainability, and performance characteristics specific to the problem at hand.

The basic syntax for a **switch** statement is:

Listing B.4 Switch statement syntax.

```
switch expression
case case1
statements
case case2
statements
...
otherwise
statements
end
```

Here's an example of using a **switch** statement to determine the day of the week based on a numeric input:

Listing B.5 Determining day of the week with a switch statement.

```
day_num = 3; % Wednesday

switch day_num
case 1
day_str = 'Monday';
case 2
day_str = 'Tuesday';
case 3
day_str = 'Wednesday';
case 4
day_str = 'Thursday';
case 5
day_str = 'Friday';
case 6
day_str = 'Saturday';
case 7
day_str = 'Sunday';
otherwise
day_str = 'Invalid day number';
end

disp(['Today is ' day_str])
```

B.4 Tips for Efficient Data Management

To make the most out of MATLAB's data structures, consider the following tips:

- **Preallocate Arrays and Matrices**: Preallocating memory for arrays and matrices can significantly improve performance, especially in large datasets or iterative computations. For example:

Listing B.6 Preallocating an array

```
n = 1000;
A = zeros(n, n); % Preallocate a 1000x1000 matrix

for i = 1:n
    for j = 1:n
        A(i, j) = i + j; % Fill the matrix with some
            values
    end
end
```

- **Use Vectorized Operations**: Vectorized operations are typically faster than loops in MATLAB. For example, instead of using a for-loop to add two vectors element-wise, use the vectorized addition:

Listing B.7 Vectorized addition

```
A = rand(1, 1000);
B = rand(1, 1000);

% Vectorized addition
C = A + B;
```

- **Efficient Data Import and Export**: Use MATLAB functions for efficient data import and export to avoid bottlenecks. For example, use 'readtable' to import data from CSV files:

Listing B.8 Reading data from a CSV file

```
% Read data from a CSV file into a table
data = readtable('data.csv');

% Display the first few rows of the table
head(data);
```

- **Use Structures and Cell Arrays Appropriately**: Choose structures or cell arrays when dealing with heterogeneous data. Structures are useful for named fields, while cell arrays are beneficial for indexed collections of mixed-type data.

Listing B.9 Using structures and cell arrays

```
% Using a structure
student.name = 'John Doe';
student.age = 21;
```

```
student.grades = [90, 85, 92];

% Using a cell array
data = {'John Doe', 21, [90, 85, 92]; 'Jane Smith',
    22, [88, 90, 95]};
```

- **Efficiently Handle Large Data Sets**: For large datasets, consider using MATLAB's capabilities for memory mapping, or use the 'tall' arrays which allow you to work with data that does not fit into memory.

Listing B.10 Using tall arrays for large data

```
% Create a tall array from a large CSV file
ds = datastore('largefile.csv');
t = tall(ds);

% Perform operations on the tall array
meanValue = mean(t.SomeColumn);

% Gather results into memory
meanValue = gather(meanValue);
```

B.5 Difference Between Arrays and Vector

What Is The Difference Between Arrays and Vector ?

- **Dimensions**: A **vector** is a one-dimensional array, while an **array** can be multi-dimensional (two-dimensional, three-dimensional, etc.).
- **Representation**: Vectors are typically represented using a single row or column, whereas arrays are represented using multiple rows and columns.
- **Element Access**: Elements in a vector are accessed using a single index, while elements in an array are accessed using multiple indices (one for each dimension).
- **Memory Storage**: Vectors are stored contiguously in memory, making them more memory-efficient for storing and processing one-dimensional data. Arrays, on the other hand, may require more memory due to their multi-dimensional nature.
- **Operations**: While both vectors and arrays support various mathematical operations, some operations may be more efficient or optimized for vectors due to their simpler structure.
- **Reshaping**: Vectors can be easily reshaped into arrays, and vice versa, using MATLAB's built-in functions like reshape and squeeze.

In general, vectors are well-suited for representing and manipulating one-dimensional data, such as sequences or lists of values, while arrays are more suitable for representing and manipulating multi-dimensional data, such as matrices, images, or higher-dimensional datasets.

B.6   MATLAB AI Chat Playground

The MATLAB AI Chat Playground is ready for you to experiment with Generative AI, answer questions, and write initial draft MATLAB code [7, 8].

B.7   MATLAB on Github

A collection of curriculum materials for educators using MATLAB and Simulink [9].

Here is a summary of the top 30 contents on it:

- **Virtual-Controls-Laboratory**: Virtual labs and mechanisms for studying controls using MATLAB.
- **Fourier-Analysis**: Learn Fourier analysis using live scripts and apps in MATLAB.
- **Thermodynamics**: Interactive examples teaching thermodynamics concepts for Mechanical Engineering using MATLAB.
- **Fluid-Mechanics**: Introductory fluid mechanics concepts taught with interactive MATLAB courseware.
- **Robotic-Manipulators**: Interactive examples teaching fundamental robotics manipulator concepts using MATLAB.
- **Numerical-Methods-with-Applications**: Teaching numerical methods like ODE solving using MATLAB.
- **Density-Functional-Theory**: Fundamentals and applications of density functional theory using Jupyter Notebooks.
- **Intro-To-Engineering**: Introducing engineering concepts like signals, data analysis and IoT using MATLAB and Arduino.
- **Machine-Learning-for-Regression**: Workflow, setup and considerations for solving regression problems with machine learning and MATLAB.
- **Machine-Learning-Methods-Clustering**: Theory and application of clustering methods using MATLAB.
- **Regression-Basics**: Fundamentals of regression analysis taught with interactive MATLAB courseware.
- **Applied-Linear-Algebra**: Teaching linear algebra applications like chemistry and mechanical engineering using MATLAB.
- **Calculus-Derivatives**: Introducing derivatives and calculus concepts with interactive MATLAB module.
- **awesome-matlab-students**: Helpful resources, tips, tutorials and opportunities for students learning MATLAB.
- **Calculus-Single-Variable**: Interactive module introducing single variable calculus concepts using MATLAB.
- **Dynamics-and-Vibrations**: Teaching fundamentals of dynamics and vibrations with interactive MATLAB examples.

- **Network-Analysis**: Interactive electrical circuits and network analysis curriculum module using MATLAB.
- **Algebra-Trig-PreCalculus**: Foundational algebra, trigonometry and pre-calculus concepts taught interactively with MATLAB.
- **Data-Acquisition-and-IOT**: Teaching concepts like sampling, sensors, data acquisition and IoT using MATLAB and Arduino.
- **matlab-getting-started**: Resources to help students get started with MATLAB.
- **Ordinary-Differential-Equations**: Introductory ordinary differential equations concepts with MATLAB interactive module.
- **Statistical-Learning**: Interactive introduction to statistical learning theory and methods using MATLAB.
- **matlab-cheat-sheet**: MATLAB cheat sheet covering basic commands and syntax.
- **System-Modeling-Control**: System modeling, analysis and control concepts taught interactively using MATLAB and Simulink.
- **Signals-and-Systems**: Interactive curriculum covering signals and systems theory using MATLAB.
- **matlab-for-engineering-students**: Curated resources for engineering students learning MATLAB.
- **Introduction-to-Programming-with-MATLAB**: Introductory programming concepts taught with MATLAB.
- **Probability-Statistics**: Introduction to concepts of probability, statistics and hypothesis testing using MATLAB.
- **Aerospace-Blockset**: Interactive aerospace examples using MATLAB Aerospace Blockset.

B.8 What Is MATLAB, and What Are Its Primary Applications?

- **MATLAB** (short for **Matrix Laboratory**) is a high-level programming language and numerical computing environment widely used for scientific and engineering applications.
- Primary applications include **data analysis**, **algorithm development**, **simulation**, **visualisation**, and **prototyping**.
- It is extensively used in fields such as **engineering**, **science**, **finance**, and **research**.

B.9 What Are the System Requirements and Installation Process?

- MATLAB is available for various operating systems (**Windows, macOS**, and **Linux**).

- System requirements vary depending on the version, but generally include a compatible processor, sufficient RAM, and disk space.
- Installation can be done by downloading the software from the **MathWorks** website and following the on-screen instructions.
- For students and educators, **academic licenses** may be available at discounted rates.

B.10   What Are the Different Components of the MATLAB Desktop Environment, and What Are Their Functions?

- The **Command Window** is used for interactive computations and executing commands.
- The **Editor** is used for writing and editing MATLAB scripts and functions.
- The **Workspace** displays variables and their values.
- The **Command History** keeps track of previously executed commands.
- The **Current Folder** shows the current working directory.

B.11   How Do I Write and Execute MATLAB Scripts and Functions?

- **Scripts** are text files with a .m extension containing a sequence of MATLAB commands.
- **Functions** are reusable code blocks that can accept input arguments and return outputs.
- Scripts and functions can be written in the Editor and executed using the **Run** button or keyboard shortcut (**F5**).

B.12   What Are the Different Data Types in MATLAB, and How Do I Work with Variables, Vectors, and Matrices?

- **Numeric** data types include **double** (default), **single**, **integer**, and more.
- **Character** and **string** data types are used for text data.
- **Variables** are assigned values using the = operator.
- **Vectors** and **matrices** are created using square brackets ([ ]) and can be manipulated using various operations.

B.13   How Do I Import and Export Data in MATLAB, and What File Formats Are Supported?

- Data can be imported into MATLAB using functions like **readtable**, **readmatrix**, and **load**.
- Supported file formats include **CSV**, **Excel**, **text**, **MAT** (MATLAB's binary format), and more.
- Data can be exported using functions like **writetable**, **writematrix**, and **save**.

B.14   How Do I Create and Customize Plots and Visualisations in MATLAB?

- The **plot** function is used to create basic 2D line plots.
- Additional plotting functions like **bar**, **scatter**, and **histogram** are available for different visualisation types.
- Plots can be customized using functions like **title**, **xlabel**, **ylabel**, **legend**, and more.

B.15   What Are the Different Control Flow Statements (If-Else, For Loops, While Loops) in MATLAB, and How Do I Use Them?

- The **if-else** statement is used for conditional execution of code blocks.
- The **for** loop is used for iterating over a specific range of values.
- The **while** loop is used for executing a block of code as long as a condition is true.
- These control flow statements can be combined with logical operators (&&, ||, ~) for more complex conditions.

B.16   How Do I Handle Errors and Debug MATLAB Code?

- MATLAB provides detailed **error messages** and **warning messages** to help identify issues.
- The **debug** function can be used to pause execution and step through code line by line.
- Setting **breakpoints** in the Editor allows pausing execution at specific points.
- The **profiler** tool can be used to analyze code performance and identify bottlenecks.

B.17 What Are the Available Resources for Learning MATLAB, Such as Documentation, Tutorials, and Online Communities?

- The **MATLAB Documentation** provided by MathWorks is a comprehensive resource covering all aspects of the software.
- **Tutorials** and **examples** are available on the MathWorks website and within the MATLAB environment.
- Online communities like the **MATLAB Central** forum and **Stack Overflow** provide a platform for asking questions and sharing knowledge.
- **Books**, **online courses**, and **video tutorials** from various sources can also aid in learning MATLAB.

B.18 What Are the Differences Between Scripts and Functions in MATLAB?

- **Scripts** are sequences of MATLAB commands executed from top to bottom, useful for automating tasks or analyses.
- **Functions** are reusable code blocks that can accept input arguments, perform operations, and return outputs, promoting code modularization and reusability.

B.19 How Do I Perform Basic Arithmetic Operations in MATLAB?

- MATLAB supports standard arithmetic operators (+, -, *, /) for scalar and array operations.
- Additional operators like ^ (exponentiation) and **sqrt()** (square root) are available.
- MATLAB follows the standard order of operations (**PEMDAS**) unless modified by parentheses.

B.20 How Do I Work with Matrices in MATLAB?

- Matrices are created using square brackets ([ ]) with elements separated by spaces or commas.
- Basic operations like **addition**, **subtraction**, **multiplication**, and **division** can be performed on matrices.
- Special functions like **transpose**, **inverse**, **determinant**, and **eigenvalues** are available for matrix operations.

B.21   How Do I Access and Manipulate Elements in Vectors and Matrices?

- Elements in vectors and matrices are accessed using **indexing** with parentheses (( )).
- Colon notation (:) can be used to access a range of elements or entire rows/columns.
- Logical indexing using boolean expressions can be used to selectively access or modify elements.

B.22   What Are the Different Ways to Create Arrays in MATLAB?

- Arrays can be created explicitly by entering values within square brackets ([ ]).
- The **linspace** and **logspace** functions can be used to create arrays with linearly or logarithmically spaced values.
- The **ones** and **zeros** functions create arrays filled with ones or zeros, respectively.
- The **eye** function creates an identity matrix.

B.23   How Do I Perform Operations on Arrays in MATLAB?

- MATLAB supports element-wise operations on arrays using arithmetic operators (+, -, *, /).
- The .* and ./ operators are used for element-wise multiplication and division, respectively.
- Functions like **sum**, **mean**, **max**, and **min** can be used to perform aggregate operations on arrays.

B.24   How Do I Concatenate and Reshape Arrays in MATLAB?

- Arrays can be concatenated horizontally using the [ ] operator or vertically using the ; separator.
- The **cat** function can be used to concatenate arrays along a specified dimension.
- The **reshape** function can be used to change the shape or dimensions of an array without changing its data.

B.25   What Are Cell Arrays in MATLAB, and How Are They Used?

- **Cell arrays** are flexible data structures that can store different data types, including arrays, strings, and other cell arrays.
- They are created using curly braces ({}) and can be accessed and manipulated using indexing and cell array operations.
- Cell arrays are useful for storing heterogeneous data or data with varying sizes.

B.26   How Do I Work with Strings and Text Data in MATLAB?

- MATLAB supports **character arrays** and **string arrays** for working with text data.
- String concatenation can be performed using the + operator or the **strcat** function.
- Functions like **length**, **substring**, **replace**, and **split** are available for string manipulation.

B.27   How Do I Read and Write Data to Files in MATLAB?

- The **fopen** function is used to open a file for reading or writing.
- Data can be read from files using functions like **fscanf**, **fgetl**, and **textscan**.
- Data can be written to files using functions like **fprintf**, **fwrite**, and **fprinf**.
- The **fclose** function is used to close the file after reading or writing operations.

B.28   What Are Structures in MATLAB, and How Are They Used?

- **Structures** are data containers that allow you to group related data elements of different types using **field names**.
- Structures are created using the **struct** function or by providing field-value pairs within parentheses.
- Fields in structures can be accessed using the dot notation (.) or curly braces ({}).

B.29   How Do I Create and Use Functions in MATLAB?

- Functions are defined using the **function** keyword followed by the function name, input arguments, and code body.
- Input arguments are passed to the function when it is called, and outputs can be returned using the **return** statement.
- Functions can be saved as separate .m files and called from other scripts or functions.

B.30   What Is the Purpose of Anonymous Functions in MATLAB?

- **Anonymous functions** are unnamed functions that can be defined and used inline without creating a separate function file.
- They are useful for passing functions as arguments to other functions or for simple, one-time operations.
- Anonymous functions are defined using the @ symbol followed by the input arguments and function body.

B.31   How Do I Work with Dates and Times in MATLAB?

- MATLAB provides the **datetime** data type for working with dates and times.
- **datetime** objects can be created from various input formats using the **datetime** function.
- Functions like **datenum**, **datevec**, and **calendarDuration** are available for date and time manipulation and calculations.

B.32   What Are the Different Ways to Handle Missing Data in MATLAB?

- MATLAB represents missing data using the **NaN** (Not a Number) value.
- The **isnan** function can be used to identify missing data elements in an array.
- Functions like **nansum**, **nanmean**, and **nanstd** can be used to perform operations while ignoring missing data.
- The **rmmissing** function can be used to remove rows or columns with missing data from a matrix or table.

B.33　How Do I Integrate MATLAB with Other Programming Languages or Software?

- MATLAB provides interfaces and toolboxes for integrating with various programming languages and software.
- The **MATLAB Engine API** allows you to call MATLAB functions from external programs written in languages like C, C++, Java, and .NET.
- The **MATLAB Compiler** can be used to generate standalone applications or libraries from MATLAB code.
- MATLAB supports data exchange with other software through file formats like CSV, Excel, and MAT-files.

B.34　How Do I Create and Work with Tables in MATLAB?

- The **table** data type in MATLAB allows you to store heterogeneous data in a tabular format with row and column labels.
- Tables can be created from arrays, cell arrays, or data files using functions like **table**, **readtable**, and **struct2table**.
- Rows and columns in tables can be accessed and manipulated using indexing, logical indexing, or column names.
- Functions like **sortrows**, **unique**, and **join** can be used to perform operations on tables.

B.35　What Are the Different Types of Plots and Visualisations Available in MATLAB?

- MATLAB provides a wide range of plotting functions for creating 2D and 3D visualisations.
- Common plot types include **line plots**, **scatter plots**, **bar charts**, **histograms**, **surface plots**, and **contour plots**.
- Specialized plotting functions are available for visualizing data structures like **tables**, **images**, and **geographic data**.
- Plots can be customized with labels, legends, color maps, and other formatting options using various plotting functions and properties.

B.36   How Do I Work with Images and Image Processing in MATLAB?

- MATLAB provides the **Image Processing Toolbox** for working with images and performing various image processing operations.
- Images can be imported into MATLAB using functions like **imread** and displayed using **imshow**.
- Operations like **filtering, edge detection, segmentation,** and **morphological operations** can be performed on images using toolbox functions.
- MATLAB also supports video processing and computer vision applications through additional toolboxes.

B.37   What Are the Different Techniques for Data Analysis and Machine Learning in MATLAB?

- MATLAB provides a comprehensive set of tools and functions for data analysis and machine learning tasks.
- Techniques like **regression, classification, clustering,** and **dimensionality reduction** are supported through toolboxes like **Statistics and Machine Learning Toolbox**.
- MATLAB integrates with deep learning frameworks like **TensorFlow** and **PyTorch** for building and training neural networks.
- Functions for **data preprocessing, feature extraction,** and **model evaluation** are also available.

B.38   How Do I Parallelize Computations in MATLAB to Take Advantage of Multiple Processors or GPUs?

- MATLAB provides the **Parallel Computing Toolbox** for parallelizing computations on multi-core systems or computing clusters.
- The **parfor** loop can be used to distribute loop iterations across multiple MATLAB workers or cores.
- Functions like **spmd** (single program, multiple data) and **codistributor** can be used for data parallelism.
- GPU computing is supported through integration with CUDA [10] and AMD GPUs, allowing computationally intensive operations to be offloaded to the GPU.

B.39 What Are the Different Techniques for Optimisation and Solving Equations in MATLAB?

- MATLAB provides a range of functions and toolboxes for optimisation and solving equations.
- The **optimisation Toolbox** includes algorithms for **linear programming, non-linear optimisation**, and **constrained optimisation** problems.
- The **Symbolic Math Toolbox** allows for symbolic computing, including solving **algebraic equations, differential equations**, and **calculus operations**.
- Functions like **fsolve, lsqnonlin**, and **fmincon** can be used for solving specific types of equations and optimisation problems.

B.40 How Do I Create and Work with Objects and Classes in MATLAB?

- MATLAB supports **object-oriented programming** (OOP) through the use of classes and objects.
- Classes are defined using the **classdef** keyword, followed by properties and methods.
- Objects are instances of classes and can be created using the **constructor** method defined in the class.
- Objects can interact with each other and access or modify their properties and methods.
- OOP principles like **inheritance, polymorphism**, and **encapsulation** are supported in MATLAB.

B.41 What Are the Different Options for Deploying and Sharing MATLAB Applications?

- MATLAB provides several options for deploying and sharing applications with others.
- The **MATLAB Compiler** can be used to create standalone applications or libraries from MATLAB code.
- The **MATLAB Web App Server** allows hosting MATLAB applications as web services accessible through a web browser.
- MATLAB code can be packaged into **MATLAB apps** or **toolboxes** for distribution and sharing with other users.
- Integration with cloud platforms like **MATLAB Online** and **MATLAB Production Server** enables sharing and running MATLAB applications in the cloud.

B.42 How Do I Integrate MATLAB with Version Control Systems Like Git or SVN?

- MATLAB provides built-in support for integrating with version control systems like **Git** and **Subversion (SVN)**.
- The **Source Control Integration** feature in MATLAB allows connecting to remote repositories and managing code changes.
- Common version control operations like **commit**, **push**, **pull**, **merge**, and **resolve conflicts** can be performed within the MATLAB environment.
- Version control best practices, such as branching and merging, can be followed to collaborate on MATLAB projects effectively.

B.43 MATLAB Plot Cheat Sheet

A MATLAB Plot Cheat Sheet covering some of the most common plotting functions and customizations [11]. Remember, many of these functions have additional optional parameters for further customisation. You can use the help command in MATLAB (e.g., help plot) to get more detailed information about each function and its options (Fig. B.2).

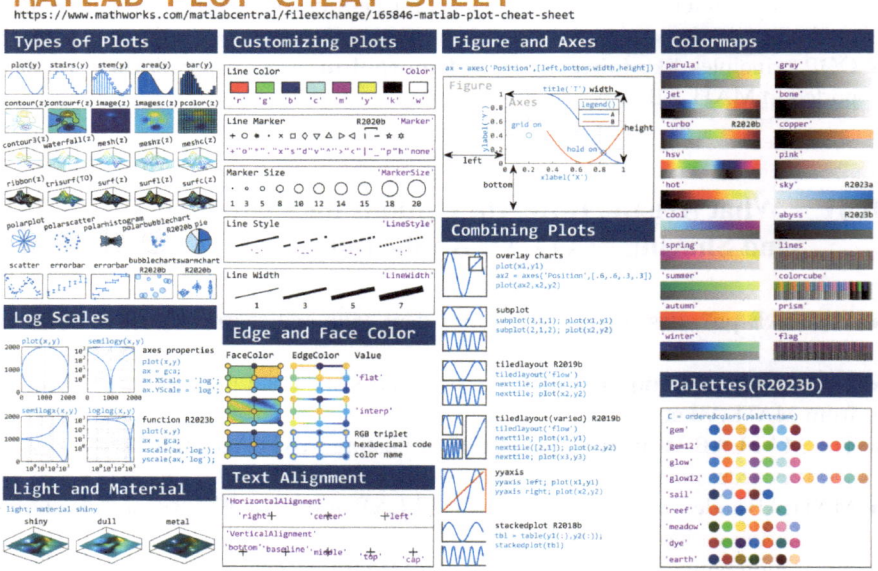

Fig. B.2 MATLAB Plot Cheat Sheet

B.44 MATLAB Resources and Online Courses

MATLAB is a versatile programming language widely used in various fields, including data analysis, numerical computation, and algorithm development. For beginners seeking to learn MATLAB, there are numerous resources and online modules available that provide a solid foundation in using the software.

- One of the primary resources for MATLAB beginners is the official MATLAB website [12]. The website offers comprehensive documentation, tutorials, and examples that cover a wide range of topics and functionalities of MATLAB. Beginners can access these resources to learn the basics of MATLAB syntax, programming techniques, and data visualisation.
- The MATLAB File Exchange [13] is another valuable resource for MATLAB beginners. It is a community-driven platform where MATLAB users can share and download MATLAB code, functions, and applications. Beginners can explore the File Exchange to find code examples, scripts, and functions that they can utilize to learn and enhance their MATLAB skills.
- Online learning platforms such as Coursera [14] and edX [15] also offer MATLAB courses for beginners. These courses provide structured learning materials, video lectures, and hands-on assignments to help beginners grasp the fundamentals of MATLAB programming. Some of these courses are self-paced, allowing learners to study at their own convenience.
- For those who prefer interactive learning, MathWorks, the company behind MATLAB, provides MATLAB Online [16]. MATLAB Online is a web-based version of MATLAB that allows beginners to practice MATLAB programming directly in a browser without the need to install any software. It provides a user-friendly interface and access to various MATLAB toolboxes and functionalities.

- Identify and utilise **authoritative resources** for learning and mastering MATLAB and its applications.
- Navigate and leverage **online documentation** and **community forums** to find solutions and troubleshoot issues.
- Explore and enrol in **online courses** and **tutorials** to enhance your MATLAB skills and knowledge.
- Locate and access **code repositories** and **examples** to learn from existing implementations.
- Understand the importance of **continuous learning** and **professional development** in the field of MATLAB and related domains.
- Develop a **learning plan** and **strategy** tailored to your specific goals and needs.

- **Authoritative Resources**: Reliable and trustworthy sources of information, typically produced or endorsed by recognised experts, institutions, or organisations in the field.
- **Online Documentation**: Digital documentation, manuals, and reference materials available on the internet, providing comprehensive information about a software, tool, or concept.

- **Community Forums**: Online platforms where users can ask questions, share knowledge, and engage in discussions with other members of the community related to a specific topic or product.
- **Online Courses**: Educational courses offered over the internet, often self-paced or instructor-led, covering various topics and allowing learners to acquire knowledge and skills remotely.
- **Tutorials**: Step-by-step guides or instructional materials designed to teach specific concepts, techniques, or processes related to a particular subject or tool.
- **Code Repositories**: Online platforms or databases where developers can store, share, and collaborate on code, often maintained using version control systems like Git.
- **Continuous Learning**: The ongoing process of acquiring new knowledge and skills throughout one's personal and professional life to stay up-to-date and adapt to changing circumstances.
- **Professional Development**: Activities and experiences that enhance an individual's skills, knowledge, and competencies relevant to their professional growth and career advancement.
- **Learning Plan**: A structured approach to identifying learning goals, strategies, and resources, often accompanied by timelines and milestones, to guide and facilitate the learning process effectively.

B.45 Official MATLAB Resources

B.45.1 MathWorks Documentation

- MATLAB Product Documentation [17], Function Reference [18], and Examples and Tutorials [19] provide comprehensive information on using MATLAB effectively.
- MATLAB supports various hardware platforms, including Parrot drones [20] and Ryze Tello drones [21], enabling users to control and acquire data from these devices.
- MathWorks organises student competitions like the MathWorks Minidrone Competition [22] and the competition for autonomous mobile robots [23, 24], encouraging students to develop innovative solutions using MATLAB and Simulink.
- The MathWorks Excellence in Innovation repository on GitHub [25] showcases innovative projects across various domains, including drones [26].
- MATLAB offers a variety of project ideas [27] and live script examples [28] to inspire and guide users in their engineering and scientific endeavours.
- Advanced MATLAB for Scientific Computing [29] is a resource that delves into advanced MATLAB features and techniques for scientific computing applications.
- The MATLAB Plot Gallery [30] provides a collection of visualisation examples to help users create informative and appealing graphics.

- MATLAB Academy offers various online courses, including MATLAB Onramp [31], MATLAB Fundamentals [32], MATLAB for Data Processing and visualisation [33], and Optimisation Onramp [34], catering to different skill levels and application areas.
- Cleve Moler's MATLAB Labs [35] and MathWorks Teaching Resources on GitHub [36] provide valuable materials for learning and teaching MATLAB
- AI with MATLAB: Tutorials and Examples [37].

B.45.2 MATLAB Community and Support

- **MATLAB Central**
- **MATLAB Answers**
- **File Exchange**

B.46 Third-Party Resources

B.46.1 Books and Textbooks

- **Introductory MATLAB Textbooks**
- **Domain-Specific MATLAB Books**
- **Advanced MATLAB Programming Books**

B.46.2 Online Courses and Tutorials

- **Massive Open Online Courses (MOOCs)**
- **Video Tutorials**
- **Interactive Online Tutorials**

B.46.3 Blogs and Forums

- **MATLAB-related Blogs**
- **Online Forums and Discussion Groups**
- **Social Media Communities**

B.47   Code Repositories and Examples

- **GitHub**
- **MATLAB Central File Exchange**
- **Domain-Specific Code Repositories**

B.48   Professional Development and Certifications

- **MATLAB Training and Certification Programs**
- **Industry-Specific Professional Development Opportunities**
- **Continuing Education and Lifelong Learning**

B.49   Learning Strategies and Planning

- **Setting Learning Goals and Objectives**
- **Creating a Personalised Learning Plan**
- **Time Management and Study Habits**
- **Blended Learning Approaches**
- **Hands-on Practice and Project-Based Learning**

Glossary

See Table B.1.

Table B.1 Comprehensive glossary of MATLAB terms and concepts

Term	Definition
Algorithm	A finite sequence of well-defined, computer-implementable instructions, typically used to solve a class of problems or perform a computation
Array	A fundamental data structure in MATLAB, consisting of a collection of elements arranged in rows and columns. Arrays can be one-dimensional (vectors), two-dimensional (matrices), or multi-dimensional
Artificial Neural Network	A machine learning model inspired by biological neural networks, implemented in MATLAB for various applications including pattern recognition and function approximation
Automation	The use of MATLAB to create systems or processes that operate automatically, often applied in industrial settings and robotics
Block Diagram	A graphical representation of a system in Simulink, consisting of interconnected blocks that represent mathematical operations or system components
Cell Array	A data structure in MATLAB that can contain elements of different data types and sizes, useful for storing heterogeneous data
Command Window	The primary interface for entering MATLAB commands and viewing results. It allows for interactive execution of MATLAB functions and scripts
Control System Toolbox	A MATLAB toolbox for analysing, designing, and simulating control systems, particularly relevant for robotics and autonomous systems
Data Acquisition	The process of sampling signals that measure real-world physical conditions and converting the resulting samples into digital numeric values that can be manipulated by MATLAB
Data Type	The classification of data which determines its possible values, operations, and storage method in MATLAB, including numeric, character, and logical types

(continued)

© The Editor(s) (if applicable) and The Author(s) 2025
Y. Chen and L. Huang, *MATLAB Roadmap to Applications*,
https://doi.org/10.1007/978-981-97-8788-3

Table B.1 (continued)

Term	Definition
Deep Learning	A subset of machine learning based on artificial neural networks with multiple layers, implemented in MATLAB for tasks such as image and speech recognition
Digital Manufacturing	The use of MATLAB and Simulink in computer-integrated manufacturing systems, often involving simulation and optimisation of manufacturing processes
Digital Twin	A virtual representation of a physical object or system, often created and analysed using MATLAB and Simulink, used in Industry 4.0 applications
Discrete-Event Simulation	A method of simulating the behaviour and performance of a real-world process or system, often used in MATLAB for modelling complex systems
Embedded Coder	A MATLAB tool for generating readable, compact, and fast C and C++ code for use on embedded processors, often used in robotics and autonomous systems
Function	A program file that can accept input arguments and return output arguments. Functions operate on variables within their own workspace, separate from the base workspace
Fuzzy Logic	A form of many-valued logic in which the truth values of variables may be any real number between 0 and 1, implemented in MATLAB for control systems and decision-making processes
GPU Computing	The use of a graphics processing unit (GPU) to perform computations in MATLAB, often resulting in significant speedups for certain types of parallel computations
Handle Graphics	MATLAB's graphics system, which provides a flexible and powerful way to create and manipulate graphical objects
Image Processing	The manipulation and analysis of digital images using MATLAB functions, often applied in computer vision and robotics
Industry 4.0	The fourth industrial revolution, characterised by the integration of digital technologies in industrial processes, often modelled and simulated using MATLAB
Internet of Things (IoT)	A network of interconnected devices that collect and exchange data, often analysed and controlled using MATLAB in industrial and smart systems contexts
Live Editor	An interactive development environment in MATLAB that combines code, output, and formatted text in a single interface, facilitating literate programming and reproducible research
Live Script	An interactive document that combines MATLAB code with formatted text, equations, and visualisations in a single environment
Machine Learning	A subset of artificial intelligence that provides systems the ability to automatically learn and improve from experience, implemented in MATLAB through various toolboxes and functions
Matrix	A two-dimensional array of numbers or expressions arranged in rows and columns. Matrices are fundamental to MATLAB's computational capabilities
M-file	A text file containing MATLAB code, typically used to store functions or scripts. M-files have a .m extension and can be executed in the MATLAB environment

(continued)

Table B.1 (continued)

Term	Definition
Model Predictive Control	An advanced method of process control that uses dynamic models to predict and optimise process performance, often implemented using MATLAB's Control System Toolbox
Object-Oriented Programming	A programming paradigm based on the concept of "objects", which can contain data and code. MATLAB supports object-oriented programming for creating complex, modular systems
Optimisation	The selection of a best element, with regard to some criterion, from some set of available alternatives. MATLAB provides various tools and functions for optimisation problems
Parallel Computing	The simultaneous use of multiple compute resources to solve a computational problem in MATLAB, often applied to large-scale simulations and data processing tasks
Predictive Maintenance	A technique used to predict when equipment failure might occur, often implemented using MATLAB's machine learning and signal processing capabilities
Profiler	A MATLAB tool that helps identify performance bottlenecks in code by measuring the execution time of functions and lines of code
Reinforcement Learning	A type of machine learning where an agent learns to make decisions by taking actions in an environment to maximise a reward, implemented in MATLAB for various control and optimization problems
Robotics System Toolbox	A MATLAB toolbox that provides tools and algorithms for designing, simulating, and testing robotic systems
Robust Control	A branch of control theory that explicitly deals with uncertainty in its approach to controller design, often implemented using MATLAB's Robust Control Toolbox
Simulink	An add-on product to MATLAB that provides a graphical programming environment for modeling, simulating, and analysing dynamic systems
Singular Value Decomposition	A factorization of a real or complex matrix, widely used in signal processing and statistics, efficiently computed in MATLAB
State-Space Model	A mathematical model of a physical system as a set of input, output, and state variables related by first-order differential equations, commonly used in control systems and implemented in MATLAB
Stateflow	A graphical programming environment in Simulink for modeling and simulating combinatorial and sequential decision logic based on state machines and flow charts
Statistical Analysis	The collection, analysis, interpretation, presentation, and organization of data, facilitated by MATLAB's Statistics and Machine Learning Toolbox
Symbolic Math Toolbox	A MATLAB toolbox that provides tools for solving, manipulating, and analyzing symbolic math equations
System Identification	The process of developing or improving a mathematical representation of a physical system using experimental data, often performed using MATLAB's System Identification Toolbox
Toolbox	A collection of specialised MATLAB functions, classes, and applications designed for solving particular classes of problems

(continued)

Table B.1 (continued)

Term	Definition
Unit Testing	A software testing method by which individual units of source code are tested to determine whether they are fit for use, supported in MATLAB through the Unit Testing Framework
Vectorisation	The process of rewriting loop-based code to operate on entire arrays at once, often resulting in more concise code and improved performance in MATLAB
Wavelet Analysis	A method for analyzing localized variations of power within a time series, implemented in MATLAB's Wavelet Toolbox and often used in signal and image processing
Workspace	The set of variables and their values that are currently accessible in the MATLAB environment. The base workspace contains variables created at the command prompt or through running scripts
xPC Target	A solution for prototyping, testing, and deploying real-time systems using standard PC hardware, often used in rapid control prototyping and hardware-in-the-loop simulation

References

1. Chen Y, Zhang G (2013) Exchange rates determination based on genetic algorithms using Mendel's principles: investigation and estimation under uncertainty. Inf Fusion 14(3):327–333
2. Chen Y, Li Y (2018) Computational intelligence assisted design (In the Era of industry 4.0). CRC Press (ISBN 978-1-4987-6066-9)
3. Chen* Y, Li Y, (2019) Intelligent autonomous pollination for future farming - a micro air vehicle solution with artificial intelligence and human-in-the-loop. IEEE Access 7(1):119706–119717
4. Chen Y, Zhang G, Jin T, Wu S, Peng B (2014) Quantitative modelling of electricity consumption using computational intelligence aided design. J Cleaner Prod 69:143–152. (15 April 2014) https://doi.org/10.1016/j.jclepro.2014.01.058
5. Seaborn, "Timeseries plot with error bands," https://seaborn.pydata.org/examples/errorband_lineplots.html, accessed May 2024
6. MathWorks, "Creating the MATLAB Logo," https://www.mathworks.com/help/matlab/visualize/creating-the-matlab-logo.html, accessed on Feb. 17, 2024
7. MathWorks, "MATLAB AI Chat Playground," [Online]. https://www.mathworks.com/matlabcentral/playground. [Accessed: Feb. 17, 2024]
8. MathWorks, "The MATLAB AI Chat Playground Has Launched," MATLAB Community. [Online]. https://blogs.mathworks.com/community/2024/11/07/the-matlab-ai-chat-playground-has-launched. [Accessed: Feb. 17, 2024]
9. MathWorks, "MathWorks Teaching Resources - A collection of curriculum materials for educators using MATLAB and Simulink.," Github. [Online]. https://github.com/MathWorks-Teaching-Resources. [Accessed: Feb. 17, 2024]
10. MathWorks, "MATLAB GPU Computing Support for NVIDIA CUDA-Enabled GPUs - Perform MATLAB computing on NVIDIA CUDA-enabled GPUs," [Online]. https://www.mathworks.com/solutions/gpu-computing.html. [Accessed: Feb. 17, 2024]
11. Zhaoxu Liu / slandarer (2024) MATLAB-PLOT-CHEAT-SHEET (https://github.com/slandarer/MATLAB-PLOT-CHEAT-SHEET), GitHub. Retrieved June 26, 2024. https://www.mathworks.com/matlabcentral/fileexchange/165846-matlab-plot-cheat-sheet
12. MathWorks Home Page, www.mathworks.com. [Accessed: Feb. 17, 2024]
13. MathWorks. (n.d.). MATLAB file exchange. https://www.mathworks.com/matlabcentral/fileexchange

14. Coursera. (n.d.). MATLAB programming for engineers and scientists. https://www.coursera.org/learn/matlab
15. edX. (n.d.). MATLAB and Octave for beginners. https://www.edx.org/professional-certificate/matlab-and-octave-for-beginners
16. MATLAB Live Editor, https://matlab.mathworks.com. [Accessed: Feb. 17, 2024]
17. MathWorks, "MATLAB Product Family," https://www.mathworks.com/products.html
18. MathWorks, "MATLAB Basic Functions Reference," https://www.mathworks.com/content/dam/mathworks/fact-sheet/matlab-basic-functions-reference.pdf, accessed on Feb. 17, 2024
19. MathWorks, "MATLAB and Simulink Examples," https://www.mathworks.com/academia/examples.html, accessed on Feb. 17, 2024
20. "Parrot Drone Support from MATLAB – Control Parrot drones from MATLAB and acquire sensor and image data," https://uk.mathworks.com/hardware-support/parrot-drone-matlab.html?s_tid=srchtitle, accessed on Feb. 17, 2024
21. "Ryze Tello Drone from MATLAB – Control Ryze Tello drones from MATLAB and acquire sensor and image data," https://uk.mathworks.com/hardware-support/tello-drone-matlab.html, accessed on Feb. 17, 2024
22. "MathWorks Minidrone Competitions," https://uk.mathworks.com/academia/student-competitions/minidrones.html, accessed on Feb. 17, 2024
23. "RoboNation Competitions," https://uk.mathworks.com/academia/student-competitions/robonation.html, accessed on Feb. 17, 2024
24. "Korea Autonomous Mini-Drone Aviation Competition," https://www2.mathworks.cn/academia/student-competitions/krmdr.html, accessed on Feb. 17, 2024
25. "MathWorks Excellence in Innovation repository on GitHub," https://github.com/mathworks/MathWorks-Excellence-in-Innovation, accessed on Feb. 17, 2024
26. "Matlab for Drone Examples on Github," https://github.com/mathworks/MathWorks-Excellence-in-Innovation/blob/main/megatrends/Drones.md, accessed on Feb. 17, 2024
27. "MathWorks Challenge Projects Program," https://ww2.mathworks.cn/academia/matlab-engineering-project-ideas.html?s_tid=ln_acad_programs_projec, accessed on Feb. 17, 2024
28. "Live Script Gallery," https://ww2.mathworks.cn/products/matlab/live-script-gallery.html, accessed on Feb. 17, 2024
29. "Advanced MATLAB for Scientific Computing," https://ww2.mathworks.cn/matlabcentral/fileexchange/106675-advanced-matlab-for-scientific-computing?s_tid=srchtitle, accessed on Feb. 17, 2024
30. MathWorks, "MATLAB Plot Gallery," [Online]. https://www.mathworks.com/products/matlab/plot-gallery.html. [Accessed: Feb. 17, 2024]
31. "MATLAB Onramp," https://matlabacademy.mathworks.com/details/matlab-onramp/gettingstarted, accessed on Feb. 17, 2024
32. "MATLAB Fundamentals," https://matlabacademy.mathworks.com/details/matlab-fundamentals/mlbe, accessed on Feb. 17, 2024
33. "MATLAB for Data Processing and Visualization," https://matlabacademy.mathworks.com/details/matlab-fundamentals/mlbe, accessed on Feb. 17, 2024
34. "Optimisation Onramp," https://matlabacademy.mathworks.com/en/details/optimisation-onramp/optim, accessed on Feb. 17, 2024. MathWorks Teaching Resources [70] https://github.com/MathWorks-Teaching-Resources/
35. Cleve Moler (2024) Experiments with MATLAB (https://www.mathworks.com/matlabcentral/fileexchange/37977-experiments-with-matlab), MATLAB Central File Exchange. Retrieved January 30, 2024
36. MathWorks, "MathWorks Teaching Resources-A collection of curriculum materials for educators using MATLAB and Simulink.," [Online]. https://github.com/MathWorks-Teaching-Resources/. [Accessed: Feb. 17, 2024]
37. MathWorks, "AI with MATLAB: Tutorials and Examples," https://www.mathworks.com/solutions/artificial-intelligence/tutorials-examples.html, accessed on Feb. 17, 2024